CHRISTOPHER COLUMBUS, COSMOGRAPHER

**A History of
Metrology, Geodesy, Geography, and
Exploration from Antiquity to the Columbian Era**

CHRISTOPHER COLUMBUS, COSMOGRAPHER

A History of
Metrology, Geodesy, Geography, and
Exploration from Antiquity to the Columbian Era

Fred F. Kravath
Captain, Civil Engineer Corps
U.S.Navy (Retired)

LANDMARK ENTERPRISES
10324 Newton Way
Rancho Cordova CA 95670

ISBN 0-910845-32-8

Cover design by Total Graphic Services
Printed in USA by Edwards Brothers, Ann Arbor, Michigan

Dedicated to those, worldwide,
who can identify with Columbus'
"Project of the Indies," and
my own effort to demonstrate the
plausibility of Columbus'
rationale.

Christopher Columbus painted by an anonymous Italian artist in the 16th century.

There is no evidence that any portraits or sculptures of Columbus were ever executed during his lifetime. This copy of an oil painting in the possession of SCALA/Art Resources, N.Y., seems to fit the descriptions of the great discoverer provided by his son Ferdinand and the best of the other contemporary chroniclers, Bartolome de las Casas. The sober, sad eyed, ascetic, but determined Columbus shown here is the favorite of many pro-Columbian scholars. This print in based on a color portrait in the possession of SCALA/Art Resource, New York, which has granted permission for its use in this book.

TABLE OF CONTENTS

APPENDICES

Table of Contents (continued)

LIST OF ILLUSTRATIONS

Chapter 2

Chapter 3

Chapter 4

Chapter 5

Chapter 6

Chapter 7

Chapter 8

Appendix A

Appendix B

LIST OF TABLES

Chapter 1

Chapter 2

Chapter 3

Chapter 4

PREFACE

Christopher Columbus, Cosmographer is a critique of the great navigator's geographic and geodetic views of the earth formed during the period 1472-1484 when he gained his initial experience as a seaman and navigator, and enunciated with increasing conviction in the period 1484-1492 when he was trying to obtain authorization and financial backing for his "project of the Indies" - to sail westward from Europe to the Orient. These "cosmographic" views, which Columbus held tenaciously through four voyages of discovery and exploration to the New World and up to the time of his death in 1506, have been ridiculed by many negatively motivated writers (good news is no news; bad news pays!) These views seem, also, to have embarassed some of Columbus' staunchest supporters.

That Columbus grossly underestimated the distance westward from Iberia to the Asian continent is not contested. The point at issue is the oft-repeated charge that his geodetic and geographic views were revolutionary, without solid foundation, and unshared by noted cosmographers and geographers of his time. This is incorrect as the text will show.

Estimates of the westward distance from Europe's west coast to the Asian coast, before Columbus' voyages revealed the existence of the American continent and the Magellan-El Cano circumnavigation of the earth (1519-1522) revealed the enormous expanse of the Pacific Ocean and the great difficulty of gaining access to it by rounding the southern tip of the American continent, were based on the size of the earth and the length of the Eurasian continent. Subtracting the length of the Eurasian continent at some latitude from the circumference of a parallel circle at the same latitude yielded the distance westward along the same parallel circle from Europe to Asia. For a variety of reasons it was quite a while after the Columbian and Magellan voyages before European geographers and cartographers began to show the airline distance from Europe westward to the Orient significantly different from Columbus' own view.

Over the millenia, people of many disciplines and in many places contributed to the sum total of what was said and became known about the size and shape of the earth, its physical features, and, in particular, the distribution of its lands and seas. Not all of it was correct, nor was the correct information always accepted in preference to the incorrect. A further complication resulted from the proliferation of measurement systems. These tended to vary from time to time and from place to place so that the correct interpretation of a measurement, a distance, or a dimension became increasingly precarious as the interpreter and the source were separated by time, language, and political barriers.

Precisely what did Eratosthenes, in the 3rd century B.C. mean when he set a sexagesimal of a meridian at 4200 stades and how did he determine this? (See pp. 78-81, incl.) Hipparchus (2nd century B.C.) made the degree 700 stades and Ptolemy (2nd century A.D.) gave it a length of 500 stadia. Strabo, at the dawn of the Christian era, said 8 Greek stades made the Roman mile, but was this always the case? Alfragan, reporting on the Arab measurement of a degree, in 830 A.D., said it had a length of 56 2/3 Arabian miles. How was the Arabian mile related to the Roman miles and the various stades?

Because of the problems introduced by various systems of linear measurement used by writers over the millenia to express geodetic measurements, Chapter 2 is devoted to *Ancient Linear Metrology* so as to provide a basis for understanding and relating their views, delineated in the remainder of the book.

Strangely, some of the very earliest geodetic measurements and estimates were the most nearly accurate. What events and which personalities "conspired," over the millenia, to create a scenario by Columbus' time in which the earth's circumference was generally considered to be some 20% less than its actual length and the Eurasian continent almost 75% too long? And what were Columbus' reasons for believing the earth was even smaller and the Eurasian continent even larger? Of equal importance, what specific events, in the wake of Columbus' discovery of the New World and stretching over a period of almost 200 years after Columbus, finally led to more realistic views of both the size of the earth and the distribution of its lands and seas?

Christopher Columbus, Cosmographer examines the views of the great philosophers, astronomers, geographers, and writers of antiquity and the middle ages through to the Columbian era and beyond. Every known significant attempt through the mid-16th century to determine the size of the earth - including Columbus' own effort - is described, usually illustrated, and the results interpreted. Tables are employed to explain the many different concepts as to the size of the earth and the distribution of its lands and seas. Maps illustrate the varying concepts of the earth over a period of almost 2000 years; eighteen of these demonstrate how these concepts changed during the period 1490 to 1595, in the wake of the wave of global exploration given impetus by Columbus' four voyages of exploration and discovery to the New World.

The appendices permit a treatment in greater depth of Columbus' four voyages of exploration and the contemporary voyages of other explorers and the almost unbelievable variation in the length of one of the most widely employed units of itinerary measurement during Columbus' time and for hundreds of years beyond - the league.

The book profiles, from a cosmographic standpoint, many of the great characters of recorded history from Aristotle through Erastosthenes, Posidonius, Strabo, Pliny, Ptolemy, Macrobius, Alfragan, Bacon, Marco Polo, d'Ailly, and Toscanelli to Columbus. Then it parades before the reader a succession of 16th century world maps produced by the great cosmographers, geographers and cartographers of that century- - such men as Waldseemuller, Ruysch, Apianus, Finaeus, Gerardus Mercator, Ortelius, Rumold Mercator, and Barentszoon. The maps are augmented by tables which interpret the cosmographic views of such 15th-16th century authorities as Ferrer, Enciso, Duran, Juan Vespuccius, Sebastian Cabot, and Diego Riberio.

Far from depicting Columbus as revolutionary in his cosmographic concepts and without support among his contemporaries, it shows him to be a man of his times and aggressive in the promotion of those ideas of illustrious predecessors and contemporaries which supported his proposal to sail westward from Europe to the Orient. Going beyond that, the book shows conclusively that the cosmographic views of Columbus and his contemporaries were not easily abandoned even after the wave of exploration which characterized the "Age of Discovery." Not till the 17th century developments of the telescope by Galileo, the pendulum clock by Huygens, and the publication of Cassini's ephemerides of the satellites of Jupiter, did the accurate determination of longitude on land become possible. And it would take still longer before Harrison developed a reliable, spring wound, chronometer so that the determination of longitude at sea finally became a practicality.

While the book is more concerned with Columbus' ideas, their origins, and the degree to which they were accepted by authorities of this time, the reader cannot help but be impressed by the acute sensitivity of the man to the wonders of nature all around him. Alexander Von Humboldt in his classic *Cosmos* (Potsdam, 1844) dwells upon "the penetration and acuteness with which . . . he could sieze and combine the phenomena of the external world . . ."

My own effort to correct the distorted characterizations of this truly great man would have been impossible were it not for the efforts of the many writers - mostly of the 19th century and before - whom I have credited in my copious reference notes. Closer to home, I am indebed to the following, as indicated: to Dr. David Dearborn, formerly Director, Steward Observatory, Univ. of Arizona, for review and suggestions for improvement of those phases of the book involving astronomy and celestial mechanics; to Brother (Dr.) Nicholas Sullivan, Prof. of Geology, Manhattan College, for overall review of accuracy and fluidity of expression; to Professor Emeritus Benjamin Keen, translator and annotator of *The Life of the Admiral Christopher Columbus by His Son Ferdinand* ", Rutgers Univ. Press, New Brunswick, N.J., 1959, for his excellent work and for permission to reproduce for publication 9 illustrations therefrom; to Bjorn Landstrom, author and illustrator (supreme) of *Columbus, the Story of Don Cristobal Colon, Admiral of the Ocean Sea . . .*", MacMillan Co., New York, 1966, for permission to reproduce 4 of his own illustrations from that superb work; and to the late Professor (and Rear Admiral, USNR) Samuel Eliot Morison for his incomparably detailed biographical coverage of the Great Navigator and his *Project of the Indies* (see Index). Indeed, it was Morison's rather extravagant criticism of Columbus' cosmography - a criticism this writer could not fully accept - which sparked the long search to achieve more rational explanations of Columbus' ideas and their origins.

Deepest gratitude must go to Dover Publications, New York, for permission to reproduce from A. E. Nordenskiöld's *Facsimile Atlas to the Early History of Cartography with Reproductions of the Most Important Maps Printed in the XV and XVI Centuries* no less than 20 maps drawn (or inspired) between c. 423 and 1595 A.D. These maps are of critical importance in proving the validity of the basic thrust of *Christopher Columbus, Cosmographer.*

Chapter 1

CHRISTOPHER COLUMBUS, THE MAN, THE NAVIGATOR, AND THE COSMOGRAPHER

INTRODUCTION

Von Humboldt is credited[1] with the cynical remark, "There are three stages in the popular attitude toward a great discovery: first, men doubt its existence, next they deny its importance, and finally they give the credit to someone else." He had Christopher Columbus' discovery of America in mind when the remark was made. And for a fact the lands Columbus found and explored in four voyages over a period of twelve years were named for another whose contacts with the New World were later and immeasurably shallower.

Columbus' rewards were quite substantial, both during his lifetime and to his heirs, although disappointing to him (and to us, in retrospect, when we consider the immensity of his achievement). However, the penalties of greatness were heart-breaking. The barbs started to fly early and they have never stopped. Bjorn Landstrom[2] succinctly summarizes some of the charges which have been leveled against Columbus by "mudslinging historians":

> . . . a charlatan, a foreign immigrant, a Jew, an imposter who stole other men's ideas, a canting humbug, a hopelessly incompetent navigator, a pious fraud, a blood-sucking murderer, a man who had no idea where he touched land, and worst of all, a man who did not even discover America.

Landstrom then balances the ledger by adding:

> Other people have been anxious to see him as their countryman, as the greatest mariner and navigator in history, as the true Bearer of Christ, worthy enough to be numbered among the saints."

Up to about the 400th anniversary of Christopher Columbus' landing on San Salvador (arguably Watling Island) in the Bahamas, his initial landfall in America, no-one doubted that he was looking for some part of "The Indies[3]" - Japan or China, or both. "La Empresa de las Indias", as Columbus himself called his project in later years, was simply to reach Asia by sailing westward. He expected to find one or more islands on the way and gold, pearls, and spices when he reached Asia. However, he had no suspicion that he would encounter the continent we now call America.

Even after four voyages of exploration and discovery which covered a period of twelve years and resulted in the discovery of the Bahamas, all of the Greater Antilles, most of the Lesser Antilles and South America, and the exploration of most of Central America, he still thought he had reached parts of Mangi (southern China) and certain offshore islands hitherto unknown to Europeans. Why was he so obstinate when, seemingly, everyone else recognized that the places he had found represented a new continent? Because these places were just about where he had determined beforehand the Asian coast and offshore islands should be.

Bartolome de Las Casas' *Historia de las Indias*, perhaps the most important single work ever written on the discovery of America, was started in Hispaniola about 1527, but most of the work on this lengthy tome appears to have been done between 1550 and 1563, upon the author's return to Spain. Still the document was not printed until 1875. Why it was not printed before is open to conjecture. That it was finally printed and thus made readily available to an army of scholars all over the world appears to have been influenced by the fever which developed in anticipation of the 400th anniversary of the discovery of the New World.

In 1871, Henry Harrisse, a noted French scholar and prolific author on the Columbian and contemporary voyages of discovery to the New World, discovered in the renowned Seville Library in the collection of books known as the Columbina[4] the Latin text of a famous letter previously believed to have existed only in Spanish and Italian versions. In fact, it was because of Harrisse's familiarity with the Spanish version contained in Las Casas' *Historia . . .* that he recognized the letter, written in Latin, as that sent from Florence, dated 25 June 1474, by the great physician/astronomer/cosmographer Paolo dal Pozzo Toscanelli to a canon of the Cathedral of Lisbon named Fernao Martin. Harrisse subsequently published the Latin text.

Martin, a confidante of the then reigning Portuguese monarch, Alonso V, had been importuned by his king to obtain a written expression of Toscanelli's views as regards the feasibility of a sea voyage westward from Lisbon to the Indies. The cosmographer's reply had been quick and positive.

Toscanelli was one prominent cosmographer of his day who believed two of the basic geographic theses in Marco Polo's book, that Asia extended considerably farther eastward than the geographers had indicated and that there was a fabulously rich land called Chipangu which lay some 1500 miles off the coast of Cathay. Chipangu, also translated as Cipanju, Zipangu, and several other variations, is, of course, Japan. Cathay is North China. Toscanelli believed that the shortest way from Spain or Portugal to the east coast of Mangi (South China) or Cathay was by sailing due westward. Stops at Antilia and Chipanju would reduce the length of each leg of the journey to that which was feasible for the sailing ships of the mid-fifteenth century. Antilia was a mythical or grossly misplaced island which had managed to find its way onto many of the Atlantic Ocean portolanos (sailing charts) of that period.

Toscanelli's letter to Martin appended a chart which, he said, would make clearer the route to be followed from Portugal, the places that ships could stop for refitting and re-provisioning, and the distances involved in each leg of the journey. Besides dwelling upon the wealth of the countries in the East, the great commercial and maritime activities of the coastal cities, and their enormous size, Toscanelli mentions with praise the Great Kahn, or King of Kings, who rules the entire region and whose residence is chiefly in the province of Cathay.

The most important part of the letter is that pertaining to the distances involved in going from Portugal to the China coast. From Lisbon to Quinsay in the province of Mangi, there are, going directly west, 26 spaces marked on the map, each space comprising 250 miles, or 6500 miles in all, making about one-third the circumference of the globe. But, as from the isle of Antilia to the famous isle of Cipangu, there are only 10 spaces (2500 miles).

Sometime in the period 1480-81, Columbus heard of the existence of the Toscanelli-Martin letter and, since the gist of what he heard about it fitted in with his own ideas, wrote Toscanelli to ask his advice. It is then he is supposed to have received from Toscanelli a copy of the cosmographer's letter and map of 1474 to Martin. We say "supposed to have received" because Columbus, himself, never mentioned the incident in any of his writings or correspondence. The correspondence was first mentioned, and a Spanish version provided, in Las Casas' *Historia de las Indias,* and Las Casas does not say how he came by the letters. Columbus wrote several more letters to Toscanelli and received a second response from the cosmographer. Neither of the two letters from Toscanelli to Columbus were dated or contained any geographical information. Nor have they survived. The letters dwelt only on the worthiness of Columbus' desired undertaking and the wealth of the lands he would find. The only geographical information was that contained in the letter to Martin, a copy of which had been provided Columbus in Toscanelli's first response.

In 1571, there was published at Venice, in Italian, a life of Columbus, attributed to his son Ferdinand, translated from a Spanish manuscript. This work contained a version of the Toscanelli letters to Martin and Columbus and a statement to the effect that the first response[5] had been written in Latin. For three centuries the Toscanelli letters were known, essentially, by this Italian version, although scholars like Harrisse had access to and were familiar with Las Casas' manuscript containing the Spanish version. A few years after Harrisse published the Latin text of the Toscanelli letter, the entire text of Las Casas' manuscript, until then unpublished, was made available to the public.

As Ferdinand himself puts it, the Toscanelli letters ". . . played a large part in encouraging him (Christopher Columbus) to undertake his voyage."

Following the Harrisse publication of the Latin text of Toscanelli's letter to Martin, he became embroiled in a controversy as to who really had made the discovery of the Latin text, he or the custodian of the Columbina at the Seville Library, one Don Jose Fernandez y Velasco. The fuss that was stirred up attracted the attention of Henry Vignaud who thought there was something apocryphal about the Toscanelli-Martin-Columbus correspondence and began to research the matter furiously.

In the meanwhile, Harrisse, published a series of books and articles dealing with Columbiana, one titled *Colomb et Toscanelli*, Paris, 1893, and another *Encore La Bibliotheque Colombine,* Paris, 1897. A keen critic, Harrisse, nevertheless, was not as suspicious of the letters we have described as Vignaud. The latter, sometime around the turn of the 20th century, published, in French, a book highly critical of Columbus, his brother Bartolomeo, his son Ferdinand, and the historian Las Casas. This was followed by an edition in English published in 1902 which took note of the furor created by the French edition and sought to counter at least the most serious criticisms.

Among the charges Vignaud levelled at Columbus were " . . . (he) . . . never spoke one word of truth on what related to himself personally . . . Throughout his letters and writings he has sprinkled incorrect statements, skillfully devised, with the object either of obscuring certain portions of his life or of hiding traces of his origin." Vignaud claims Columbus lied about his age, claiming to have been born in 1436, not 1451, so as to make more credible a long list of fictitious accomplishments, helpful in selling his "project" and himself as its leader.

Specifically,

> . . . there was no famous Admiral in his family . . . , . . . he was never at the University of Pavia. The only education he ever got was such as he obtained from the schools founded and maintained by the Genoese weavers and what he was able to procure for himself. He never commanded a galley for King Rene, nor fought in a campaign for that monarch, as he claimed to have done . . . At the time in question, Columbus was only nine years old. He had neither overrun the seas as he boasted to have done, nor sailed for forty years, as he wrote in 1501 to the Catholic kings. Forty years earlier than 1501 bring us to date 1461, . . . when Columbus was 10 years old.

Vignaud disputes other historians as to the date Columbus first arrived in Portugal, claiming it was not 1470 as Las Casas, Ferdinand Columbus, and later, Humboldt, Washington Irving, Fiske and others had indicated, but " . . . 1476, when he was twenty-five years of age." And, on the subject of why Columbus left Portugal, "It was untrue that . . . (he). . . left Portugal secretly because King Jaoa II wished to rob him of his secret, as Las Casas and Ferdinand Columbus allege. He fled the country because through family connections he belonged to the Braganza party which King Jaoa pursued with hatred, and because, therefore, his life was no longer safe in Portugal."

Vignaud brands as untrue the Columbus-inspired stories that he had been invited by King Jaoa to return to Portugal; or that he had made proposals to Genoa, England, and France. More important is Viganud's statement that " . . . we must consider as absolutely disproved the assertion that he (Columbus) had long prepared himself for his great discovery by the study of those authors who could enlighten him on this subject." It was Vignaud's contention that Columbus' vocation" . . . dates only from the confidence made him by the pilot who by accident had discovered . . . unknown islands or lands, and who had convinced Columbus of the reality of his discovery."

Further, " when he (Columbus) sailed from Palos in 1492, (he) had no intention of opening a new route to the Indies . . His sole object was to go whither the pilot had been . . ." And, says Vignaud, " . . . he does not appear to have proposed anything more to the Catholic kings, the text of their capitulations mentioning neither the Indies nor the land of spices, or of any country to the East, but solely, . . . islands and lands he may discover in the ocean . . ."

The most shocking charge that Vignaud made concerned the Toscanelli-Martin-Columbus letters. He said they were frauds, pure and simple! There never had been a canon, Fernam Martin, therefore there could not have been any letter sent by Toscanelli to him. Since there had been no Toscanelli-Martin correspondence, a copy could not have been sent to Columbus. Nor had there been any "second letter" sent by Toscanelli to Columbus, or any contact at all between the two! The chart which had been appended to the Martin letter, and which Las Casas avers he had seen with his own eyes, had been drawn by Columbus himself, for it had incorporated all of the explorer's cosmographical ideas. Well then, did Columbus also write the fraudulent letters?

Now here the reader detects Vignaud wrestling with himself. The thinly disguised animus toward Columbus which pervades his coverage of the man's life and accomplishments would seem to impel him towards such a conclusion. After all, had not Harrisse concluded that the Latin text of the Toscanelli-Martin letter which he had discovered in the Colombina had been in Columbus' own handwriting?

Surprisingly, and it seems almost reluctantly, Vignaud comes to another conclusion. "The forger was, to all appearance, Bartholomew Columbus, who was a good cosmographer, but a bad Latin scholar; he was also very devoted to his brother (Christopher). The copy . . . (of the Latin text of the Toscanelli-Martin letter found by Harrisse). . . is in a hand as much resembling his as it does his brother's."

"The invention of the story of Columbus' correspondence with Toscanelli dates probably from Bartholomew's arrival in Spain, the period when it was rumored his brother had been instructed by a pilot; but the documents themselves were only written later . . ."

"These documents, composed on the opinions of Columbus, must have been fabricated after the discoverer's death . . . They were first produced between 1547 and 1552, the time when Las Casas, who is the first to record them, was revising his book, and was placed in possession of all the Columbus family papers. It was also the time when the story of the pilot who had instructed Columbus was revived by the publications of Oviedo and Gomara."

"The person who gave them to Las Casas can only have been the same as placed him in possession of the Columbus family papers. At the period mentioned, Luis Colon, the Third Admiral of the Indies, was sole proprietor of these papers. He was a reckless and unscrupulous person . . . he alone could then dispose of . . . (the papers)."

" . . . These letters and map attributed to Toscanelli . . . which no one ever used, and which no one ever knew except he who produced them seventy years after their inscribed date, have never served any other purpose than to create the impression that Columbus had a scientific idea, and that it was this idea which led him to his great discovery."

As to a motive for his complicated hoax, Vignaud points out that the stories about the pilot " . . . did harm to the heirs and successors of Columbus, to whom the Crown refused to continue the honours and extraordinary privileges granted to the great discoverer, contesting, if not indeed the discoveries, at least the importance of the personal part he had played therein." The Columbus family contested this "Indian giving" on the part of the crown and the correspondence between Columbus and Toscanelli--showing that Columbus had the basic idea for his great discovery at least before 1481, the year Toscanelli died, and that the great cosmographer approved of it--was, probably helpful during this litigation. Thus, the letter written before Bartholomew's death in 1514 had the desired effect in protecting ,to a degree, the substantial Columbus family interests in the New World.

Vignaud achieved some distinction as a historian even if his findings in *La Lettre et la Carte de Toscanelli* (1901) and *Toscanelli and Columbus* (1902) have not been generally accepted. After the two books cited, he brought out three volumes on Columbus and one on Amerigo Vespucci, and a year before his death at the age of 92, he summarized his views on Columbus in a small book, *La vrai Christophe Colomb et la Legende* (1921)[6]

In the rest of this chapter, the charges touched on by Landstrom and particularized by Vignaud will be responded to, as a matter 500 years old is susceptible to response. It is believed best to segregate this treatment into three parts, "Christopher Columbus, the Man", " . . . the Navigator", and " . . . the Cosmographer."

CHRISTOPHER COLUMBUS, THE MAN

Origin

Morison, on the subject of Columbus' origin, says "There is no mystery about . . . (his) birth, family, or race. He was born in the ancient city of Genoa sometime between August 25 and the end of October, 1451, the son and grandson of woolen weavers who had been living in various towns of the Genoese Republic for at least three generations. As there was a good deal of moving about along the shores of the Mediterranean in the Middle Ages, some of the Discoverer's remote ancestors doubtless belonged to other races than the Italian. His long face, tall stature, ruddy complexion and red hair suggest a considerable share of 'barbarian' rather than 'Latin blood, but do not prove anything; and he himself was conscious only of Genoese origin."[7]

It was far from rare, during the period of Columbus' four voyages for the appellation of "Jew" to be leveled at a person if it was desired to cause him trouble. This was the period of the Inquisition when flimsy charges of heresy and religious insincerity were hurled at Christianized Jews and non-Jews alike. It will be recalled that Columbus' departure from Palos on his First Voyage had been delayed until the 3rd of August, 1492, because of the difficulty in finding the three ships he needed, most Spanish ships having been assigned to take the 250,000 Jews (who refused conversion) to Moslem North Africa and, to a much lesser degree, Christian European havens.

During the over 800 years of gradually declining Islamic domination of the Iberian peninsula, which ended with the capitulation of Granada in January, 1492, the Jews had been exceptionally well-treated by the Moslem overlords. Many had attained high positions in the government and in the universities. Hence, the reflexive feeling of distrust for them among many in Christian Castile and Aragon when the Islamic yoke was finally cast off. And, since the official policy of the Crown was "convert or leave," Jew and Moor alike were considered little better than potential political enemies.

Whether Columbus was descended from an ancient Phoenician Jewish family, as some have claimed, is not known. After all, all of the first Christians were Jews, before other Palestinians, Syrians, Phoenicians, and Greeks of Europe, Asia Minor, and Africa, and still later Romans and the peoples they had conquered. Las Casas has said:

> In matters of the Christian religion, . . . he was a Catholic and of great devotion . . . He observed the fasts of the Church most faithfully, confessed and made communion often, read the canonical offices like a churchman . . . , hated blasphemy and profane swearing, was most devoted to Our Lady and to the seraphic father St. Francis; (and) seemed very grateful to God for benefits received . . . He was extraordinarily zealous for the divine service; he desired and was eager for the conversion of. . . (the Indians), and that in every region the faith of Jesus Christ be planted and enhanced . . .

Education

Ferdinand Columbus, one of the prime biographers of the great discoverer, is responsible for the claim that his father had attended the University of Pavia. This appears to have been disproved in that the meticulous and well-preserved matriculation records of that ancient institution fail to support this claim. Still, it is not inconceivable that the young Columbus, of meager financial resources, may have attended the university on a part-time basis, or for a limited period, without having this event indelibly recorded for all time.

There is no question, however, that Columbus could read and write in Spanish, Portuguese, Latin, and Italian - the last oddly enough the poorest of all. As Morison points out, this was not at all unusual among middle class Genoese of Columbus' day who spoke a dialect very different from Tuscan or classical Italian. It was a language of common speech that was never written. He may have learned some Italian and Latin in the weaver's schools in Genoa. During the years he was in Portugal, he learned to speak and read Portuguese, but Castilian appears to have been the first language he learned to write proficiently, this with a Portuguese slant.

Columbus achieved a good, basic understanding of astronomy, geography and the mathematics incidental to both, as well as to navigation at sea. And, he was a journeyman cartographer. He was well read, particularly of those in Scripture and ancient and medieval history and geography who had offered ideas parallel to, or supportive of, his own, over-riding cosmographic thesis - that the shortest way to the Indies was by going west and that this was well within the capabilities of a well planned, outfitted, and directed expedition.

Without question, Columbus was self-educated. But in his deductive methods of reasoning, his lively curiosity, his accurate and unusually sensitive observation of natural phenomena, and his ambition to win wealth and recognition, he achieved a level of knowledge possessed by very few more formally educated of his contemporaries. After all, he was certainly put to the test in both Portugal and Spain when the reigning monarchs in each country had him argue his theories with the most learned cosmographers of the day. If he could not convince them, it is also clear that they could not shake him from his own convictions. In any event - and this point appears to have been forgotten, overlooked, or purposely shelved - any assessment of Columbus' ultimate intellectual achievements must be made against the general level of knowledge in his day - not ours.

Stealing Other Men's Ideas - The Story of The Pilot

In the earliest days after the establishment of the first, ill-fated Spanish settlement at Navidad, on the north coast of Haiti, soon after Christmas 1492, members of Columbus' expedition began to voice a rumor which seemed to grow in detail and adherents with the retelling and the degree to which Columbus found it necessary to impose discipline and work assignments among them. The rumor was simply that an obscure pilot, some say his name was Alberto or Francisco Sanchez, others that his name is not known; hailing from Huelva, Audalusia, Portugal, or the Basque country, or again, from parts unknown, had told Columbus precisely the course to follow to reach an island unmarked on any existing chart, far to the west, where the pilot had seen strange looking inhabitants running around naked.

As the story goes, sometime in 1483 or 1484, the pilot and a crew of seventeen had sailed from a Spanish or Portuguese port bound for England and Flanders. An easterly gale which developed soon after departure, and which persisted for 28 or 29 days, drove the ship and its complement far off course. When the storm abated the pilot and crew beached their battered craft on a nearby island and performed emergency repairs, while determining the latitude of the island and guessing at its longitude. Then they refloated their ship, re-provisioned it as best they could, and set sail eastward as the winds would allow.

The journey home was long and painful. With the provisions exhausted, the greater part of the crew fell sick and died before the ship at last fetched Madeira. Columbus, then a resident there, gave them refuge, caring for the pilot in his own home. However, worn out by their privations and the labor of their return, the ship's complement, including the pilot, soon died. But, the "secret" of where they had been and what they had encountered, did not die with them. Touched by Columbus' kindness toward him, the pilot had Columbus record all of his observations as to the course to be followed and the wind directions likely to be encountered in reaching the island. Some chroniclers say this island was Hispaniola, others that it was one of the Lesser Antilles, while the majority say no one knows for sure.

The earliest chroniclers of the story of the pilot were Oviedo, Las Casas, Gomara, and Garcilaso.[8] Oviedo relates the story as he heard it and then says flatly he doubts its authenticity. Las Casas indicates he heard the story from several of Columbus' companions of his first voyage and that he believes it. He doubts, however, that it played much of a role in persuading Columbus because there were so many more important incidents in Columbus' life, all well documented, which the great explorer himself credits. Gomara tells the story about the same way as Oviedo and Las Casas and goes on to say he not only believes it but the credit for the discovery of the "New World" belongs to the pilot, not to Columbus. Garcilaso, known as the Inca (for he was descended from that Peruvian dynasty) and also known for embellishing a good story, tends to agree with Gomara.

Over the years, since the time of the foregoing, there have been many who have retold the story. Suffice it to say that most serious biographers of Columbus such as Washington Irving, Henry Harrisse, and Samuel Eliot Morison consider the story not improbable, but doubt that it had any significant effect on Columbus' subsequent actions, siding with Las Casas' assessment.

A Murderer ?

The charge of murderer can only refer to the treatment accorded the natives of Hispaniola by the Spaniards under Columbus and his successors. Historians have accorded Columbus, the first administrator for the Spanish crown, the brunt of the blame for a policy which resulted in the practical extinction of the native population.

The second voyage of Columbus marked the beginning of European empire-building in the Americas. The seventeen ship armada of 1493 carried a thousand colonists and the supplies and materials necessary for their support. Tools and implements for the mining of gold and arms and horses for the "reduction" of the island permitted prompt initiation of both activities. The following years brought more colonists and artisans. Seeds and plants and livestock were introduced and as "reduction" proceeded towns began to spring up. The groundwork for the erection of a civilization had been laid.

As Bannon[9] puts it:

The early years on Española were hectic and troubled. The lust for gold and quickly won wealth, the restless character of the first colonists, the lack of administrative competence in the Columbus brothers (Christopher and Bartolomeo) led to disorder, rebellion, and practical anarchy.

Neither the gentle Arawak nor the revolting Carib could command the Spaniards' respect. Indian customs and habits and morals seemed to offer nothing for the credit side of the ledger. The early impulse was to range the native with the higher animals and treat him accordingly. The prospect of his enslavement awakened no scruples in the hearts of the Spanish conquistador, and no compunction. Columbus, as early as 1494, on his return to Española, had proposed the enslavement of the Caribs as a source of wealth to compensate for the disappointing yield of gold.

As a result of an Indian uprising in 1495 Columbus condemned the natives to pay tribute, which, when they were unable to meet the terms in gold, was commuted into personal services. From that point the Spanish had gone on to make more drastic demands.

Following the Indian uprising of 1495 during which several Spaniards were killed, Columbus mounted a punitive expedition which killed some but captured many of the Arawaks involved. Five-hundred of "the best males and females" from among the captives were loaded on a ship and sent to Spain. Many did not survive the voyage. Those that did were sold on the slave market at Seville.

The Spaniard who came to the New World, regardless of previous position, considered himself an hidalgo, a gentleman, upon arrival. According to the standards of the time, work was beneath the dignity of a gentleman. Fighting and governing were his professions. The growing of food, the building of towns and roads, the mining of gold, these fell to the lot of the Indians. While the Indians were accustomed to the sub-tropical climate, they were not used to heavy labor and the combination of the two took a heavy toll among them.

There is some evidence that syphilis was extant in Europe before it was spread by infected Spanish conquistadores and seamen returning from the New World. It is, however, certain that many common diseases of the white man, against which he had built up, over centuries, a partial immunity, exacted a frightful toll among the Indians.

By no means did all this happen during the direct administration of Española by the Columbus brothers. The administration of Francisco de Bobadilla, from 1499 to 1501, was even worse than that of his predecessors. Nicolas de Ovando, who succeeded Bobadilla is credited with putting the colony on a sound footing administratively, but he, too, and his successors were faced with essentially the same Indian problems, revolt and a disinclination to work for the white man. Outright slavery, "repartimiento"[10], or "encomienda"[11], were all uniformly despised by the Indians and resisted. Indian mothers killed their young, lest they grow up to know hateful Spanish slavery. Many Indians committed suicide rather than submit to it.

For all of the foregoing reasons, the Indian population of Española, variously estimated as somewhere between 300,000 and one million at the time of Columbus' discovery, had been reduced to 60,000 by 1508 and 14,000 by 1514. In 1548 Oviedo doubted whether 500 Indians remained. Columbus has received the greatest share of the blame for this first example of practical genocide in the New World, and for the introduction of the more docile, but hardier, black African slaves to do the work the Indians couldn't and the Spaniards wouldn't.

CHRISTOPHER COLUMBUS, THE NAVIGATOR

Nautical Experience and Competence

That Columbus exaggerated his nautical experience both in quality and quantity, especially (but not exclusively) during the period he was trying to sell his "project" to the sovereigns of Portugal, Castile, and Aragon, appears well established. But that he became a good navigator and highly intuitive sailor during the period of his four voyages of exploration and discovery appears just as certain. In attempting to justify this conclusion, it is desirable to split the review into two periods: that preceding his first voyage of exploration and discovery, which may be considered the preparation phase, and that which encompassed his four voyages to the New World.

The Preparation Phase, c. 1472 to 3 August 1492

Not a great deal is known about Columbus' nautical experience prior to his first voyage of exploration and discovery of the New World. What is known would indicate that it was not very extensive.

He did not go to sea at an early age as he represented to Isabella and Ferdinand during the period he was trying to sell his "Project of the Indies" and even later, but evidently made his first voyage from Genoa to Marseilles and Tunis in his very early twenties, and at least one to Chios[12] and the Levant in 1474, when he was 23. The latter was an important one for the young Columbus for it initiated an association with the powerful trading and banking houses in Genoa which was to last a lifetime. His employer on the voyage to Chios was the firm of Spinola and Di Negri which had considerable interests in Portugal.

Following the voyage to Chios, the ships refitted at Noli[13] and set sail for European trading ports on the Atlantic. However, off Cape St. Vincent, the convoy of which Columbus' ship,[14] the Bechalla, was a part, was attacked by the French corsair de Cazanove (also known as Colombo, but no relative of Columbus'). Some Genoese ships, including the Bechalla were sunk. While accounts differ, it would appear that Columbus and some of his shipmates swam ashore - a good trick, in itself - and were given refuge by members of the Genoese colonies at Cape St. Vincent and later Lisbon.[15]

Not long thereafter, the firm of Spinola and Di Negri dispatched another fleet with orders to pick up the survivors and proceed to England. The port of call was evidently Bristol, a major English port, particularly for trading with Ireland and Iceland. Columbus claimed later that he had gone on to Galway, Ireland, and, in 1477, to Thule (Iceland), in fact sailing some 100 leagues beyond. This claim has been discounted by some Columbian historians. However, it is clear that he made at least one round trip between Genoa and Lisbon, a number of voyages between Lisbon and the Madeiras, and probably two round trips to West Africa.[16] It was on one of these trips that he claims to have checked Alfragan's determination of the length of a degree of arc along a meridian of the earth's surface - enraging many scholarly Columbophobes in the latter half of the 19th century and to this day.

In 1479, while resident in Lisbon, Columbus courted and married Felipa Perestrello e Moniz, the daughter of the late Captain-General of Porto Santo, one of the Madeira group. The family had important connections, for Perestrello, in his youth, had been a squire of Prince Henry (the Navigator). Perhaps as important, the Captain-General had accumulated an imposing library containing works by the most highly respected cosmographers, geographers, and astronomers. His widow, impressed by her son-in-law's ambition and zeal to learn, allowed him access to the library. No doubt, among the works in the library was Sacrobosco's *De Sphaera*[17] which had become one of the chief textbooks of 14th and 15th century Portuguese mariners. This work treated on the terrestrial globe, of circles great and small,[18] of the rising and setting of the stars, and of the orbits and movements of the planets.

Admiral A. Teixeira da Mota [19] reports that the employment of astronomy in navigation occurred first (among Europeans) in Portugal, midway through the 15th century. In due course there emerged techniques for the determination of latitudes by observing the North Star and, later, the sun. The development of the techniques involved, which took place over a long period of time, made it possible for many to learn piloting by practice at sea in the company of competent pilots. This appears to have been the main school of navigation for the Portuguese until the middle of the 16th century.

Admiral da Mota relates that ". . . Columbus had become familiar (in Portugal) with this phase (technique) of navigation by astronomy, but he failed to assimilate it properly and did not transmit it to the Spaniards who came into contact with him."

Whether or not he assimilated navigation by astronomy properly, Columbus made every effort to employ what he learned about the subject during his various trips with Portuguese pilots during the period 1479 to 1484. During the same period, Columbus came into frequent contact with a number of pilots and navigators who had ventured, or been blown, far to the west of the Portuguese island groups of the Azores, Madeiras, and the Cape Verde Islands. Thus, he learned what to expect in the way of prevailing winds (and ocean currents propelled by those winds) in the vicinity of the various islands as well as the Atlantic coasts of Portugal, Spain and West Africa.

On his voyages to Guinea, which occurred during the period 1482-1484, he was struck by the force of the easterlies coming off the African continent and figured there was at least a chance they might carry all the way across the Atlantic to the Indies. He had already become well aware of the persistent westerlies in the latitude of the Azores which made any attempt at a westward crossing in these latitudes highly precarious. The ease with which northeasterlies propelled ships from Lisbon, Cape St. Vincent, and the Spanish Atlantic ports of Cadiz southwestward towards the Madeiras and the Canaries where the easterlies became dominant was not lost on Columbus either.

On his several voyages, Columbus made it a point to observe the work of the pilots, to question them, and on occasion to carry out observations and participate in the dead-reckoning process.

The courses to be followed on a proposed voyage were laid down on a mariner's chart beforehand. Dead reckoning was the practice of trying to follow the courses laid out and the recording of the actual courses taken, as estimated. Each course had a direction, or bearing, and a length, or distance sailed on that course. The course was obtained from the mariner's compass which had been in use for at least three centuries before Columbus' time. Distance on a course was determined by estimating the speed with which the bubbles in a ship's wake receded and recording the time spent on each course. From the dead-reckoning data entered on his chart, a pilot was able to estimate the number of leagues or Roman miles an island, coastal promontory, or position at sea lay, and in what direction, from his starting (or other known) point. This data could be converted[20] to give the latitude and longitude of a new position and latitudes could be checked with astrolabe or quadrant. However, getting a reliable reading of altitude of a heavenly body from the deck of a small ship subject to six "degrees of freedom" [21] was extremely difficult, if not impossible. This was best done after arrival at an island, where the astrolabe or quadrant could be hung from the limb of a tree or from an improvised tripod.

At night, the altitude of the pole star had been assumed for centuries to give latitude directly and during the day the sun, at noon, provided an indication of latitude. To obtain precise latitude from the sun at noon, it was necessary to have a set of tables which would indicate what the declination of the sun would be at noon on a particular day. This was not yet available to Europeans although it had been to the Islamic world for centuries. However, everyone who had ever gone to sea was aware of the sun and the path it followed across the heavens during the course of the daylight hours as well as the manner in which this path changed during the course of the year and with change in latitude. Maps of Columbus' day based on Ptolemy (see Figure 5, Chapter 3) generally showed the number of hours of daylight on the longest day of the year, the summer solstice, for the various "klimata," or zones of latitude. Most experienced Portuguese pilots had developed, or acquired from other pilots, home-made tables or formulae for approximating the latitude of a ship's position from the altitude of the sun at noon on a particular day of the year. Until the advent of published tables of the sun's declination throughout the year (about 1509 in Portugal and at least ten years later in Spain) most attempts to determine latitude from the sun were crude approximations at best and highly inaccurate at worst.

When the heavens were clear and the seas not too violent, observations of Polaris were made at night to provide a check on the compass' north and to give latitude directly. However, although not known to 15th century European navigators, Polaris at this time was not fixed but significantly circumpolar. (This will be given further treatment in the pages ahead.)

Thus, as a practical matter, dead reckoning was the means by which the 15th century pilot navigated, although - for the reason given in the foregoing - we cannot go as far as Morison who claims that "celestial navigation formed no part of the professional pilot's or master's training in Columbus' day, or for long after his death" adding:

> ...it was practiced only by men of learning such as mathematicians, astrologers, and physicians, or by gentlemen of education like Antonio Pigafetta who accompanied Magellan, or D. Joao de Castro, who on India voyages in the 1530's and '40's had everyone down to the ship's caulker taking meridional altitudes of the sun. Mathematics was so little taught in common schools of that era and the existing ephemerides (compiled largely for astrologers) were so complicated, that even the best professional seaman could do nothing with them.

Nor was dead-reckoning free from error. Course was determined by a magnetic compass with a pivoted needle and no means of damping against the motion of a ship so that the compass needle did not always point to magnetic north. There is also variation in which each magnetic meridian stretching from one magnetic pole to the other follows an independent course, usually irregular, and sometimes highly so. In the open Atlantic, in the 15th century and for a long time thereafter, variation might be suspected but there was no means for making corrections for it.

The determination of distance followed on a particular course required an estimation of the average speed and an accurate record of the elapsed time. In a rough sea, attemping to gauge the rapidity with which bubbles in the wake of a ship recede, can be a frustrating operation for the wake disappears rapidly. As for time, the ampoletta, or half-hour sand glass, could be quite erratic in rough weather and if the ship's boy in the watch, whose duty it was to reverse it promptly, was sleepy, slack, or lost count, a reduced estimate of distance covered would result. Happily, this was rarely fatal, for it only served to counter the invariably inflated estimate of speed (and therefore distance) of which pilots and navigators over the millenia have been guilty.

Contributing to the precariousness of dead-reckoning was the problem of inaccurate steering and the consequent difficulty of assessing the mean course. The existence of currents and no means of detecting them added to the problems.

Navigation, ship-handling in shallow water and at sea, and the art of sailing - getting from point A to point B somehow, regardless of wind direction and force, currents, weather, season of the year, or time of day - were the arts which Columbus picked up during the years from 1472 to 1484 as a passenger and, at best, part-time crew member. Also during this period, in between his voyages on behalf of his Genoese employers, he is known to have been employed as a cartographer, a field in which his younger brother, Bartholomeo, seems to have preceded him. Producing "portolani," sea charts, undoubtedly was a great training ground for the budding cosmographer as well as the aspiring mariner and explorer.

In the period 1480-1481, Felipa bore Columbus a son Diego on the island of Porto Santo, where they had been resident, and died shortly thereafter. In the same period, Columbus corresponded with the great Florentine physician/astronomer/cosmographer Paola Toscanelli and received encouragement as regards the feasibility of a westward voyage to the Indies.

Following Columbus' voyages to Guinea, which appear to have taken place in the period 1482-1484, he presented his plan for a westward voyage from Lisbon to the Orient to King John II of Portugal. The king was interested and referred the project to a Royal Commission for review. That body ruled Columbus' plan infeasible, far less likely to succeed than Portugal's own effort to reach India by rounding Africa. Columbus went to Spain with his son, Diego.

In the period 1484-1485, Columbus was a member of the household of the Duke of Medina Celi from whom he attempted to obtain financial backing for his plan. In early 1486, Columbus journeyed to Cordova and in May of that year presented his plan to Ferdinand and Isabela. The reigning monarchs of Aragon and Castile were then deeply involved in a war with Granada, attempting to overturn the last Islamic stronghold in the Iberias. A commission appointed to investigate the project, like the Portuguese commission before it, ruled the project infeasible (essentially because the distances involved in a westward voyage to the Orient were far too great for the ships of that day). However, Columbus, convinced of the soundness of his approach, persisted in his campaign to win backers wherever he could find them.

In the latter part of 1488, Columbus journeyed to Lisbon to again present his plan for a westward voyage to the Indies to John II of Portugal. In December of that year, before he was able to gain John II's approval, Bartolomeu Dias returned to Lisbon having successfully rounded the Cape of Good Hope. The way was now clear for the next leg of Portugal's southern route to India, up the east coast of Africa and across the Arabian Sea and Indian Ocean. Further consideration of Columbus' project now clearly out of the question, he returned to Spain. In the ensuing four years, he labored mightily to drum up support for his project, but is not known to have gone to sea again until--after the fall of Granada--the Spanish monarchs approved his first voyage of exploration and discovery.

Period of the Four Voyages, August, 1492-November, 1504

If little was known about Columbus' preparation for his "project of the Indies," by contrast documentation of his four voyages of exploration and discovery is so extensive that it has fostered in the almost 500 years since the first voyage a veritable mountain of literary effort. Our concern, here, is to extract from this mountain specific evidence of Columbus' nautical competence.

Figures 1A and 1B illustrate the trans-Atlantic routes taken by Columbus on his four voyages and the specific areas in the New World which were explored. Appendix A provides a chronological brief of "European Discoveries and Exploration in the Americas Through the Year, A.D. 1504." (1504 was the year that Columbus returned from his fourth and last voyage of discovery and exploration).

Landstrom, who rates Columbus' competence as a mariner highly, says " . . . His one revolutionary innovation was his route across the ocean, out with the trade winds and home with the westerlies." Morison is at once more critical and more complimentary.

Columbus' Discovery of Westerly Compass Variation and Circumpolarity of Polaris

Morison reports that on Columbus' first voyage, the pilots, on checking the compass needles against the polestar just after dusk and prior to dawn (using the "pilot's blessing" method) [22], obtained some highly unusual results. On September 13, the needles pointed a few degrees west of the polestar at dusk and about the same variation east of the polestar at dawn. On the seventeenth, at dusk, the variation of the needle was a full point (11 1/4°) to the west. At dawn, the needles were "true" (no variation). On the 30th, the results were about the same as on the 17th.

Now, the "pilot's blessing" method of checking the compass needles against the polestar was hardly precise and Columbus realized it. But, he also knew the pilots were not likely to be off more than a few degrees. Thus, he came to the conclusion that the ships had been passing through zones of westerly variation of the compass and so reported it in his log. Morison believes this was the first report ever made by Europeans on westerly variation. To account for the different variations at dusk and dawn, Columbus concluded that Polaris was not marking true north on the celestial sphere, but was rotating about that point. He had discovered for himself the diurnal rotation of Polaris, a fact which was denied by many late medieval and renaissance astronomers. Morison indicates further that ,"practical seamen had assumed for centuries that the North Star marked true north."

The following explanation of the events related in the foregoing is best understood by referring to Figure 2.

Actually, the ships had passed through a zone of no compass variation on the 13th, but Polaris, in 1492, described a circle with a radius of 3° 27' about the celestial pole. At nightfall, on 13 September 1492, Morison reports that Polaris was this full distance east of the celestial north pole. The time span between the dusk reading on 13 September and the dawn reading on the 14th, at latitude 28° N is about 11-1/3 hours. Thus, Polaris, at dawn, should have rotated in a counter-clockwise direction 11-1/3 x 15° = approximately 170°[23] from its initial position (due east of the celestial north pole). This would appear to observers on the ship to be about 3 1/2° west of the compass needle.

On the 17th, dusk having arrived 5 minutes earlier than on the 13th, Polaris would have rotated 95.92 (hrs.) x 15°/hr. = 1438.8° past, or 1.2° short of its position at dusk on the 13th. This would have appeared to observers on the ship as being about 3 1/2° east of the compass needle, if there had been no deviation of the earth's magnetic field. However, the ships had entered a zone of 2° westerly variation. This should have put the compass needle about 5 1/2° west of Polaris. Instead, the pilots read this as a full point (11 1/4°) west - not a good reading. The dawn reading should have yielded a variation of about 1 1/2° east. It was read as "true", not bad for the method used.

On the 30th of September, when the pilots reported a full point westerly variation of the needle at dusk, the ships had reached the isogonic line of 7° westerly variation. Dusk on this date arrives at about 5:45 p.m., some 35 minutes earlier than on the 13th, so that the total elapsed time since dusk on the 13th was 407.42 hours. Thus, Polaris would have rotated 407.42 x 15° = 6111.3°, which is equivalent to saying that Polaris was 8.7° short of its position at dusk on the 13th. This would have been perceived as about 3 1/2° east of the compass needle had there not been magnetic variation of the compass. With the 7° westerly variation of the compass, Polaris should have been seen as 10 1/2° east of the compass needle. Thus, the reported difference of a full point was not bad.

Dawn, 1 October, arrived about 12.083 hours after dusk the previous day and 419.503 hours after dusk on the 13th, so that Polaris would have rotated a total of 419.503 x 15° = 6292.5°, or 172.5° counter-clockwise from its original position at dusk on the 13th. With no compass variation, this would have been perceived as a bit less than 3 1/2° west of the compass needle. However, with 7° westerly compass variation, the needle should have pointed 3 1/2° west of Polaris. The pilot, however, read "true" - not a good observation.

That Columbus came to the conclusion he did when the concept of Polaris being circumpolar was denied, the position of Polaris vis-a-vis the magnetic compass was ascertained by the "pilot's blessing" method, and westerly compass variation had not yet been experienced by European navigators - would seem to indicate that he was, at the very least, highly intuitive.

Credits for Figure 1a: Routes for the 2nd, 3rd, and 4th voyages have been superimposed on a map titled "The Voyage of Discovery - from Palos de Moguer, August 3, 1492, to San Salvador, October 12, 1492", pp 62 and 63, "The Life of the Admiral Christopher Columbus by his son Ferdinand", translated by Benjamin Keen, Rutgers University Press, 1958. The approximate routes for the 2nd, 3rd, and 4th voyages have been drawn from a variety of sources, but chiefly Keen's translation and Samuel Eliot Morison's "Admiral of the Ocean Sea, A Life of Christopher Columbus."

Use of the base map has been authorized by Professor Emeritus Benjamin Keen.

For details of voyages and exploration within the New World, see Appendix A, which also outlines and illustrates the first voyage approach routes, landfalls, and routes through the Bahamas proposed by authorities other than Morison or Keen.

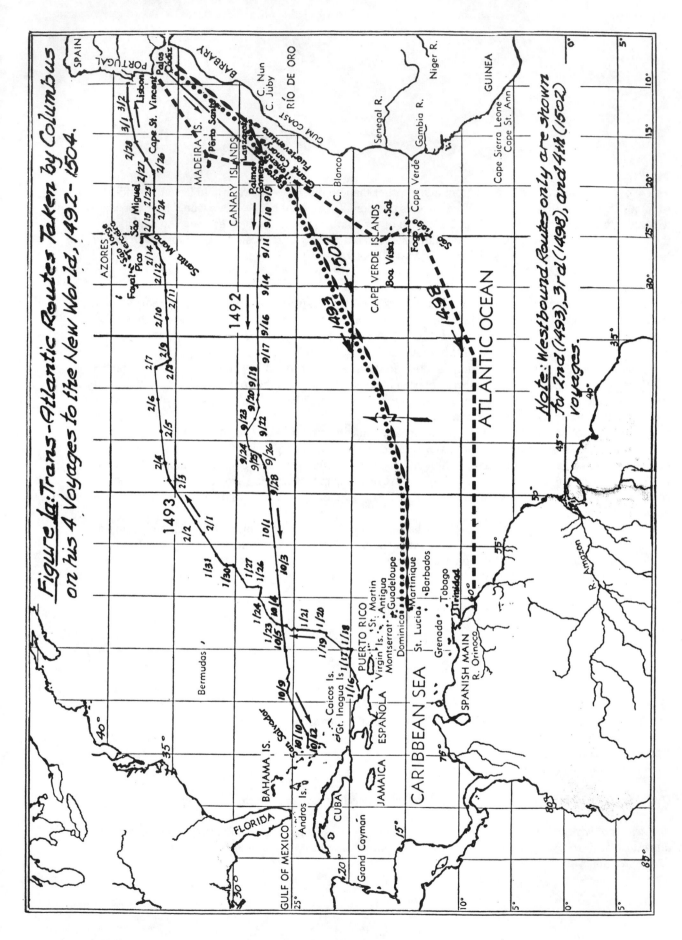

Figure 1a. See page 15 for credits.

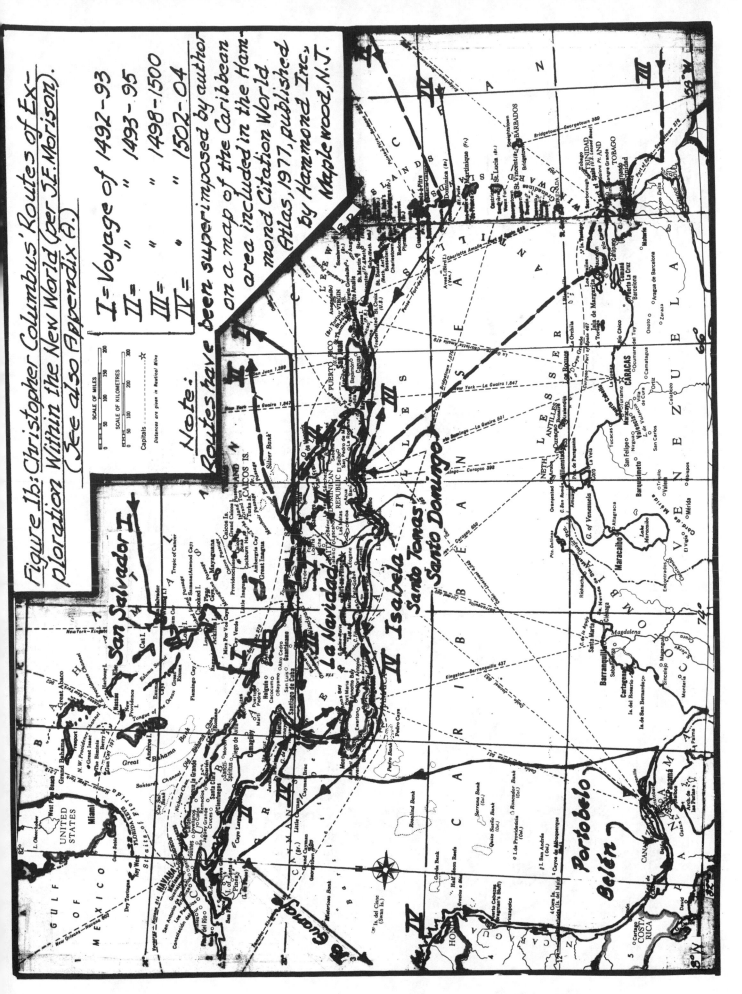

Figure 1b: Christopher Columbus' Routes of Exploration Within the New World (per J.E. Morison). (See also Appendix A.)

I = Voyage of 1492-93
II = " " 1493-95
III = " " 1498-1500
IV = " " 1502-04

Note:
Routes have been superimposed by author on a map of the Caribbean area included in the Hammond Citation World Atlas, 1977, published by Hammond Inc., Maplewood, N.J.

SCALE OF MILES
0   50   100   150   200
SCALE OF KILOMETRES
0   50   100   200   300

Capitals are given in Nautical Miles.
Distances are given in Nautical Miles.

A Comparison Between the Position of Polaris and What the Compass Should Have Shown, in 1492 at:

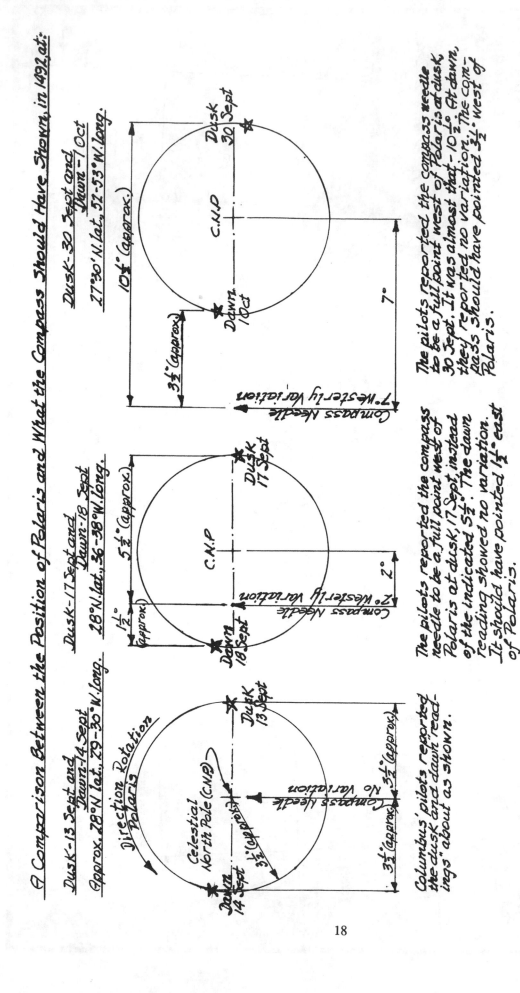

Figure 2.: Columbus Discovers Polaris is Circumpolar and Encounters Zones of Westerly Compass Variation on First Voyage of Exploration and Discovery, Outbound from the Canaries.

18

Columbus' Use of Polaris and the Sun in Navigation

Despite Morison's high regard for Columbus' skills as an intuitive navigator and seaman, this did not extend to Columbus' attempts to utilize the heavens in his work.

Morison claims that [24] "the testimony of . . . (Columbus') . . . own journals proves that the simple method of finding latitude from a meridional observation of the sun, long used by the Arabs in 'camel navigation' of the desert, was unknown to (him). Polaris observations for latitude he made not infrequently on his last two voyages, but these observations though 'not too bad' were of no use to his navigation, because he never knew the proper correction to apply.

Without taking direct issue with Admiral Morison on this point, it would appear that his position is too strong. Morison, himself reports that [25], " . . . in. . . (Columbus') . . . Book of Prophecies there is a calculation of latitude that is very nearly correct." Quoting Columbus, "In the harbor of Santa Gloria (Montego Bay) in Jamaica, the altitude of the Pole was 18° when the guards were on the Arm." Morison then continues, "Columbus, therefore made an error of less than half a degree. Considering that he now had a perfectly stable platform to 'shoot' from, and a whole year to make repeated observations and average them (Columbus' ships had been beached, and he and his crews were marooned for a year in Santa Gloria harbor) this was nothing to boast about; but it does show that the Admiral's technique with the quadrant had improved since his last voyage. After all, it was one of the best latitude observations on record for the early years of the sixteenth century."

The reference to " . . . the Guards on the Arm," simply means that Kochab, the Beta star in the constellation Ursa Minor (the Little Dipper) was, at the time of observation, either due east or due west (Columbus did not specify which arm) of Polaris and that Polaris was probably at the same altitude as the celestial north pole, although about 3 1/2° east or west of the pole. (See Figures 3 and 7 and the section which follows, titled "On Telling Time from the Guards.")

Morison's deprecatory remark about the stability of the platform and the time available to Columbus for checking and averaging his observations, should not detract from Columbus' achievement. The same opportunities were available to geographers and cartographers of Columbus' time, including the vaunted Portuguese, who nevertheless showed some of the best known places at incorrect latitudes.[26]

By the time of his Santa Gloria observation, Columbus had become at least as good as his Portuguese contemporaries and better than most Spanish navigators in reading the heavens.

The proposition that Polaris was worthless as a navigational aid in 1492 without precise corrections for circumpolarity, or that the sun could not be used to determine latitude without a formal table of its declinations throughout the year, is extreme. Figure 3 illustrates the positions of Polaris with respect to the celestial north pole at dusk and dawn on eight representative dates throughout the year 1492 as might be observed from a fixed platform at latitude 28° N. and longitude 45° W. (The principle holds, however, in any year, at any site, fixed or moving, provided Polaris is visible at dusk and dawn and allowance is made for any movement of the observing platform). The positions of Polaris shown in Figure 3 are for the days of the vernal and autumnal equinocti (when nights are of approximately 12 hours duration), the Summer Solstice (the shortest night, about 9 hours), the Winter Solstice (the longest night, about 14 1/2 hours) and four dates intermediate to the foregoing - May 6, August 7, November 8, and February 7. As indicated, averaging the dusk and dawn positions of Polaris results in a) elimination of all error both in azimuth and altitude, on March 21 and September 23; b) effective elimination of altitude error on June 22 and Dec. 22 while azimuth error averages less than 1.32°; and c) reduction of errors in azimuth and altitude on May 6,

Aug. 7, Nov. 8, and Feb. 7 to 0.75° or less.

Averaging was obtained in practice by following compass courses adjusted at dusk and then again at dawn to conform with azimuth readings of Polaris as obtained by "Pilot's blessing method." Determining latitude by observing Polaris' altitude with the quadrant, a less vital operation, also benefited from averaging. However, in this case, dusk and dawn readings were simply recorded in the ship's log and on the chart of the ship's voyage.

Figures 4a and 4b illustrate, "The Relationships between a Ship's Position at Sea and the Heavenly Bodies Available for Determining Latitude." The method of averaging dusk and dawn quadrant readings is a rough approach to providing the corrections for Polaris circumpolarity. An approach for the determination of an approximation to the Sun's declination is described in the following.

Figure 5 titled, "The Declination of the Sun, at Noon, Throughout the Year," shows that the declination of the sun, from day to day, at noon, varies in a rather regular manner, actually a sine curve. Table 1 shows that the daily change in declination also varies in a rather regular manner, being a maximum of 24 minutes in the 14 days before and after each of the equinoxes and diminishing to zero at the solstices. There were, without question, dozens - if not hundreds - of ways of approximating the daily change in declination so that with an observation of the sun's altitude at noon an approximation of the latitude of the observer could be made.

Hipparchus (fl.160 B.C.), who invented trigonometry, could predict the declination of the sun precisely and his methods and tables were handed down via Strabo (c. 63 B.C.-c. A.D. 19) and Ptolemy (fl. A.D. 140) to the Arabs who over-ran all of North Africa and most of the Iberian peninsula by the middle of the 7th century. The Islamic influence in Spain and Portugal was to remain powerful, although diminishing, through the period of Charlemagne, the Viking raids, the Holy Roman Empire, the Crusades, the Empires of the Mongol Khans and Tamerlane, until the 14th century by which time Islamic control of territory had been reduced to about 13,000 square miles in southern Spain comprising the Kingdom of Granada. This last bastion of Islam in Europe was to fall to Ferdinand and Isabella in the "crusade" initiated in 1484 and successfully completed in 1492.

Arab astronomical knowledge began to be passed on to Europeans as early as the first conquest in the 7th century and to re-appear in Latin via such 13th, 14th, and 15th century writers as Roger Bacon, Sacrobosco, Pierre d'Ailly, and Toscanelli. The astrolabe, a sophisticated astronomical instrument (which provides the time of day or night by observing the altitude of the sun or a star above the horizon and which shows graphically the positions of the sun, the stars, and, with the aid of an almanac, the moon and the planets at any moment of the year) was perfected by the Arabs between the 7th and 10th centuries although its Greek origins go back as far as Hipparchus. The astrolabe became the most important tool of the astronomer, the astrologer, and the surveyor in Europe until the advent of telescopic instruments in the 17th century[27].

Sometime in the late 15th century, a much simplified version of the astrolabe-the mariner's astrolabe, illustrated in Figure 7a-was developed in Europe for use at sea. The first European seamen to utilize the astrolabe were the Portuguese in the latter half of the 15th century, followed by the Spanish in the first quarter of the 16th century.

The mariner's astrolabe was a portable metal circular disc, varying from 4 to 8 inches in diameter, with a suspending ring at the top, so it could be hung vertically. The disc was inscribed near the perimeter to read degrees or multiples thereof (depending on the diameter of the disc) between the horizon and the zenith.

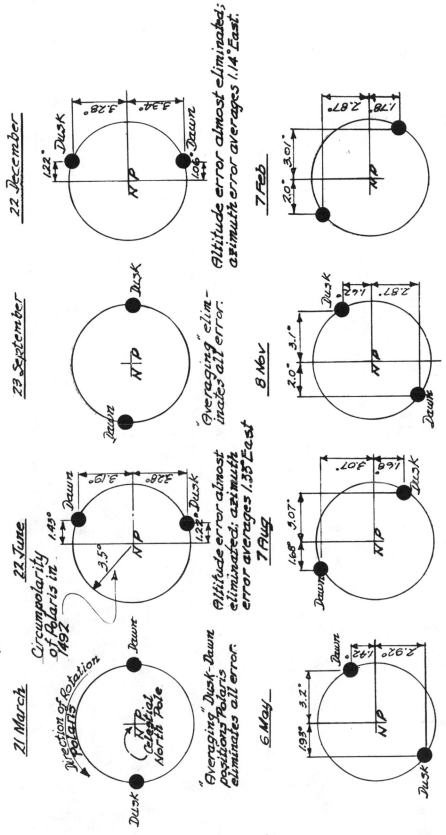

Figure 3: On the Usefulness of Polaris in 1492 (or any time), without Correction as a Navigational Aid. Shown above are the positions of Polaris with respect to the Celestial North Pole at dusk and dawn on 8 representative dates throughout the year. "Averaging" dusk and dawn positions of Polaris (in essence, considering each to be at Celestial North Pole) is an effective way of minimizing aggregate 24 hour error in azimuth and altitude due to circumpolarity.

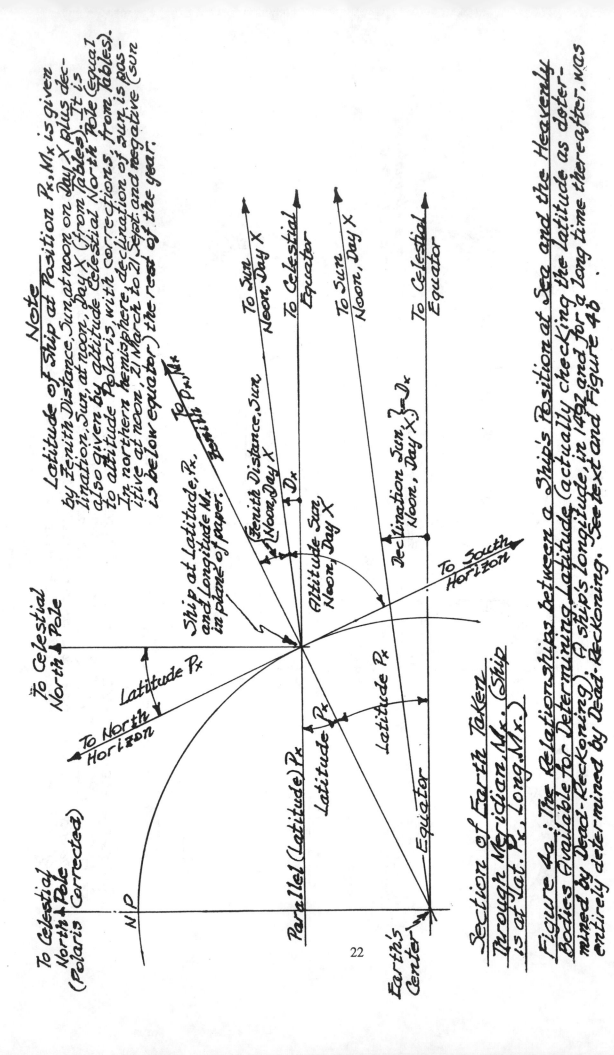

Note

Latitude of Ship at Position P_x, M_x is given by Zenith Distance, Sun at noon on Day X plus declination, Sun, at noon, Day X (from Tables). It is also given by altitude Celestial North Pole (equal to attitude Polaris with corrections, from Tables). In northern hemisphere declination of Sun, is positive at noon, 21 March to 21 Sept. and negative (Sun is below equator) the rest of the year.

To Celestial North Pole

To Sun Noon, Day X

To Celestial Equator

To Sun Noon, Day X

To Celestial Equator

Zenith P_x, M_x

To P_x, M_x

Zenith Distance, Sun (Noon, Day X)

D_x

Altitude Sun, Day X

Declination Sun, Noon, Day X } = D_x

Ship at Latitude P_x, and longitude M_x in plane of paper.

To South Horizon

To Celestial North Pole

Latitude P_x

To North Horizon

Latitude P_x

Latitude P_x

Parallel (Latitude) P_x

Latitude

Latitude

Equator

To Celestial North Pole (Polaris Corrected)

N P

Earth's Center

22

Section of Earth Taken Through Meridian M_x. (Ship is at Lat. P_x, Long. M_x.)

Figure 4a: The Relationships between a Ship's Position at Sea and the Heavenly Bodies Available for Determining Latitude (actually checking the Latitude as determined by Dead-Reckoning). A ship's longitude, in 1492 and for a long time thereafter, was entirely determined by Dead-Reckoning. See text and Figure 4b.

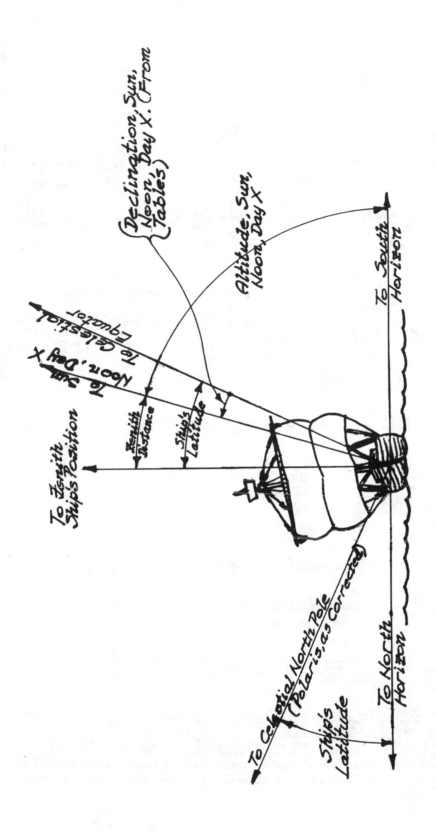

(Declaration, Sun, Noon, Day X. (From Tables))

Altitude, Sun, Noon, Day X

To Sun, Noon, Day X

To Celestial Equator

To Zenith, Ships Position

Zenith Distance

Ship's Latitude

To South Horizon

To Celestial North Pole (Polaris, as Corrected)

Ship's Latitude

To North Horizon

Figure 4b: Another Way of Presenting the Relationships between a Ship's Position and the Heavenly Bodies Available for Determining Latitude. (Ship is proceeding due west in sketch - towards the reader. Plane of paper is the plane containing meridian of ships position.)

23

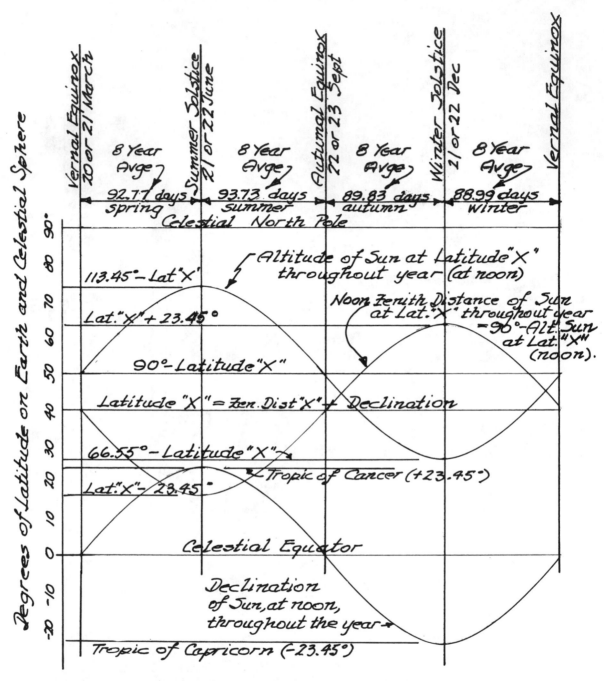

Figure 5: The Declination of the Sun at Noon Throughout the Year and Its Effect upon the Altitude and Zenith Distance of the Sun as observed at Noon at any Latitude "X"
(Note: The declination of sun at noon is approx. given by 23.45° sin "y" where "y" = 0.9856° × N, and N = number of days after vernal equinox.)

24

In the period between the winter and summer solstices dawn arrives somewhat earlier each successive day, the days become a bit longer, and - most important to the mariner before electronic navigational aids became available - the sun crosses the meridian a bit higher in the sky than it did the day before. The process is reversed in the period between summer and winter solstices. The change in the sun's declination (altitude at the equator) from day to day (at noon) is small, varying from a maximum change of 24' per day in the 28 days which bracket the vernal and autumnal equinocti (each) to a minimum of 1' or less during the 5 days which bracket each of the solstices. The following listing shows - as an example - how the sun's declination changes during the spring, between the vernal equinox and the summer solstice:

Days after Vernal Equinox	Increase in Declin. Per Day	Days after Vernal Equinox	Increase in Decl per Day
0 - 14	24'	64 - 65	11'
15 - 23	23'	66 - 68	10'
24 - 29	22'	69 - 70	9'
30 - 33	21'	71 - 73	8'
34 - 38	20'	74 - 76	7'
39 - 41	19'	77 - 78	6'
42 - 44	18'	79 - 80	5'
45 - 48	17'	81 - 83	4'
49 - 51	16'	84 - 85	3'
52 - 54	15'	86 - 88	2'
55 - 56	14'	89 - 90	1'
57 - 60	13'	91 - 93	0
61 - 63	12'		

Since latitude was given by the sum of the zenith distance and the declination at noon (equal to 90° - altitude of sun at noon + declination) and the declination at noon on the day of the vernal equinox is zero, each pilot had his own method of approximating the declination so as to determine his latitude. One simple method was to add 23' for each of the first 31 days after the vernal equinox, 17' minutes for each of the next 31 days, and 5½' for each of the last 31 days. This would always be within 6½' of the correct declination, equivalent to 6½ nautical miles, 8.12 Roman (Italian) miles, or (in 1492) 2.03 Spanish leguas (leagues) - a tolerable error.

Table 1: Demonstrating the Daily Increase in the Sun's Declination at Noon in the Period between the Vernal Equinox and the Summer Solstice (Pilots at the turn of the 15th-16th Century, before Tables were available giving the computed position of heavenly bodies for each day of a period, had developed various methods of obtaining the approximate declination of the sun at noon.

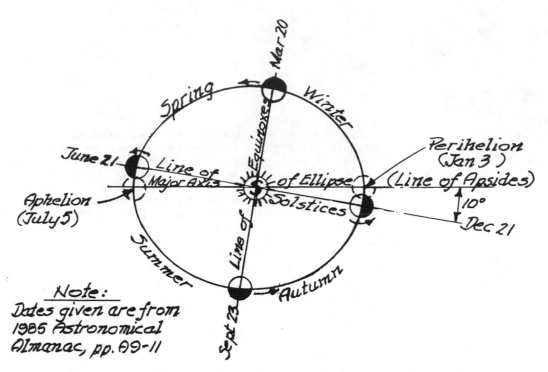

Note:
Dates given are from
1985 Astronomical
Almanac, pp. A9-11

Figure 6a: The Seasons Are Not of Equal Length.
The four seasons are not of equal length owing to the
eccentricity of the earth's orbit about the sun and
to the angle of 10° which the line of solstices makes with
the major axis of the orbital ellipse. Kepler's Law of
Equal Areas (the radius vector from the sun to any planet
sweeps out equal areas in equal times). This requires the
earth to take a different time to cover each of the seasons.

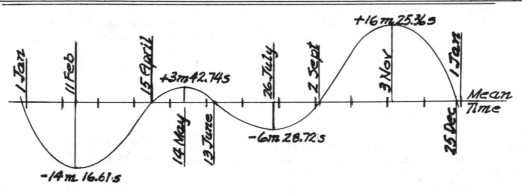

Figure 6b: The Equation of Time, the time interval
between mean solar time (clock time) and solar time
at noon shown for the entire year. Sun crosses meridian at noon, solar time. Plus sign means sun is early,
Minus sign, sun is late. (See note on Figure 6c.)

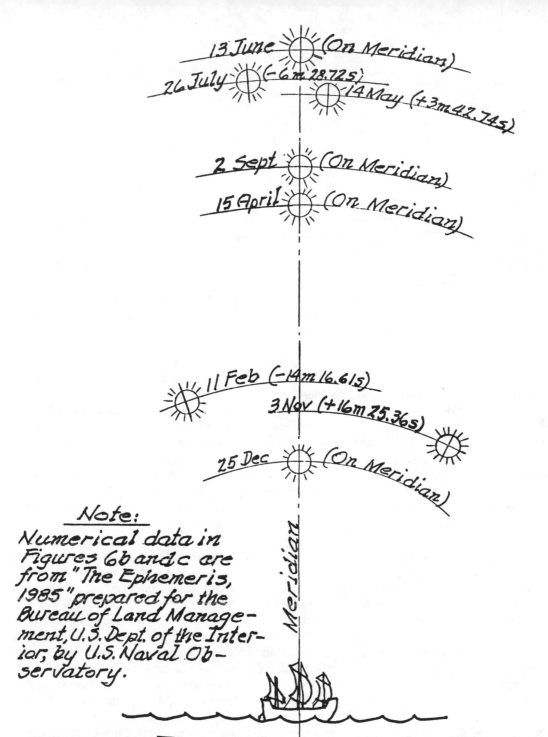

13 June ☀ (On Meridian)
26 July ☀ (- 6m 28.72s)
☀ 14 May (+3m 42.74s)

2 Sept ☀ (On Meridian)

15 April ☀ (On Meridian)

11 Feb (-14m 16.61s) ☀

3 Nov (+16m 25.36s) ☀

25 Dec ☀ (On Meridian)

Note:
Numerical data in
Figures 6b and c are
from "The Ephemeris,
1985" prepared for the
Bureau of Land Manage-
ment, U.S. Dept. of the Interior, by U.S. Naval Ob-
servatory.

Meridian

Figure 6c: The Sun Rarely Crosses the Meridian Precisely at Noon, Clock Time (Mean Solar Time)

Owing to the differences in the earth's orbital velocity around the sun, the sun rarely crosses the
meridian at mean solar time (clock time). There
are 4 times during the year when solar time and
mean solar time at noon are essentially the same.
On all other days, the sun crosses the meridian
either before or after noon, clock time. The difference
between mean solar time and solar time, both taken
at noon, is called the Equation of Time.

An alidade, a device pivoted at the center of the disc with peep holes near both end pointers, permitted sighting on an elevated object, usually but not necessarily a heavenly body. The angle between the object and the zenith (zenith angle or distance) was indicated by the pointer.

It appears that the circumpolar nature of Polaris was known to the Arabs well before Al-Biruni's time (fl. c. A.D. 1000) and, as a result, several authors suggested using the mean altitude at upper and lower culmination of a number of circumpolar stars for the determination of latitude (If latitude determination is to be made during a single night, the method is limited to latitudes and seasons of the year where nights have a length significantly greater than 12 hours and stars whose upper and lower culminations, both, occur during this night period). Eventually, the Arabs developed tables giving the correction to be applied to Polaris' altitude to obtain the altitude of the celestial north pole, equivalent to geodetic latitude. Ibn al-Haitham (965-1038 A.D.) devoted a separate work to the calculation of latitude and Al-Biruni applied the method of circumpolar stars to the sun (see Chapter 4).

The astrolabe was, however, used for a variety of earthly purposes, besides its basic astral function, e.g., the calculation of the distance of an inaccessible place, the height of a building or mountain, the depth of a well whose diameter could be measured and the bottom seen, even the length of a degree of arc of the earth's surface (see Al-Biruni, Chapter 4).

The quadrant, an outgrowth of the astrolabe, seems, however, to have preceded the use of the astrolabe at sea. This instrument, illustrated in Figure 7b, was invented in the 13th century. Columbus attained some proficiency in its use, but not until he was able to spend a whole year with it ashore, while marooned in Jamaica, as has already been described.

The quadrant was simply a thin plate of metal or wood, a quarter of a circle in shape, marked in degrees around the arc, with pin-hole sights fixed to the 90° straight edge. A weighted plumb line, hung from the intersection of the straight edges, passed over the degree markings so that if a heavenly body was sighted through both pinholes, the line indicated the angle of elevation from the horizon, or altitude, of the body. On a ship in any kind of weather except a dead calm, the quadrant was almost as difficult to use as the astrolabe. In the case of both instruments the problem arose from the near impossibility of finding any place on a ship, underway, from which the instruments could be suspended, which was motionless. Thus, the plumb line was rarely plumb, the plane of the quadrant rarely vertical, and the eye of the observer still more rarely, if ever, able to align the peep holes and the heavenly body at the instant the plumb line was vertical. Morison's criticism of Columbus' inaccuracy in the use of the astrolabe on the first voyage and the quadrant on succeeding voyages is intended to offset Columbus' own claims to proficiency in the use of these instruments, rather than point to an inferior level of competence vis-a-vis other navigators. Both the astrolabe and the quadrant proved so difficult to use at sea that their use was replaced in the 16th century by the Cross-staff (illustrated in Figure 8).

a. MARINER'S ASTROLABE

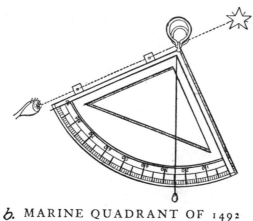

b. MARINE QUADRANT OF 1492

Altitude is read from the point where thread cuts Arc

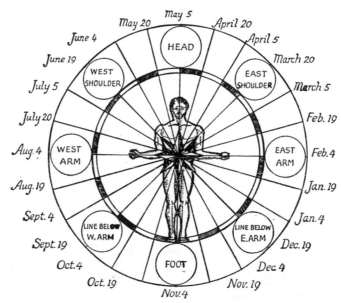

c. DIAGRAM FOR TELLING TIME FROM
POLARIS, 1942

*Kochab (β Ursae Minoris) moves counter-clockwise one line per hour.
Its position relative to Polaris at midnight is indicated for each date*

Figure 7: Astronomical Instruments Used with Varying Degrees of Success by Columbus.

Note: *While the Mariner's Astrolabe shown in Figure 7a was manufactured in 1632, it is scarcely more elaborate than that used by Columbus and other Spanish, Portuguese, and Italian navigators since at least the early 14th century. The Astrolabe, an astronomical instrument of greater sophistication, was invented (some say) by Ptolemy (c. 2nd Cent., A.D.) and perfected by the Arabs.*

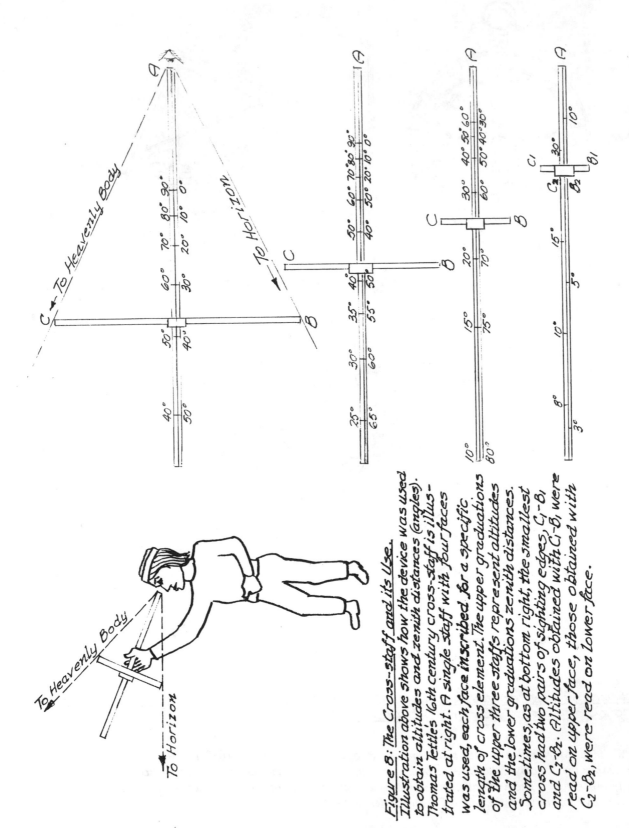

Figure 8: The Cross-staff and its Use.

Illustration above shows how the device was used to obtain altitudes and zenith distances (angles). Thomas Tuttle's 16th century cross-staff is illustrated at right. A single staff with four faces was used, each face inscribed for a specific length of cross element. The upper graduations of the upper three staffs represent altitudes and the lower graduations zenith distances. Sometimes, as at bottom right, the smallest cross had two pairs of sighting edges, C_1-B_1 and C_2-B_2. Altitudes obtained with C_1-B_1 were read on upper face, those obtained with C_2-B_2, were read on lower face.

Telling Time From "The Guards"

Morison indicates that "Columbus could tell approximate clock time from the "Guards of the North Star . . ." A device such as that illustrated in Figure 7c was utilized. Polaris was sighted through a hole in the center and a movable arm representing the constellation was rotated until it hit Kochab. The illustration indicates how the diagram on the device would have been graduated in 1942 (when Morison was completing work on his book, *Admiral of the Ocean Sea, A Life of Christopher Columbus*). Dates disposed about the circumference of the diagram indicate where Kochab should be at the zero hour of that date (midnight on the preceding date). Thus, if in the year 1942, a reading were taken on Dec 19, and Kochab found to be pointing to half-way between "East Arm" and "East Shoulder" (or 4 1/2 lines counter-clockwise beyond "Line below East Arm") the time would be 4:30 a.m.

While the ampolleta, the sand hour-glass, was the prime device for telling time and changing the watches at sea, the "Guards of the North Star" at night, and the sun crossing the meridian at noon (maximum altitude during the day), provided checks on the occasionally erratic performance of this instrument and its operator.

Columbus' Use of Lunar Eclipses to Determine Longitude

During the period of the four voyages of exploration and discovery there were several partial and total eclipses of the moon. On two of these occasions, Columbus attempted to calculate the longitude of his position by noting the local (sun) time of total or maximum eclipse and comparing this with the time the same eclipse was predicted to occur at Nurnberg, each hour of difference in the two times being equal to 15° of longitude. The data for the eclipses at Nurnberg came from Regiomontanus' astronomical ephemerides which first became available in printed form in 1490, fourteen years after Regiomontanus' death. Morison who reports on both of Columbus' calculations and the errors in the results does, however, indicate that no one else taking observations from points in the New World for another century did any better.[28]   First, however, a few words about Regiomontanus.[29]

Johann Muller, better known as Johannes de Monte Regio and, after his death, as Regiomontanus, was born in 1436 at Konigsberg, Franconia, the son of a miller. A student of the astronomer George Peurbach, he inherited from his mentor a tremendous respect for the works of Ptolemy and a conviction that a thorough knowledge of Greek was an absolute prerequisite to the accurate and complete understanding of that master. Peurbach and he both had concluded that translations of Ptolemy, made into Arabic and from Arabic into Latin, contained errors not in the original Greek.

After Peurbach's death in 1461, Regiomontanus, with the financial assistance of Cardinal Bessarion, a Greek by birth, spent six years in Italy visiting the principal cities and collecting every Greek astronomical manuscript he could get his hands on. Some years later, he settled in Nurnberg where he erected an observatory and commenced publishing on a prolific scale. Certainly the most popular of his works were his astronomical ephemerides which were to render invaluable services to Portuguese and Spanish navigators. Columbus had a copy of Regiomontanus' ephemerides.

Even more important than the ephemerides was his treatise on trigonometry, the first systematic work on this subject, and his "Tabulae Directionum" containing a table of sines for every minute and a table of tangents for every degree. The renown he gained (probably as a result of his ephemerides) resulted in Regiomontanus being summoned to Rome by the Pope in 1475 to carry out the long contemplated reform of the Calendar. However, he died the following year, at the age of 40, before he could get started on his task.

Returning now to Columbus, on September 14, 1494, while his fleet lay anchored off Saona Island at the southeastern tip of Hispaniola, he observed an eclipse of the moon. The Regiomontanus ephemerides he carried indicated the time of the eclipse in Nurnberg. Allowing for the time difference between Nurnberg and Cape St. Vincent, the southwestern tip of Europe, Columbus came up with a time difference of 5 1/2 hours between Saona and Cape St. Vincent, or 82.5° of longitude. The actual difference in longitude between these points, as Morison points out,[30] is only about 60°, or 4 hours. Thus, Columbus' error was a whopping 22.5° of longitude, or 1 1/2 hours. Maps of that era available to Columbus[31] showed Europe stretched out in the east-west direction, the length of the Mediterranean being shown as 61°, instead of about 41° (see Figure 5, Chapter 3) and the difference in longitude between the vicinity of Nurnberg and Cape St. Vincent being shown as about 38° instead of 20°. The mapping error actually served to reduce Columbus' error in determining his longitude at Saona. An allowance of 20° for the Nurnberg-Cape St. Vincent longitudinal distance would have made his Cape St. Vincent-Saona longitudinal distance 18° greater, or 100.5°, for an error of 40.5°.

The other occasion when Columbus attempted to determine longitude from an eclipse of the moon occurred on the night of February 29, 1504, while marooned in the harbor he called Santa Gloria (St. Ann's Bay, Jamaica) during his fourth voyage. Columbus has recorded the incident in his *Book of Prohecies*[32]:

> Thursday, 29 February, 1504, I being in the Indies on the Island of Jamaica in the harbor called Santa Gloria which is almost in the middle of the island on the north side, there was an eclipse of the moon, and as the beginning thereof was before the sun set, I could only note the end of it, when the moon had just returned to its light, and this was certainly two hours and a half after the night (fell), five ampolletas most certainly. The difference between the middle of the island of Jamaica in the Indies and the island of Cadiz in Spain is seven hours and fifteen minutes, so that in Cadiz the sun sets seven hours and fifteen minutes earlier than in Jamaica (see almanac).

Columbus' calculation of seven hours and fifteen minutes difference between Cadiz and Jamaica was equivalent to a difference of 108.75°. The correct difference in longitude between these sites is 70.91°, so that his calculation was too great by 37.84°, or 2.523 hours. Had Columbus used the correct longitudinal distance between Cadiz and Nurnberg instead of that shown on the distorted maps of Europe, his estimate of the longitudinal distance between Cadiz and St. Ann's Bay would have been about 18° still greater in error.

The two attempts at determining longitude were not only crude, but biased. The events predicted in Regiomontanus' almanac were probably "mid-eclipse", when the moon is halfway through the earth's shadow. It would take far better timing devices than Columbus' ampolettas to determine the precise local time of each of these events. In his second attempt, he did not witness the entire eclipse but only " . . . the end of it, when the moon had just returned to its light . . . ". Thus, his estimate of the time of mid-eclipse was, at best a guess (all total lunar eclipses are not of equal duration).

More important, Columbus had long considered the island of Zipangu (Japan), which since his first voyage in 1492-3 he identified with Hispaniola, to lie some 90° of longitude west of Lisbon or Cape St. Vincent (see Chapter 7). St. Ann's Bay on the north coast of Jamaica he estimated to be about 15° of longitude west of Saona Island in Hispaniola. Where Columbus got his distorted geographical ideas and whether these ideas were exclusive to him is the "business" of this book and just what the major elements of his "cosmography" were will be discussed in the following section. It is clear, however, that Columbus' attempts at determining longitude were intended to reinforce - not refute - his cosmography, about which "his mind was made up". It must be noted, however, that Columbus was not the only one who erred significantly in determining New World longitudes. Morison describes a mighty effort made by the intelligentsia of Mexico City in 1541 to determine the longitude of that place by timing two lunar eclipses. Their result, based upon ephemerides prepared at the observatory in Toledo, Spain, was that Mexico City was 120° 38', or 8 hours and 2.5', west of Toledo. The correct difference in longitude is 95° 4.8', equivalent to 6 hours and 20.3'. Thus, the efforts of the most sophisticated scientific group in the New World at that time yielded results little better than Columbus' efforts some 37-46 years earlier.

A Summary on Nautical Competence

Considering the relative meagerness of Columbus' professional experience at sea prior to his four voyages of exploration and discovery, the competence, resourcefulness, intuitiveness, and range of nautical skills he developed during his four voyages is amazing. Despite some claims that his attempts at celestial navigation left something to be desired, he appears to have developed at least as much competence in this branch of the mariner's arts as any of his contemporaries. He became exceptionally skillfull in every phase of ship handling and sailing, in deep water on trans-Atlantic and trans-Carribbean voyages, and in treacherous coastal exploration of the Lesser and Greater Antilles, Venezuela, Honduras, Nicaragua, Costa Rica, and Panama. He rode out countless storms successfully and was afterwards able to effect sufficient repairs to resume his voyage or, if he had beached his ship for repairs, to put to sea again. In an age when dead-reckoning was the prime means of navigation over great expanses of ocean, he became exceptionally skillfull at it.

Morison sums it up this way " . . . Columbus was a great navigator. He took his fleet to sea not as an amateur possessed of one big idea, but as a captain experienced in 'el arte de marear' . . . Over and above his amazing competence as a dead-reckoning navigator, he had what a great French seaman, Jean Charcot, recognized and named 'le sens marin,' that intangible and unteachable God-given gift of knowing how to direct and plot 'the way of a ship in the midst of the sea.' "

CHRISTOPHER COLUMBUS, COSMOGRAPHER

Cosmography has been defined[33] as "a science that describes and maps the main features of the heavens and the earth, including astronomy, geography, and geology." Whether Christopher Columbus was ever considered to be a cosmographer by his contemporaries, or even by himself, there can be no question that he had very well-developed ideas on the subject.

During most of his adult life he is known to have considered the earth to be a sphere, although in the last years of his life Columbus came to believe that it "was not round in the way that is usually written, but has the shape of a pear that is very round, except in the place where the stem is, which is higher."[34]

During the greater part of his adult life, when he had no reservations as to the sphericity of the earth, he considered a degree of latitude, and of longitude along the equator, to measure 56 2/3 Roman (Italian) miles holding with the Arab astronomer and writer Alfragan, as reported by Cardinal D'Ailly and other cosmographers of the 14th and15th centuries. He even claimed to have checked the length of a degree of latitude on one of his voyages to Guinea and found Alfragan to be correct. (See Chapter 7.)

On the distribution of land and ocean in the east-west direction, he held with Marinus of Tyre, as modified by Marco Polo, rather than with Ptolemy, as did most learned cosmographers of the day. However, Columbus was not alone in this. Toscanelli, the celebrated physician/astronomer/cosmographer, resident in Florence, and Behaim, the Nurnberg cartographer/cosmographer, had remarkably similar, but not identical, views. The thrust of these views was to stretch the length of the Eurasian continent very significantly and thus narrow the width of the Atlantic commensurately. With the use of certain islands in the Atlantic as way stations, for refitting and repairing ships and taking on water and provisions, a crossing of the ocean from Europe to Asia became feasible, so Toscanelli, Behaim, and Columbus believed.

As to Columbus' views of the relationship between the earth, sun, moon, planets, and stars, he was a man of his times, believing in the geocentric theory where all the heavenly bodies revolved about the earth, but in highly individual, yet predictable, patterns. He was reasonably familiar with the paths of the sun and moon throughout the year, with eclipses, solar and lunar, partial and total, and with the use of Regiomontanus' Ephemerides calculated for the years 1474-1506, as has already been explained in the section on "Christopher Columbus, Navigator."

It is with Columbus' specific views on the size of the earth and the distribution of land and ocean, particularly in the east-west direction and in where he got these views, that we are most interested. We approach this subject by quoting three distinguished authors.

Samuel Eliot Morison had this to say in his *Admiral of the Ocean Sea, A Life of Christopher Columbus.*[35]

> The important thing that Columbus obtained from Toscanelli, apart from the prestige of having an eminent scholar approve his enterprise, was the Florentine's approval of Marco Polo. For the Venetian traveler had added some 30° of longitude to the easternmost point of China described by Ptolemy. And beyond Mangi, Cathay, Quinsay and Zaitun, 1500 miles out to sea, Marco Polo placed the fabulously wealthy island of Cipangu (Japan) with its gold-roofed and gold-paved palaces. Even at that, Toscanelli predicted a sail of some 5000 nautical miles from Spain to China, although the voyage could be broken at the mythical island of Antilia ("well known to you," he wrote to Martins) and at Japan. Columbus, however, thought he knew better, and that the ocean was even narrower than Toscanelli supposed.

The circumference of the globe can easily be figured out by multiplying the length of a degree by 360. But how long was a degree? That problem had been bothering mathematicians for at least eighteen centuries. Eratosthenes around 200 B.C. made a guess at it that was very nearly correct: 59.5 nautical miles instead of 60. Columbus, however, preferred the computation of Alfragan. That medieval Moslem geographer found the degree to be 56 2/3 Arabic miles, which works out at 66.2 nautical miles; but Columbus, assuming that the short Roman or Italian mile of 1480 meters was used by Alfragan, upon that false basis computed that the degree measured only 45 nautical miles, roughly 75 per cent of its actual length, and the shortest estimate of the degree ever made. Arguing from this faulty premise, Columbus concluded that the world was 25 per cent smaller than Eratosthenes, 10 per cent smaller than Ptolemy, taught.

Not content with whittling down the degree by 25 per cent, Columbus stretched out Asia eastward until Japan almost kissed the Azores. The way he figured it was something like this; and you can follow him on any globe, however small. Ptolemy taught that the known world covered half the globe's circumference, 180° from the meridian of Cape St. Vincent (long. 9° W of Greenwich) to "Catigara" in Asia. That was already a 50 per cent over-estimate, but Columbus insisted it was all too small. He preferred the estimate of Marinus of Tyre, who stretched out the known world to 225°.

To that Columbus added an additional 28° for the discoveries of Marco Polo, and 30° for the reputed distance from eastern China to the east coast of Japan. The total width of Europe and "The Indies" thus measured 283°, and as Columbus proposes to start west from Ferro in the Canaries, which is 9° west of the "beginning of Europe" at Cape St. Vincent, he has only 68° of ocean to cross before hitting Japan.

Columbus, moreover, had two more corrections to be taken into account, and applied them in such a way as to give him all the breaks: First, assuming that Marinus of Tyre's already exaggerated linear distance from Cape St. Vincent eastward to the end of Asia was correct, the distance in degrees was too small, because Marinus's degree (so Columbus thought) was oversize. So, instead of 68° of open water to be crossed between the Canaries and Japan, there were only 60° of longitude to cover. Second, as Columbus estimated a degree of longitude on the equator to be 45 nautical miles, it would measure only 40 miles on latitude 28°, which he proposed to follow for his ocean crossing. Therefore he had only 60 x 40 or 2400 nautical miles (750 leagues) to sail. As we shall see, he expected to hit land at exactly that distance from the Canaries on his First Voyage. In other words, his calculations placed Japan about on the meridian of the Anegada Passage, Virgin Islands.

A brief table will exhibit the colossal errors of these fifteenth-century optimists, the distances being reduced to nautical miles, and assuming Behaim's length of a degree to be the same as that of Columbus:

	Toscanelli	Martin Behaim	Columbus	Air-line*
Canaries to Cipangu (Japan)	3000	3080	2400	10,600
Canaries to Quinsay (Hangchow)	5000	4440	3550	11,766

* Between the respective meridians, measured on latitude 28°

Of course this calculation is not logical, but Columbus' mind was not logical. He knew he could make it, and the figures had to fit. To anticipate a bit, the Portuguese king's committee of mathematicians will have no difficulty seeing the flaw in the reasoning; for even if he were right and Ptolemy wrong about the length of a degree (which they would hardly be disposed to admit), he had applied the corrections both ways in order to narrow down the ocean as much as possible.

David B. Quinn, in his *North America from Earliest Discovery to First Settlements, the Norse Voyages to 1612*,[36] some 35 years after Morison, put it this way:

It was Paulo Pozzi Toscanelli who, in 1474, produced a world map (now lost) and a descriptive letter which provided a(n) . . . attractive western projection. Toscanelli had been impressed both by the alleged discovery of Antilia and by reading in Marco Polo that Cipango (Japan) lay 1000 or 1500 miles to the east of southern China, and in similar latitudes, as he understood (or misunderstood) Polo. Moreover, Polo had indicated that the eastward land extension of Asia was greater than that accepted though not dogmatically insisted on, by Ptolemy. Ptolemy's "Terra Incognita" which was the name given on some versions of his world map for the unterminated land, that lay farthest east, could well be Polo's Cathay. Therefore the sea route from Europe to Asia might plausibly be shorter than had hitherto been envisaged, while both Antilia and Cipango could act as stepping stones on the way. If, in addition, it could be argued that 24,000 miles or thereabouts for the circumference of the earth was much too great (the classical Greek estimates were now known to differ rather widely and the unit of measurement was uncertain), then a voyage to Asia across the Atlantic became feasible.
Toscanelli finally settled on a figure for the gap between Europe and Asia of some 6000 to 6500 miles. He reckoned that a ship might be able to sail continuously through the ocean for some 2500 miles, when she could put in to refit at Antilia which

he evidently regarded as a civilized country. She would then make a further 2500 miles to Japan, and again refit, and so be able to travel the final 1500 miles at the most to Polo's Quinsay (Hang-chow), from which all other parts of Asia would be accessible to her. This controlled and documented fantasy had a strong appeal to the Portuguese and was soon to appeal equally to an Italian guest of theirs, Christopher Columbus.

Somewhat later in the text, after describing the elements of Columbus' nautical experience and what he had learned from others at Galway, Bristol, and Madeira about voyages westward into the Atlantic, Quinn says:

> . . . it seems by 1485, or perhaps a little earlier, he had copied the Toscanelli letter, studied Pierre d'Ailly's great work, and steeped himself both in the current cosmographical theories and in the maps which were being produced of the western European shores and the ocean that lay off them.
>
> His basic theory rested on that of Toscanelli but was farther refined. Columbus seems to have claimed that he would find Antilia some 1500 miles westward from the Canaries on the 28th parallel, and Cipango 1000 miles further on, with Cathay only one further 1000 miles beyond, so making a mere 4,500 miles from Spain to Asia. His arguments on these points were easily resisted by those whom King John II consulted, since on an orthodox view some 10,000 miles lay between Europe and Asia.
> . . .

Finally, we excerpt some informative passages from R.A. Skelton's *The Cartography of Columbus' First Voyage*:[37]

> According to Ferdinand Columbus, his father's speculation on the possibility of discovering the Indies by a westerly voyage sprang from the Portuguese navigation to Guinea. 'It was in Portugal that the Admiral began to think that, if men could sail so far south, one might also sail west and find land in that quarter.'
>
> Columbus' eager search for authorities to support his hypothesis is exemplified in his annotations (postille) to the works of classical and medieval cosmographers, notably Seneca[38] and Pierre d'Ailly.[39] Both in the postille and in Ferdinand's Historie we find the basic concept underlying Columbus' plan for the enterprise of the Indies, namely his estimate of the width of the ocean between Europe and Asia, expressed in degrees and converted, by a calculation peculiar to himself, into miles and sailing time. From his reading, supplemented by the correspondence with Toscanelli between 1474 and 1481, Columbus arrived at the conclusion that Cipangu lay about 80° west of the Canaries, and the coast of Cathay some 30° further west.
>
> His interpretation of the Moslem geographer Alfragan, controlled by observations made on Portuguese voyages to Africa in 1485-8, led Columbus to evaluate the equatorial degree as 56 2/3 Roman miles or 45 nautical miles, 'the shortest estimate of the degree ever made,' in Professor Morison's words.

But Columbus was a cartographer, and it cannot be doubted that he also sought graphic expression for his theoretical concepts in globes and maps, to which indeed the Journal refers (e.g. on October 24th and November 14th, 1492). The characteristics of the cartographic sources consulted by Columbus can be readily determined. They must have reflected the views of Toscanelli (derived from Marinus of Tyre and from Marco Polo) on the longitudinal extension of Asia, and on the width of the ocean; and they must have provided a precise quantitative statement of these distances, in terms of degrees.

The world map or chart which Toscanelli sent to Lisbon in 1474, and later to Columbus, has not survived, but it is possible to reconstruct it from the data given in his letter; the most convincing reconstruction is that of Hermann Wagner. From this map Columbus obtained a figure of 26 'spaces', that is 130°, for the width of the ocean (which presumably occupied the centre of the map) between Lisbon and Quinsay, on the coast of Cathay. Toscanelli also laid down, in the ocean, the island of Antilia, which is found in many nautical charts of the 15th century; and, 10 spaces (50°) further west, Cipangu.

We cannot as yet point to any extant cartographic work executed before 1492, and embodying Toscanelli's concepts, that could have come to Columbus' notice. The globes to which he alludes in his Journal have disappeared; and no non-Ptolemaic map of the 15th century graduated in longitude has yet been brought to light. The Genoese world map of 1457, of elliptical form, in the Biblioteca Nazionale, Florence, has been claimed by Professor S. Crino as a copy or derivative of Toscanelli's map, but there are many difficulties in this identification. A chart in the Bibliotheque Nationale, Paris, has a circular inset world map in which Charles de la Ronciere saw an expression of Columbus' views before his first voyage; but this association also has failed to win general acceptance.

The celebrated globe prepared by (or for) Martin Behaim at Nuremberg in 1492, however, depicts eastern Asia and the ocean dividing it from Europe in a form which closely resembles the concepts of Toscanelli and Columbus. This parallelism has impressed all scholars, some of whom have drawn the erroneous conclusion that Columbus must have been acquainted with Behaim or his globe, or that Behaim had access to Toscanelli's papers.

Professor Morison remarks that 'the scale, the eastward extension of Asia, and the narrow ocean on this globe are so similar to the false geographical notions on which Columbus based his voyage, as to suggest that Columbus and Behaim were collaborators'; yet he is compelled to add that 'there is no positive evidence of their trails ever crossing, nor is it physically possible that Columbus could have set eyes on the Nuremberg globe. The inference is inescapable that Columbus and Behaim drew on a common map-source, and that (as G. E. Nunn suggested) 'prior to his first voyage Columbus used a map of Eastern Asia similar in concept to that Behaim presented on his globe. It is possible and even probable that this concept had its origin with Toscanelli.'

The foregoing assessment by three very distinguished scholars, while agreeing on the general proposition that Columbus' cosmographical ideas were in error, differ somewhat in tone. The theme is common, however, that Columbus was basing his conclusions as to the length of the Eurasian continent, the width of the Atlantic, the circumference of the earth, the existence of islands in the Atlantic such as Antilia and Cipangu, on the works of others - among them Marinus of Tyre, Alfragan, Marco Polo, and Pierre d'Ailly. Also given great weight is Paolo Toscanelli's general agreement with Columbus' ideas and that Martin Behaim seemed to share them. That there had been attempts to determine the circumference of a spherical earth by measuring the length of a degree of latitude and that the results differed is alluded to as is the possibility that some confusion existed as to the length of the units used to express the results. That the three writers consider the experts of Columbus' day, those to whom the monarchs of Portugal and Spain turned to evaluate Columbus' plan, superior to Columbus as cosmographers is clear, but that these mathematicians, geographers, and astronomers were themselves in error is not so clear.

What was the state of geographic and geodetic knowledge in Columbus' time and what was the basis for it? Was he so out of step with his peers and with the savants of the day? Who were Eratosthenes, Marinus, Ptolemy, Alfragan, Toscanelli and Behaim and how did they impact Columbus, the cosmographer, and the Columbus story? Why did Marco Polo's adventures in the Orient excite cosmographers, geographers, princes and Columbus? How was the size of the earth determined and expressed? It is proposed, in the following chapters, to develop the answers to these questions in an orderly, chronological manner, starting with the earliest recorded history. Nothing less will set the record straight on Christopher Columbus, cosmographer.

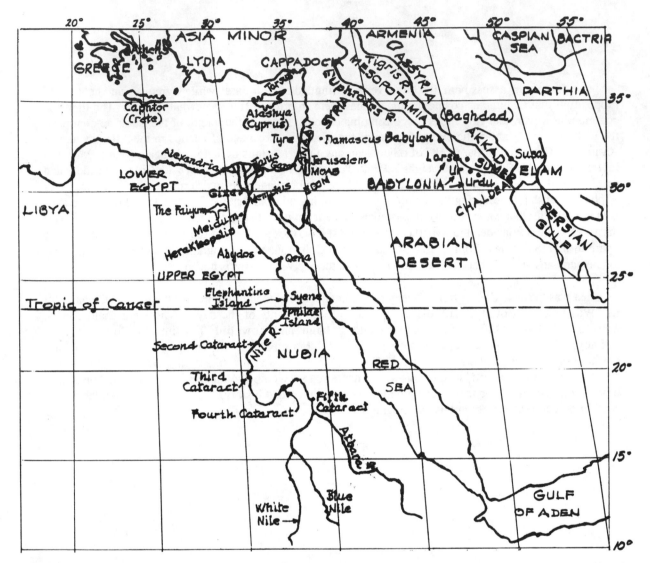

Figure F-1: Egypt, Mesopotamia and Their Neighbors, c. 3100 - c. 500 B.C.
The region of the world where the first civilizations were spawned - with irrigated agri-
culture, animal domestication, cities built around temples, government buildings
and central markets, systems of writing, weights and measures, luni-solar
calendars, transportation, and a regularized system of trade. The civilizations
shown achieved prominence or hegemony over their neighbors at different
times. Conquest and trade (as well as proximity) tended to make written
languages and systems of weights and measures similar, if not the same.

40

Chapter 2

ANCIENT LINEAR METROLOGY

FOREWORD

It is proposed in Chapters Three through Eight to examine in some detail the expression of ancient and medieval philosophers, astronomers, and cosmographers as to the size and shape of the earth and the distribution of land and water on its surface. Since such expressions are concerned with the lengths of arcs, from a degree to an entire circumference, taken along meridians or parallels of the globe, they must, inevitably, be stated in multiples of length units in use in the locale of the estimator at the time the estimate was made. It is the purpose of this chapter to examine the systems of linear measurement which came into use in the ancient world so that the true significance of the expressions of linear magnitude to be encountered in Chapter Three and beyond can be assessed. For illustrative purposes only, occasional reference is made in this chapter to medieval units. In Chapters Four through Eight, as we proceed through the Medieval into the Renaissance periods, each important geodetic event or pronouncement by cosmographers, geographers, navigators, or others which bears on Columbus' cosmographic or geographic beliefs will be explored.

The origins of weights and measures go back to the earliest days of the human race, many millenia before the time of the mature civilizations. When men first began to depend on each other, to work together, and to exchange goods, a common language of measurement became a necessity. Each social unit had its own language in this regard. As the social unit grew, in size, from tribe to tribal associations to nation, and, in sophistication, from a culture to a civilization, a consolidation and refinement of the measurement language was part and parcel of the general change in the character of the social unit.

Trade and conquest were perhaps the most important factors in promoting the use of the units of measurement of one country in another and in trending toward the standardization of measurement systems.

Proximity was also a factor of significance. Thus, people residing in border areas could be expected to know of, and use to some degree, the units of measurement of neighboring countries. Other factors contributing to change in measurement systems, as in languages, were mass immigration, re-settlement, and hostage-taking. All of these factors tended to make the measurement systems of different lands similar in many ways, if not the same.

Among the very earliest mature civilizations and power centers on earth were those spawned in Mesopotamia, the land dominated by the Euphrates and the Tigris, and that which sprang up along the Nile in ancient Egypt before 3200 B.C.

The civilizations of the borderlands of Palestine, Anatolia, Syria, Phoenicia, and Persia either never developed into power centers such as Pharaonic Egypt, Babylonia, and Assyria, or, in the case of Persia, did so much later. However, through trade, special skills, and temporary alliances - with each other, with Egypt, or with whomever was dominant in Mesopotamia, the borderlands exercised varying degrees of influence over the power centers and, frequently, much in their own right. Sumerians, Akkadians, Babylonians, Mittani, Kassites, Elamites, Assyrians, Persians, Macedonians, and Parthians succeeded one another in governing Mesopotamia, Persia, and Parthia as well as - at times - Anatolia, Syria, Palestine, and Egypt.

From 2850 B.C., when King Narmer is supposed to have united the Upper and Lower Kingdoms of Egypt, to 1700 B.C. when Egypt came under the dominance of Hyksos invaders from Syria, and again from 1550 B.C. when the Hyksos were driven out of Egypt until 945 B.C., the Egyptians were ruled by Egyptians. However, starting in 945 B.C., when Libyans came to power, foreigners governed Egypt,Nubians, Assyrians, Persians, Macedonians and Romans,except for the last of the native dynasties, which ruled from 404-343 B.C. (sandwiched between two Persian regimes). While Egypt never developed a foreign empire comparable to that of the Babylonians, Assyrians, and Persians, or later, the Macedonians and Romans, it exercised considerable cultural and economic influence over Mesopotamia, Syria-Palestine and, later Greece and Rome. Further, as havens from time to time for various other nationalities, in particular the Jews, it both influenced and was influenced by these people who, when they moved on to other places took much Egyptian culture and practice with them.[1]

Trade developed very early between Egypt, Anatolia, and Mesopotamia via Syria-Palestine. Thus, the systems of weights and measures of each became known to the other, the dominant systems generally being those associated with the older and more powerful civilizations. Somewhat later, trade was initiated between the Minoan and Mycenaean civilizations of Crete and Greece with Egypt and Anatolia. The Egyptian influence on Greece was profound and, when Greek colonies were established in Italy and Sicily, this influence "rubbed off" on the early Romans. Greek colonies were also established in the first millenium in Spain, Southern France, Africa, Asia Minor and along the European shores of the Black Sea. Whatever measuring systems were employed by the mother city-states were established in the colonies.

Another great Mediterranean colonizer and trading nation was Phoenicia which produced the best seamen and navigators of the ancient world. The city states of Tyre and Sidon founded colonies all over the North African coast, Carthage - one of the most important - dating from 814 B.C. The Carthaginians, together with the Phoenicians, founded colonies in Spain, western Sicily, Sardinia, Malta and the Balearic Islands. Phoenicia, influenced as much by Babylonia as by Egypt, spread measuring systems somewhat different than that spread by the Greeks.

When Alexander the Great conquered Egypt, there was initiated a Hellenic-Egyptian relationship which was to last 300 years. The largest Greek city became, not Athens, but Alexandria. Greek soldiers, astronomers, mathematicians became integrated into Egyptian national life. Greek trade multiplied. Even though Greek systems of measurement had originally developed on the Egyptian mold, over the years other influences caused some changes. Now, the somewhat changed Greek units and systems were impacting the much older Egyptian system of measurement.

When the Romans conquered Greece in 146 B.C., there began a relationship unprecedented between conqueror and vassal. Exhausted by hundreds of years of internecine warfare, as well as periodic bouts with the Persians and, more recently, the Romans reclaiming Italian and Sicilian territory held by Greek colonies, the Greeks became Rome's most peaceable subject state. In turn, Rome let Greece run its own affairs, by and large.[2] Rome respected Greek accomplishments in the arts and sciences, if not some of its free-wheeling schools of philosophy. As the Roman empire grew, the problem of administering the conquered territories was eased, to a degree, by Greeks in the Roman service. One of the administrative problems encountered by the Romans was the multiplicity of systems of weights and measures in use in the many colonies and conquered territories. While Roman weights and measures (similar, but not identical to Greek systems) were installed in the subject areas, it was found almost impossible to make natives stop using the systems of measurement with which they were familiar. This was the same throughout Italy itself as it was in Spain, Gaul, Britain, Syria-Palestine, Anatolia, and North Africa. Nor did all Greeks use identical units of measurement.

The problem of a diversity of measurement systems is still with us today, despite almost two centuries of the metric system. We are concerned here, however, with ancient (and, later, medieval) measurements. Let us examine their origins.

AN INTRODUCTION TO
ANCIENT LINEAR UNITS OF MEASUREMENTS

Anatomical Units
Thumb, Finger, and Digit

The earliest linear measurement units were anatomical, or body-related. The thumb, fingers, palm, hand, forearm, upper arm, full arm, and foot were favorite appendages for indicating lengths - singly, in multiples, and in combinations. At various times and places, the thumb dimensions were used as the basis for all linear measurement. King David I of Scotland (c. A.D. 1150) set the inch as the width of the thumb at the base of the thumbnail. And, to account for the difference in the size of men, he specified that the thumb width of a large, medium, and small man be added together and divided by 3 to arrive at this working length standard. The width of the thumb nail, about 3/4 inch, was the English counterpart of a most important Roman length unit, the digit (digitus) which ultimately became standardized at about 1/54th meter (18.52 mm).

The Roman system was an outgrowth of the ancient Greek system, just as the Greek system was an outgrowth of the earlier Egyptian and Babylonian systems, particularly the former. These systems were similar in many respects, but not identical. All units of measurement in these systems were relatable to the finger or digit breadth. However, the manner of determining the finger breadth or width varied, not only between the various civilizations, but within each from time to time and place to place. In some cases the width of the palm was divided by four to obtain finger width. In at least one important case, the ancient Sumerian system (c. 2500 B.C.) - which was the foundation of the Babylonian, Chaldean, and [probably] the Persian systems, and heavily influenced the Greek systems - three fingers made the palm. This leads to the supposition that by "finger," the Sumerians may have meant thumb width, or possibly, knuckle to knuckle span.

By the beginning of the Christian era, the digit or finger of approximately 1/54 metre (18.31-18.79 mm) was in use in Egypt, Assyria, Persia, Syria, Asia Minor, Greece, Italy and the Roman colonies, including Great Britain. However, while apparently the dominant size, this was not the only digit in use. A somewhat later unit adopted by the Greeks which spread wherever Hellenic influence was strong was the finger, defined as 1/16th the Attic foot of 0.3083-0.309m., or 19.27-19.31 mm. The Attic foot was used in the dimensioning of the base of the Parthenon,[3] hence, it is sometimes called the Parthenon foot. This length foot is also referred to as the Olympic foot. Hence, the finger based on this foot is also referred to as the Parthenon or Olympic finger. Even larger fingers than the Olympic came into use, some as large as 22.5 mm. in breadth, and, at least in some of the countries which became Islamic, became predominant. [4]

Palm, Fist, Hand, Lick and Span

The Greeks appear to have been the only ones to employ the knuckle as a unit of measurement. This was arbitrarily given a length of 2 fingers.

The palm, whose width was nominally four digits or fingers, meant different things in different places and in different times. In most systems of measurement, it meant the width of the hand at the knuckles. The Greeks, however, made it the width across the four fingers at the base of the nail of the smallest finger (see Figure 1a.) The Egyptians may have used the same method of determining palm width as the Greeks (or vice versa) but there is at least the possibility that the Egyptians may have measured between the centers of the knuckles of the first and fourth fingers of the hand (see Figure 1b).

In Italy during the 13th to at least the 15th centuries (A.D.), certain of the city-states used the term palm presumably to mean hand length, i.e., the length of the hand from the wrist joint to the tip of the longest finger. This unit, which had a different length in each of the city states of Florence, Venice, and Genoa, averaged about 9 3/4 inches, or 24.8 cms throughout Italy, although some have given the "palmo" a length of as much as 11 1/2 inches, or 29.2 cms. In either case, this would indeed be an enormous hand, raising the question as to what was really meant by "palmo." The ancient Greeks, for instance, put the hand length at 10 fingers, about 7 1/2 inches, or 18.5 cms. This is a more realistic figure for hand-length.

In some systems of metrology, the palm breadth came to be called the hand (this was so in Florence, Venice, and Genoa in the time frame indicated in the last paragraph). This probably stems from the fact that the Romans did not use the hand as a linear unit, preferring the palm while the later city-states needed the term palm for another purpose. The ancient Sumerian system did not use the hand either, preferring the palm, but both the Greek system and its major progenitor, the Egyptian system, used both the hand and the palm. In the Egyptian system, the hand was equal to 5 digits and the palm 4. In the Greek system, the hand equalled 6 fingers and the palm four.

In Islamic metrology, the "gabda" (fist) was equal to four "asba" (fingers), occupying the same place as the palm in ancient Egyptian, Greek, and Roman metrology.

Returning to the hand, this was used in other ways (than as a synonym for palm) to designate linear units. Thus, the lick, or small span, was the distance between the outspread thumb and forefinger, while the span was the distance on the spread hand between thumb and smallest finger, both measurements being taken to the outside of the digits involved. The lick appears to have been used only by the Greeks, in antiquity, and was equal to 8 fingers (see Figure 1c). The Egyptians had two versions of the span - the small span of 12 digits, illustrated in Figure 1e and the large span of 14 digits, presumably for the hand of heavy laborers, thousands of whom were employed in the construction of the famous tombs of the Pharoahs, the Pyramids, which constructions took place over a time span of many centuries.

Figures 1a and 1c are based on pp. 129 and 130 of A.E. Berriman's "Historical Metrology " first published by J.M. Dent & Sons. ltd., London, 1953. Figures 1b, 1d, and 1e are from an article by Louis Barnard Jr. included in the "Report of the 50th National Conference on Weights and Measures - 1965," National Bureau of Standards Miscellaneous Publication No. 272, issued April 1, 1966.

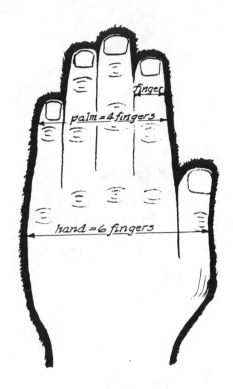

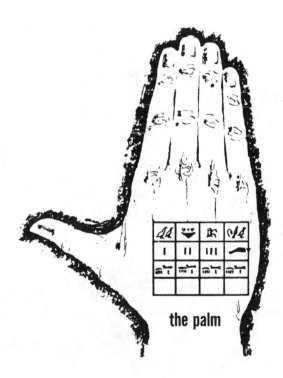

Figure 1a:
The Greek finger, palm (equal to 4 fingers), and hand (equal to 6 fingers).

Figure 1b:
The Egyptian palm (equal to 4 digits) and the hand (equal to 5 digits). The digit evidently was the width of the longest finger, as shown in Figure 1d.

the palm

Figure 1a, 1b. See page 44 for credits.

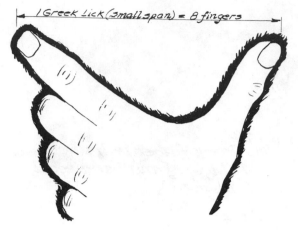

1 Greek lick (small span) = 8 fingers

Figure 1c: The Greek lick, or small span, with a length equal to 8 finger widths (see Figure 1e).

Figure 1d: The Egyptian digit (width of longest finger), equal to ¼ palm width.

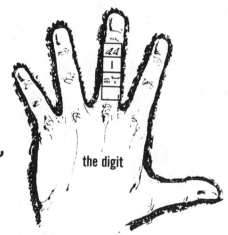

the digit

the span

Figure 1e: The 12 digit Egyptian small span. There was also a great span of 14 digits.

Figure 1c, 1d , 1e. See page 44 for credits

The Foot[5]

One of the most important of the anatomical units was the foot. In the very earliest of metrological systems, the Sumerian (early Babylonian) and that of Egypt of the Old Kingdom (in both cases c. 2500 B.C.), the foot was not as important as the digit, palm, or cubit. However, with the ascendence of Hellenic and Roman influence in the ancient world, the foot surpassed these other anatomical units in importance, i.e., in becoming the base unit for larger linear units of measurement. Later, following the advent of Islam, the foot faded in importance and utilization in those areas of the world which came under Islamic dominance.

In Egypt, as the "t'ser," it was equal to 4 palms or 16 digits and measured approximately 0.2963 metre. The foot/palm/digit size relationship of 16/4/1 was adopted by many civilizations, including the Greek and Roman, but it was never exclusive, nor immune from change where adopted. Thus, the Sumerian foot was equal to 20 shusi (digits), or 2/3 Kus (cubit), the latter the most popular foot/cubit relationship, but again, not the only one. The Sumerian foot, sometimes called the Early Babylonian (not be to confused with the Babylonian, a much later unit), measured about 0.335 metre. This unit appears to have been the predecessor of the Assyrian foot of approximately 0.329 metre.

Two prominent foot lengths utilized in the Graeco-Roman world were the Ancient Greek or Pelasgic, essentially the same size as the Egyptian t'ser, approximately 0.296 metre, and the Greek Olympic, or Attic, foot (sometimes also called the Parthenon foot). The latter unit had a length of approximately 0.309 metre. The Romans, ever interested in standardization, adopted the Ancient Greek foot as their own official foot.[6] Being also aware of the widespread use of the Olympic foot, there was a studied effort to establish an official relationship between the two feet. This emerged as 25/24 and all other units of measurement based on the foot and common to the Greek Olympic and the Roman systems became related in the same ratio.

As Roman influence grew in the ancient world, the Roman (Ancient Greek) foot became employed in Syria, Asia Minor, the Middle East, and Africa, besides Greece, Italy, and the Roman colonies, including Great Britain. The Greek Olympic foot also gained adherents, and in just about the same places.

It would be nice, indeed, if we could assume that the foregoing were the only feet of antiquity, or even classical Greece and Rome. But, we cannot. Variants of the Babylonian foot (0.3142-0.3166 m.) were widely employed throughout Greece and the Roman Empire. In Greece, as the Athenian foot of 0.316 m., it became the basis for a combination anatomical/decimal/sexagesimal system of linear measurements which overlapped a similar system based on the Greek Olympic foot. In the Roman Empire, the Babylonian foot became known as the "long" foot, the "short" being reserved for the Ancient Greek foot.

Before leaving the subject of the foot, it is necessary that mention be made of certain other ancient feet which enjoyed some reasonable acceptance or which are otherwise important to our story. The Pythic foot, in its double form measuring 0.481-0.490 m., found significant employment in Egypt, Assyria, Persia, Asia Minor, Greece, Africa, and Sardinia. The Plinian foot, equal to one-half the Talmudic cubit, measured 0.2742-0.2775 m., and was employed to some degree in Egypt, Asia Minor, and Sardinia.

Still another foot employed in varying degrees throughout Greece, the Roman Empire, Egypt, Syria, Asia Minor, the Middle East, Africa, and Sardinia was the Drusian foot (0.3342-0.3415 m.) which appears to have been based on the Sumerian foot.

Finally, there was the Philetaerian foot of 0.3529 m. which was employed over a significant period of time in the vicinity of the important island of Philae in the Nile River, just above (south of) the First Cataract. Philae was a religious and cultural center in ancient Egypt and, as that country was over-run by a succession of foreign invaders, became the last stronghold of the ancient Egyptian religion. The Philetaerian foot may have been employed in one of the most important geodetic estimates in ancient history. (See Chapter 3.) It will be noted that 1/16th of 0.3529 m. is 22.06 mm., which puts the Philetaerian digit in a special class, larger than the Egyptian, Greek, and Roman digits or fingers and more like the medieval Muslim asba related to the "Black" cubit of Caliph Al Mamoun. The "Black" cubit was used in an important geodetic measurement conducted in A.D. 830. (See Chapter 4.)

As late as the beginning of the Christian era, there was in use in Chaldea a Royal Babylonian foot rated as three digits longer than a Roman foot of 16 digits. On the assumption that the digit was the same as the Egyptian/Ancient Greek/Roman digit of 1/54 metre, or 0.01852 m., this foot had a length of 0.3519 m., just about the same as the Philetaerian foot.

Within the ancient Greek cultural orbit there appear to have been several differently sized foot measures in use at one time or another and in one place or another. In metropolitan, port, or border areas, three or more might be in use simultaneously.

Summarizing, the most important ancient feet were the following:

Name of Foot		Length in Metres
Egyptian T'ser, Ancient Greek, Short Roman	=	0.2926 - 0.300
Sumerian (Early Babylonian), Assyrian, Drusian, Long Greek	=	0.329 - 0.341
Babylonian, Athenian, Medium Greek, Long Roman	=	0.3142 - 0.3166
Greek Olympic, Attic, or Parthenon	=	0.3075 - 0.309
Double Pythic	=	0.481 - 0.490
Plinian (= 1/2 Talmudic cubit)	=	0.2742 - 0.2775
Philetaerian, Babylonian Royal	=	0.3518 - 0.3529

The Cubit (Elle) and Brachium (Pic)[7]

The next larger body-related unit used in ancient linear metrology, was, perhaps, the most ubiquitous. It was the celebrated cubit, believed to have been invented by the Egyptians about 3000 B.C. The Sumerians are known to have been using this unit, which they called the ell, as early as 2500 B.C. The cubit was the length of the forearm from elbow joint to the end of the longest finger (see Figure 2a). The Romans had an alternate name for the cubitus. It was the ulna, from which is derived the English ell and the French aulne. (Ulna, in English, is the proper name of the forearm bone on the opposite side of the thumb.)

Another unit, sometimes referred to as the cubit, was the brachium, or pic. This was the length of the upper arm. The origin of this unit is obscure, but it became quite popular among Mediterranean countries during the Middle Ages particularly, but not exclusively, for the measurement of cloth. Columbus used it to describe the height of a wave at sea, and Toscanelli, a contemporary Florentine physician, astronomer, and cosmographer, defined his mile in terms of the brachium. That the brachium enjoyed some significant use in ancient Rome is shown by a legend engraved upon a stone tablet at the foot of the monument of Cosutius, now the Museo Capitolino, in Rome. The legend identifies several units of linear measurement in use at the time. Among them are the Roman foot, the Greek foot, the braccio di tela,[8] the braccio di merc.,[9] the staiolo, and the architect's cana. As indicated in the previous section, the ratio of the length of the Greek (Olympic) foot to the Roman (Ancient Greek) foot is shown to be 25/24. The relationship of the braccio di tela/braccio de merc./staiolo is shown to be 3/4/6. The absolute lengths of the braccio di tela and the braccio di merc, are shown to be 0.636 m. and 0.8496 m., respectively.

Before ancient Egypt was over-run by a succession of foreign conquerors, there were only two official cubits in Egypt (although, as a result of immigration and trade, several other cubits had already been introduced). The small cubit of 24 digits, 6 palms, or 3/2 t'ser, measured 0.450 m. The Royal cubit of 28 digits, or 7 palms, had a length of about 0.525 m. (See Figure 2b.)

The first foreign rulers of Egypt were the Hyksos, of uncertain origin, who invaded and settled in the Nile Delta in the 18th century, B.C. Coming from the Syria-Palestine region which had been introduced to the horse and chariot by the Hittites, the Hyksos overcame the native Egyptian kings and ruled Egypt from 1670 to 1570 B.C. During this period, the Hyksos - while restrained in the changes they introduced to Egypt - permitted the expansion of immigration into the Nile Delta region, their own stronghold. The pharaoh ruling Egypt during the time that the Biblical Joseph, son of Jacob, became the chief advisor of the pharaoh,was a Hyksos. It was he who invited Jacob and his sons to come and settle in the Delta region. The first Israelites were, of course, the sons of Jacob (renamed Israel). During the period that the Israelites spent in Egypt until Moses led them into the Sinai, the Israelites had multiplied [10] and had come to be considered a foreign force by the native Egyptian pharaohs who had expelled the Hyksos in 1570 B.C. In the region peopled by Israelites, Syro-Palestinian and Mesopotamian weights and measures existed side by side with the official Egyptian measures.

Egypt was under Assyrian rule from 671 to 652 B.C.; experienced a large immigration of Jews in 587 B.C. when Judah was overcome by the Babylonians; came under Persian rule from 525 to 404 B.C.; after a brief period of native rule, was under Persian rule again from 342-332 B.C.; came under Macedonian-Greek dominance when conquered by Alexander the Great in 332 B.C. and remained so under the Ptolemaic Dynasty until 30 B.C., when it became a Roman province. By this time, the following cubits were in use somewhere or other throughout the country:

Name of Cubit		Length in Metres
Roman	=	0.444
Egyptian "short"	=	0.450
Greek Olympic	=	0.463
Assyrian "short"	=	0.494 - 0.507
Royal Egyptian and Babylonian	=	0.523 - 0.527
Talmudist, Large Assyrian	=	0.540 - 0.548
Sacred Hebrew, Palestinian, Royal Persian, and Chaldean	=	0.637 - 0.644

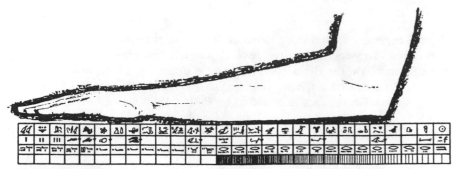

the small cubit

Figure 2a: *The Egyptian "small" cubit of 24 digits or 6 palms compared with 28 digit Royal cubit.*

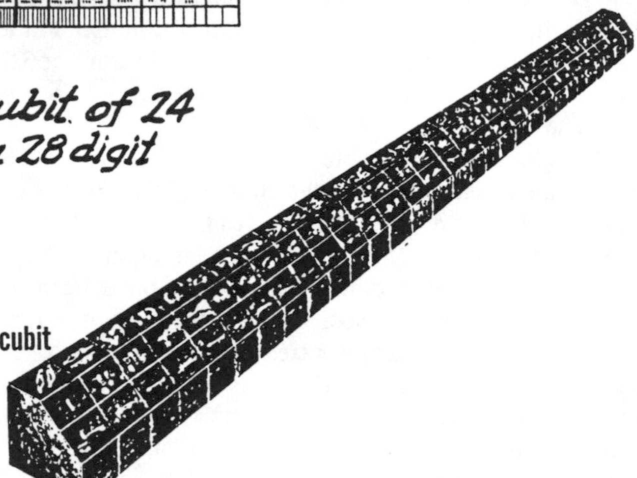

the royal cubit

Figure 2b: *The Royal Cubit of 28 digits or 7 palms was used only on royal construction projects. Shown in illustration is a black granite master (of the Royal cubit) against which all working cubit sticks, small or royal, had to be checked for accuracy each month.*

Credit
Figures 2a and 2b are from same source as Figures 1b, 1d, and 1e — Lewis Barnard's article

Figure 2a and 2b. See next page larger scale version of Egyptian Royal Cubit rod.

EGYPTIAN ROYAL CUBIT

THE CUBIT OF KING AMENHOTEP I 1559-1539 B.C. 18th DYNASTY.

ORIGINAL IN THE LOUVRE, PARIS.

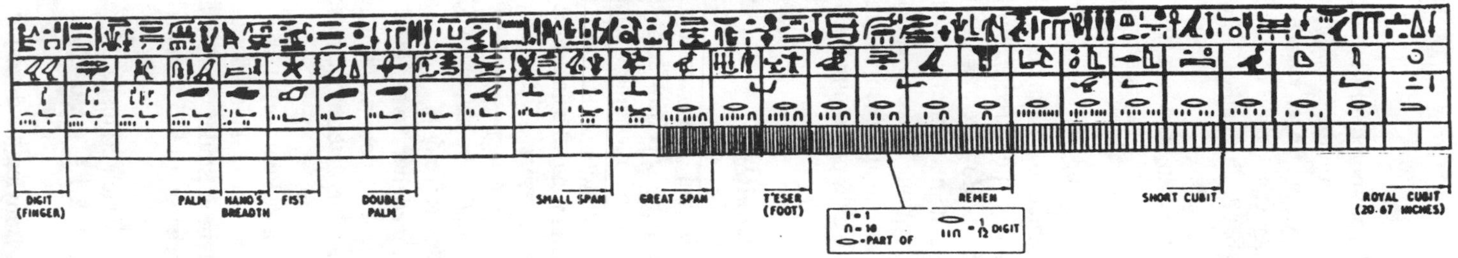

DIGIT (FINGER)	PALM	HAND'S BREADTH	FIST	DOUBLE PALM	SMALL SPAN	GREAT SPAN	TESER (FOOT)	REMEN	ROYAL CUBIT (20-67 INCHES)

SHORT CUBIT

REMEN
I = 1
II = 10
◯ = PART OF
◯ = ½ DIGIT

Figure 2c

Figure 2c: A Photograph of A Model Egyptian Royal Cubit Rod, Engraved in Granite, Constructed by the Division of Applied Physics, National Measurements Laboratory, Commonwealth Scientific and Industrial Research Organization, Australia. Mr. M.J. Puttock, formerly of the Laboratory, directed the construction, based upon a photograph in a German archeological text. This reproduction is made possible through the courtesy of Mr. Puttock and Mr. P.E. Ciddor of the Laboratory.

51

The short Egyptian cubit of 24 digits appears to have been the prototype for both the Ancient Greek and Roman cubits of 0.444 m. The Greeks enlarged this to 0.463 m. to form the Olympic cubit. There are two explanations as to how this was done. One holds that the Olympic finger was enlarged from the Ancient Greek finger of 0.01852 m. to 0.01929 m., 24 of which gave the cubit of 0.463 m. The other (propounded by Petrie) says the finger remained at 0.01852 m. but 25, instead of 24, were used to make the Olympic cubit.

While there was some proliferation of the cubit in Greek metrology, it was not as great as in Egypt. Rome, with its penchant for standardization, had even less. In any event, the cubit never enjoyed the usage in classical Greece and Rome, of the foot.

As early as the 3rd millenium B.C., the Sumerians had developed a system of linear measurement which employed three sizes of elles or cubits. The small elle, equal to 3 palms or 9 digits, had a length of 0.297-0.2988 m. The medium elle, equal to 4 palms or 12 digits, measured 0.396-0.3984 m., while the large elle, equal to 5 palms or 15 digits, measured 0.495-0.498 m. However, as a result of a succession of strong rulers from various parts of Mesopotamia and its immediate neighbors, as well as immigration and trade, other cubits (and related units) were introduced into the region. The cubits listed for Egypt when it became a part of the Roman Empire were probably also in use in Mesopotamia, although the two Assyrian, the Royal Babylonian, and the Royal Persian cubits or elles predominated.

Some Evidence on the Size of the Digit, Foot and Cubit[11]

In 1639, John Greaves, Gresham Professor of Geometry at Oxford University and subsequent Professor of Astronomy at the same institution, journeyed to Rome for the specific purpose of determining the length of the ancient Roman foot. In the Vatican gardens, he found a monument commemorating a young architect named Statilius Aper who had died in his 21st year. The monument, which had been made in the first century A.D., was in fact a large and elaborate cinerarium and the instruments of the deceased's profession which it portrayed in relief included the Roman foot-measure. Greaves reported that the Roman foot measures 1944/2000, or 0.972 of an English foot, thus 11.664 inches, or 296.27 mm. The Roman foot was rated 16 digits, thus making the Roman digit equal to 0.729 inches, or 18.52 mm = 1/54 meter.

In 1750, James Stuart, subsequent author (with Nicholas Revett) of *The Antiquities of Athens,* measured the platform of the Parthenon, in Athens, and determined that it was 100 Greek feet in width and 225 Greek feet in length. A century later, Francis Crammer Penrose confirmed these measurements. The Greek foot - in particular, the Parthenon, or Olympic foot - was thus determined to be 1.0126 English feet, 12.151 inches, or 308.64 mm.

From the relationships developed by Greaves, Stuart, and Penrose, it is evident that the Greek (Parthenon) foot was equal to 12.151/11.664, or 1.042 Roman feet. This relationship is attested to by Strabo, Pliny, and other ancient writers who said that the Olympic stade, which was equal to 600 Greek feet, was the same length as the Roman stadium which had been set to 625 Roman feet. Thus, 625/600 = 25/24 = 1.042.

The Roman digit had been set equal to the Egyptian digit of which 20 made the remen, and $20\sqrt{2}$ made the Royal cubit. Hence, the Royal cubit must have equalled 20 x 1.414 x 0.729 ins. = 20.619 inches, or 523.73 mm.

By way of corroboration, on Herodotus' visit to the city of Babylon, he determined that the Royal cubit in use there was longer by 3 finger breadths than the Greek cubit. The Greek cubit was rated 24 fingers, thus Herodotus put the Babylonian royal cubit at 27 Greek fingers. As the Greek finger measured 25/24 the Roman or Egyptian digit, the Babylonian Royal cubit at the time of Herodotus equalled 27 x 25/24 x 0.729 inches = 20.503 in. or 520.78 mm. It is interesting to note that the Harmhab cubit-rod was just about this length. It is equally interesting that the ratio of the length of this Royal cubit to the large Sumerian cubit of the time of Gudea (c. 2175 B.C.) was 25/24.

In the sixth book of Pliny's *Natural History,* the Babylonian Royal foot [12] is indicated as being 3 digits longer than the Roman foot of 16 digits. Thus, the Babylonian Royal foot was 19 digits, 19 x 0.729 inches = 13.85 inches, or 351.8 mm. As the relationship between the cubit and the foot appears to have been 3/2, almost universally, this would have made the Babylonian cubit 527.7 millimeters or 20.78 ins.

Clearly, from the foregoing, it would appear that by the time of Herodotus, the Royal cubits of Egypt and Babylon were of comparable if not identical length.

In a most unusual work, *Inductive Metrology, or The Recovery of Ancient Measures from the Monuments* (previously referred to several times), Petrie describes his procedures and results in making over 4000 measurements on more than 600 monuments, buildings, tombs and artifacts (tablets, cubit and other measuring rods, nilometers). From his own measurements, he has deduced the units of measure employed in building each structure (or artifact) by its ancient builder (or fabricator). His field work was done in Egypt, modern Iraq, Iran, Syria, Asia Minor, Greece, North Africa, Sardinia, and Italy and the ancient Roman colonies, including Great Britain.

In the simplest terms, Petrie summarizes the logic behind his approach to "Inductive Metrology," e.g. " . . . in most ages when men have to lay out or plan any work that requires to be alike on two sides, if they have a measure, they will use it; and they will probably use whole numbers in preference to fractions, and round numbers in preference to uneven ones, merely for convenience in their work."

Petrie describes his results both in his text and in two spread-sheet tables (one in English, utilizing British Inches, and one in French, utilizing millimetres). Table 1 is a condensation of the data appearing in his spread-sheet tables, deleting information not vital for our own purposes. All lengths in Table 1 are stated in metric only, to be consistent with the convention adopted for the greater part of this book. In support of Table 4, sections of Petrie's, *General Results* (pp. 144-148 of text) are quoted below:

> **3rd.** By these means the exact values of the Sacred Hebrew or Royal Persian cubit, the Royal Egyptian cubit, the Egyptian digit, the Assyrian "hu" or "U", the ancient Greek foot, the Olympic foot, the Drusian foot, the Plinian foot, and the Pythic foot, have been ascertained from the monuments, along with the probable errors of these determinations.

> **4th.** The following measures have been found in more extended use than was previously known. The Sacred Hebrew or Royal Persian cubit is found to have been used in Greece, Mohammedan Persia, and apparently by North American mound-builders.[13]

> The Pythic foot is found to be an Egyptian measure used also in Assyria, Persia, Asia Minor, Africa, and Sardinia. Its Arabian form is connected by Quiepo with the Hebraio-Persian cubit.

> The Assyrian "great U" of Oppert[14] is found in Syria, Asia Minor, Sardinia, and Roman Britain, and is very probably the basis of medieval English units, including the British inch.

> The digit has been found in Assyria, Persia, and Syria, and shown to be independent of the Royal Egyptian cubit.

> The Roman foot has been found in Syria, Asia Minor, and Greece; and its identity with the Pelasgic foot (as well as the Etruscan cubit) exhibited.

> The Olympic foot has been found in Asia Minor and Roman countries, as well as in medieval England, and its origin from the ancient digit pointed out.

> The Royal Egyptian cubit is found to have been used in Assyria, Persia, Syria, Asia

Table 1: A Condensed Version of W.M. Flinders Petrie's "Synoptic Table of the Units of Measure Found by Inductive Examination of the Monuments", in Millimetres.

(From pp. 142-143 of his book "Inductive Metrology, or the Recovery of Ancient Measures from the Monuments", Hargrove Saunders, London, 1877.

Country of Origin of Monument or Artifact Measured	Digit	Plinian Foot = Talmudic ¼ Cubit	Pelasgic Foot = Ancient Greek = Roman	Olympic Foot	Babylonian Foot	Drusian Foot	Indian Hasta	Double of the Pythic Foot	Half of Assyrian Great "U" or Su-Klu	Royal Egyptian and Babylonian or Ionian Cubit	Assyrian Cubit	Sacred Hebrew, Royal Persian and Chaldean Cubit
Egypt	18.40–18.526	274.7		308.8	344.2	334.9		480.4		524.5	543.4	637.3
Assyria	18.54							483.5	507.1	523.1	542.64	642
Persia	18.64							485.7		525.6–527.1	542.8	643.4
Syria	18.31–18.79		296.3			336.7			514.0	526.4		640
Asia Minor	18.45–18.613	277.0	295.5	307.5		339.2	454.5	490.12	507.1	524.1–525.6		
Greece	18.38–18.41		294.0–294.5	308.74	314.9	336.7	454.1	489.8		524.9		644.5
Africa			298.1			338.7		488.5				
Sardinia		274.2				336.7		481.4	509.9			
Italy and (Roman) Colonies	18.295–18.588		292.6–296.6	309.0	316.1	341.5			506.8		542.8	
Great Britain			294.5	309.0	316.6	335.7			505.5–507.7		541.9	

Note: Each entry above is supposed to represent the mean value of many derived from various sites. When -within a country- differences were noted between sites, Petrie listed the results for each site seperately. We show only the least and greatest values. Petrie made his own direct measurements of monuments and artifacts in British Inches. After obtaining- "inductively" - the value of each ancient unit of measurement employed at a site, he stated these in British Inches. Seemingly irrational entries (above) which show lengths in hundredths and even thousandths of a millimetre result from Petrie's "inductive" and averaging processes and later conversion from British measure to metric.

Minor (agreeing with what Herodotus states), and Roman Gaul.

The Babylonian foot, derived from the regular decimal division of the Egypto-Babylonian cubit, is found to have been the principal unit in Greece, to have been used by the Romans, and to have been continued in England, principally as a Saxon unit; it was also used in Mohammedan Persia.

The Assyrian cubit (as determined by Oppert) is found to be an ancient Egyptian unit, used also in Persia, Italy, and Sardinia; and it is also identical with the early Christian Irish unit, the commonest unit of pre-historic remains, and the unit of Mexico.

The Drusian foot is found to be also an Egyptian unit, and to have existed in Syria, Asia Minor, Greece, Italy, Africa, and Sardinia, and to be apparently the origin of the most usual measure in medieval England.

The Plinian foot is found to be of Egyptian origin, and to have been used in Asia Minor.

The Philetairean foot is likewise found first in Egypt.

The ancient Indian hasta has probably been determined, and is found to be an Aryan unit also used in Asia Minor and Greece.

The Step and Pace

The anatomical body unit next larger than the cubit was the step, or half-pace. In both the Greek and Roman systems, the step was equal to 2-1/2 feet or 1-2/3 cubits. And, in both the Greek and Roman systems, the step was equated to the sum of the foot and cubit, thus, was equivalent in length to 40 digits or fingers. In the Greek Olympic system this measured 0.7716 m., while in the Roman system, the length of the step was 0.74074 m.

In both systems, two steps made the double-step, or pace. This seemed to enjoy a greater usage among the Romans than among the Greeks, although all ancient peoples utilized pacers (bematistes, in Greek) to carry out itinerary measurements, no matter in which units the measurements were ultimately delineated. The Roman pace, at least up until the first century B.C., was equal, as has been indicated, to five feet. Sometime during the first century, another pace appeared upon the scene, one equal to six feet.[15] The five foot pace was used widely by the Romans, its most important use by far being to make the Roman mile: 1000 paces, or 5000 feet, made the mile. In fact, the name "mile" is a derivative of "millepassus," 1000 paces. The six foot pace, taken 1000 times, may have been the basis for an early "geographic mile", but this is conjectural. The metrological literature is singularly sparse on the six foot pace.

The uses of the five foot pace, taken in multiples, will be explained further under "Multiple and Combination Units."

The Fathom and Yard

Just as the pace was an important Roman unit, but seems not to have been accorded the same status by the Greeks, the fathom was a singularly Greek unit in origin which was not used as much by the Romans. The Greeks defined the fathom as the distance between fingertips (of the longest fingers) when a man's arms are outstretched horizontally to its greatest reach or spread. (See Figure 3a.) It was equated to four cubits or six Greek feet. This unit was introduced to Britain either during the Roman conquest or when many Belgic tribes migrated across the English channel a few centuries later.

An outgrowth of the fathom (which itself has achieved enormous acceptance throughout the world by maritime nations as the unit for indicating the depth of navigable waters) was the English yard. This was in use in England long before it received the official definition (illustrated in Figure 3b), the distance from nose to the end of the thumb of an outstretched arm, horizontal, and in line with the chest. As can be seen, the English yard was somewhat smaller than one-half the Greek fathom.

Multiple and Combination Anatomical Units, Including Arbitrary Units

We have seen that, for whatever reasons, a number of body-related units of linear measure came into being in various parts of the world. These units became related in arbitarily chosen multiples to other units of independent origin. Usually, one unit was chosen as the basic standard and all others units were then related to the standard, directly or indirectly. The definition of the basic and dependent units were so worded (or re-worded) as to lend credence to the arbitrarily chosen numerical relationship.

When the need arose for units smaller than the digit or larger than the fathom, or for units intermediate in size to existing units, they were created. Sometimes the approach was simply to multiply or divide the existing unit by a whole number, sexagesimal, duodecimal, or decimal in character. On other occasions, two existing units were added to create a new unit of a desired size.

An early example of the employment of both multiple anatomical and arbitrary units is given by the Sumerians who, around 2500 B.C., had established the palm as their basic unit of linear measurement. Obviously, they needed smaller units than the palm (which was about 100 mm in width). Thus, they divided the palm by 3 to obtain a sossus, and by 180 to obtain a line. At this time, the "line" represented the width of a clear impression made on a clay tablet with a stylus, the intent being to form human and animal figures and cuneiform characters (used in writing). One hundred eighty is a nice sexagesimal number and the line, thus obtained, was approximately 1/2 millimeter or 0.00056 meter in width.

Illustrating a mix of combination, multiple anatomical, and arbitrary units, we have the Greeks, a couple of millenia later, who added the foot, equal to 16 fingers, and the knuckle, equal to 2 fingers, to obtain the 18 finger pygme . This could also be obtained by adding the lick of 8 fingers to the hand-length of 10 fingers. Similarly, they added the foot and the palm to obtain the 20 finger pygon and the foot and the cubit to obtain the 40 finger step. They multiplied the span of 12 fingers by 6 to obtain the xylon, which could also be obtained by taking three cubits of 24 fingers. Six feet, or four cubits, made the fathom. Ten feet made the pole, also called the arkana. Six poles (arkanas) or 10 fathoms made the cable of 60 feet. Ten poles made the plethron of 100 feet. Six plethrons, 10 cables, 100 fathoms, or 400 cubits made the 600 foot stade or stadion. Four stades made the ride and, when the basic building block was the Greek Olympic foot of 0.30864 metre, two rides were exactly equal to the Roman mile based on the short Roman (Ancient Greek) foot of 0.2963 metre, i.e., 4800 Greek Olympic feet equalled 5000 Roman feet = 1481.5 metres.

Not all of the units described were in use at any single time or at any single place. When they were used, however, the proportions between units were as indicated. We see, in the foregoing a mix of sexagesimal, decimal, and binary multiples.

The Romans followed a pattern quite similar to that of the Greeks, but, ever the practical ones, managed to reduce the number of units of length employed while accentuating a duodecimal relationship. Perhaps the most famous unit of length introduced by the Romans was the "uncia," or inch, arbitrarily set at 1/12th the length of the foot. The uncia had a length of 2.47 cm or 0.972 U.S. inch.

An example of a unit which was both anatomical and arbitrary is the English rod. Some English monarch of the 16th century set the rod as the length of the left feet of the first 16 men out of church on a certain Sunday. The men were to stand toe to heel as shown in Figure 4. The rod is still in use in English speaking countries. It's current length is 16-1/2 feet.

One of the earliest examples of an arbitrary unit was the Egyptian remen, nominally equal in length to 20 digits. It may have originated as the double-remen, the length of the diagonal of a square whose side

Fathom = 6 Greek Feet

Figure **3a** (above): The fathom, the distance between fingertips of the outstretched arms, was the largest body-related unit in the ancient Greek metrological system. It was equal to 4 cubits, 6 feet, 8 spans, 12 licks, 16 handbreadths, 24 palms, or 96 fingers.

Source for this illustration is the same as for Figures 1a and 1c.

the yard

In the 12th century, English King Henry I decreed the yard to be the distance from the tip of his nose to the end of his thumb.

Figure 3b (directly above): The English yard, derived from the Greek fathom, is approximately one-half the length of the Greek unit. It will be noted that the yard stretches to the end of the thumb, not the longest finger.

Credit for this illustration the same as for Figures 1b, 1d, 1e, 2a, and 2b.

was the Royal cubit of 28 digits, i.e., 1 double remen = $\sqrt{2}$ Royal cubits = 1.414 x 28 digits = 39.6 or approximately 40 digits. Conversely, the Royal cubit was the length of the diagonal of a square whose side was the remen, i.e., one Royal cubit = $\sqrt{2}$ remen, or one remen = $\sqrt{2}/2$ Royal cubit = 0.707 x 28 digits, or approximately 20 digits. Since the Egyptian digit was approximately 1/54th metre, or 0.01852 m., the remen had an approximate length of 0.37 m.

It is believed the remen and the double remen were used in land measurement in ancient Egypt, for it permitted the ready doubling or halving of areas without altering their shapes.

the rod
In the 16th century, the rod was set as the length of the left feet of the first 16 men out of church on a certain Sunday.

Figure 4: The Rod

Source for this illustration is the same as for Figures 1 b, d, and e, 2 a, b, and 3 b.

Berriman has shown[16] that the remen was related (approximately) to other ancient units of linear measurement as follows:

Roman foot	=	4/5	remen
Greek "	=	5/6	"
Ancient Assyrian foot	=	8/9	"
Roman cubit	=	6/5	"
Greek "	=	5/4	"
Short Assyrian cubit	=	4/3	"
Royal "	=	$\sqrt{2}$	"
Talmudist "	=	3/2	"
Palestinian "	=	$\sqrt{3}$	"

Also, 500 remen made one version of the Greek itinerary unit, the stade.

Another ancient Egyptian linear unit used in land measurement as early as the period of the Rhind Mathematical Papyrus[17] (c. 1991 B.C.) was the Khet which had a length of 100 Royal cubits. The setat was a unit of area equal to one square Khet. However, it was probably intended to signify 100 parallel strips each one cubit wide and one Khet long, and was frequently referred to as a 'cubit of land.'[18]

One hundred cubits was also referred to as the "hayt" (rod or cord) and, in the time of the Ptolemies (323-30 B.C.), the "schoenia."[19] The different names accorded the unit measuring 100 cubits may have stemmed from the intent to employ short, rather than Royal, cubits in certain of the cases, but this is conjectural. It is more likely that at different times in Pharaonic Egypt (Old, Middle, and New Kingdoms, and periods of Libyan, Nubian, Assyrian, Persian, and Ptolemaic rule) and at different places (Delta/Lower Egypt, Thebes/Upper Egypt, and 1st Cataract/Border Areas) different nomenclature were employed. What has been mentioned is only a sampling of that.

Botanical Units

Another type of linear measure which developed hand-in-hand with anatomical units was botanical in origin. Originally, this was due to the requirement for units somewhat smaller or larger than the body-related units available. Since the smallest body unit was the finger (frequently called the digit), one way of subdividing the finger into smaller units was to relate it to one of several types of seeds.

In 14th century England, during the reign of King Edward II, three grains of barley, dry and round, placed end to end, lengthwise, was considered the equivalent of the average thumb width, or one inch. By the 16th century, the rule had become " . . foure graines of barley make a finger; foure fingers a hande; four handes a foote." Thus, besides the fact that 64 barley grains made a foot, the number of grains per finger of about 3/4 inch in width was, by this time, greater than the number of grains per thumb-width of one inch had been two centuries earlier. Clearly, different strains of barley were being employed as parameters, or, different dimensions of barley grain. The barley grain was also used in continental Europe and parts of Asia to provide a unit of length smaller than the digit or finger. Usually, 4 grains of barley made the digit, indicating the *length* of the barley grain, about 4.6 mm, was the parameter. However, one finds that the *width* of the barley grain, 6 to the digit (or about 3.1 mm) was also employed.[20]

In areas of the world, such as Mesopotamia and the Nile delta, where there is considerable marshland, reeds (which need little encouragement to grow) are abundant and are used for a variety of purposes. As early as 2500 B.C. the reed had found its way into the Sumerian system of linear measurement. A small reed was made equivalent to four small elles (cubits) or 12 palms; a medium reed was defined as six medium elles (cubits) or 24 palms; and a large reed was equated to six large elles (cubits) or 30 palms. The Sumerian palm was almost 100 mm, so the three sizes of reeds provided units with approximate lengths of 1.2 meters (47 inches), 2.4 meters (94) inches, and 3 meters (117.5 inches or 9.8 feet). The Egyptian measuring reed is reported[21] as having a length of 6 Royal cubits, i.e. 3.15 metres (124 inches or 10.34 feet).

An amusing story[22] descriptive of ancient anatomical, botanical, arbitrary, and related linear units of measurement found in the Orient, is contained in the writings of Hiuen Tsiang (Yuan Chwang), 603-668 A.D. A Chinese who had traveled in India, in A.D. 629, reported the following: "In point of measurements, there is first of all the yôjana (yu-shen-na); this from the time of the holy kings of old has been regarded as a day's march for an army. The old accounts say it is equal to 40 li; according to the common reckoning in India it is 30 li; but in the sacred book (of Buddah) the yôjana is only 16 li. In the subdivision of distances, a yôjana is equal to 8 krosas (keu-lu-she); a krosa is divided into 500 bows (dhanus); a bow is divided into four cubits (hastas); a cubit is divided into 24 fingers (angulis); a finger is divided into 7 barleycorns (yavas); and so on to a louse (yuka), a nit (liksha), . . . a cow's hair, a sheep's hair, a hare's down, . . . and so on for seven divisions, till we come to a small grain of dust: this is divided seven-fold till we come to an excessively small grain of dust (anu); this cannot be divided further without arriving at nothingness, and so it is called the infinitely small (paramanu)."

We note from the foregoing that the Chinese or the Indians, or both, are not above larding a good story. We also note that from the finger through the bow (dhanus), which at four cubits (hastas) is equivalent to the Greek fathom, the relationships are identical to those in the Greek system. The relationship between the finger breadth and that of the barley corn differs significantly from that of 16th century England, but to a much lesser degree from the relationship used in continental Europe where 6 barley widths of approximately 3.1mm made the digit of 18.52mm or the finger of 19.3mm. The 7 barleycorn ratio found in India could indicate that a somewhat smaller barley grain had developed there or that the finger dimensioning, if not actual size, differed from that in the western nations. Table 4, Chapter 4, indicates that finger widths of 20.78mm and 22.52 mm. were employed in several Islamic countries.

If we assume that the Indian finger of the 7th century had about the same breadth as the fingers and digits reported for earlier Egyptian, Greek, and Roman systems of measurement, which average out to be 18.8 mm., or 0.742 ins., then the yôjana, at 384,000 fingers, had a length of 7219 meters.[23] This compares reasonably well with the relationship of 16 Chinese li given in the sacred book of Buddha. The Chinese li itself varied between 536 and 650 meters at different times and places in China. Thus 16 li would be equivalent to 8576-10,400 meters.

On the other hand, the yôjana's ancient definition as a day's march for an army compares reasonably well with the 40 li specification of "the old accounts." A day's march, on a continuous basis, can vary significantly with climate, terrain, condition of the road, trail, or open country being traversed, and the number of days troops have been on the march. In moderate climes, 15 statute miles per day for troops bearing their own packs and arms is considered a reasonable figure. Major Rennel[24] reported that the mean day's march of explorers sent by Nero to find Meroe (around 17° N latitude) in the Sudan was 14.55 Roman miles, or 21,563 metres. Forty li would range from 21,440 to 26,000 metres. However, if 40 li compares well with a day's march, it does not with 384,000 fingers, for such fingers would have to measure about 2-1/4 inches, or 5-1/4 centimetres in breadth.

An important conclusion to be drawn from the preceeding story is that the length of the Indian yôjana (and probably the Chinese li, as well) varied considerably with time, and (inevitably) with place. And, at least in some places, there were more than one yôjana in use at any given time. As we shall see when we examine intinerary units of other systems in greater detail, this characteristic was not restricted to the yôjana or the li.

Itinerary Units of Measurement

The story related by Hiuen Tsiang serves as a good introduction to the topic of "itinerary measurements," the measurements of geography, geodesy, military operations, and athletic contests - particularly running.

All ancient linear systems had itinerary units related, at first, to human performance and later augmented to include the performance of animals such as the horse and camel. Some, for instance, defined the distance that troops could march in four hours without rest, or per-day, on a continuous basis, allowing for rest. Units such as the Roman mile, schoenus, parasang, and league (legua, leuga, lieue), the last of medieval rather than ancient provenance, came into being to express such distances. It was more convenient to say 12 miles than 12,000 paces or 60,000 feet. And, it was still more convenient to express a longer distance as, say, six schoeni, or parasangs, than the equivalent number of miles.

As indicated in the relation of the story about the yôjana, which unit we will encounter again in Chapter 4 (see, in particular, coverage of Aryabhata and Bhaskara Acharya), the length of itinerary units tended to vary from time to time and (probably) place to place. And, the longer the unit, the greater the variation. This was not entirely due to the relationship of the unit to human or animal performance, which had to vary in different climates, altitudes, terrain, etc. It was also due to the compound specification of the unit in terms of the smaller anatomical units such as the foot and cubit, both of which varied so widely, as has been shown.

An important itinerary unit in the ancient world was the Greek stade, or stadion. This unit was originally defined as the distance a strong man could run without stopping for breath.[25] Later, it became the standard length of the Olymic track and was equated to 100 fathoms, 500 remens, or 600 Greek Olympic feet. Alexander the Great and his armies introduced the stade (stadion) into the Middle East and (while this unit does not figure in the previous story about the yôjana) probably India. (The anguli/hasta/dhanu [bow]/krosa/yôjana relationship would be identical to that of the Greek Olympic finger/cubit/fathom/5 stades/40 stades.) The Romans, in their zeal for standarization, made their own stadium equal to the Greek Olympic stadion. The Romans had earlier standardized their foot on the Ancient Greek foot. Thus, the Roman stadium of 625 Roman feet equalled in length the Greek stadion (stade) of 600 Greek Olympic feet, at least nominally equivalent to 185.2 metres.

Strabo (63 B.C.-19 A.D.), the most famous Greek historian/geographer of the ancient world, said eight stadions (stadia, stades) made the Roman mile. Yet he reported that the earlier Greek historian/geographer, Polybius (b. 210 B.C.) had said it took 8-1/3 stades to make a Roman mile.[26] Specifically, Polybius had said that eight stadia and two plethra made the mile. A plethron equalled 100 Greek feet, so this put the mile at 5000 Greek feet. Since the original definiton of the Roman mile was 1000 paces or 5000 Roman feet, it appears more than likely that Polybius was using the Ancient Greek foot around which the Romans had standardized, measuring about 0.2963 m. Strabo, on the other hand, had set eight stades of 4800 Greek Olympic feet (0.30864 m.) equal to the Roman mile. In either case, this yields a mile with a length of 1481.5 metres.

Now, the foregoing explanation is our own, not Strabo's, who, seemingly was unaware of the variation, in length, of such commonly used anatomical units as the finger, foot, cubit and pace. Thus, Strabo also reports that the "ater," which the Greeks called "schoenus," varied in length throughout Egypt

61

from 30 to 60 stadia.[27] He also reports that the "parasang" was reckoned by some writers at 60 stadia, by others at 40, and others at 30. [28] While some early Christian era writers saw in this a rough equivalence of the schoenus and parasang, [29] a more likely conclusion is that the stade, schoenus, and parasang all varied in length from time to time and place to place. (The variation in the length of the stade is explored in Chapter 3.) Aubrey Diller, one of the most sensitive and sophisticated twentieth century writers on ancient metrology, has summarized the situation on the ancient Greek stade well in the following:

"The fact is . . . that the Greek stade was variable and in particular instances almost always an uncertain quantity. The most problematical aspect of the ancient measurements of the earth is the length of the respective stades. Some light can be thrown upon it, but the matter requires circumspection, and those who blithely convert in casual parentheses or footnotes are usually unaware of the difficulties and mistaken in their statements.[30]"

Terrestial, Gravitational, and Angular Units

Terrestrial units are based on the dimensions of the earth and are usually designed to be of such a length that, when multiplied by 10^x, where x is a positive or negative whole number exponent, equals some geodetic dimension. Clearly, a knowledge or firm opinion about earthly dimensions is required for the creation of a terrestrial unit. Thus, the metre was intended to be one ten-millionth of the length of a quadrant of a meridian on the earth's surface taken between the North Pole and the Equator. When the metre was originally envisaged, in 17th century France, the earth was conceived as being spherical. Measurements of a degree of latitude close to the Equator, within the Arctic Circle, and at several temperate zone sites, revealed the difference in curvature of a great circle at different latitudes. Before this, and going back to ancient times, the earth was generally held to be spherical by astronomers, geographers and philosophers.

An apparent counterpart of the metre in ancient times was the Greek Olympic foot of 0.30864 m. It represents 1/100th of a second of arc of a great circle of a spherical earth taken as 40 million metres in circumference (i.e., 0.30864 m x 100 x 60 x 60 x 360 = 40,000,000 m). No one has yet suggested that the Greeks had this in mind when the Olympic foot was created.[31] Successive measurements or estimates of the size of the earth by ancient Greek geographer/astronomers were always stated in stades and no one can be sure how many Olympic feet made the stade in which the measurement was given. (Greek measurements of the earth's size are taken up in the next chapter.)

Already mentioned is Vitruvius Pollio's, 1st century B.C., six foot pace. Now, 1000 such paces based upon the Ancient Greek and adopted official Roman foot of 0.2963 m. yield a mile of 1777.8 metres, not particularly significant. But, 1000 paces of six Olympic feet, equivalent to 1000 Greek fathoms based on the Olympic foot, yielded a unit of 1851. 84 metres. Now this is significant, for it is precisely the length of the Miglio Geografico or geographic mile of 18th century Italy which is equal to the length of one minute of latitude or of longitude at the Equator. Further, what could be more logical than that 1000 fathoms should make the nautical mile of the same basic length as the geographic mile?

Another example of a unit with terrestrial significance is the sacred cubit of the Hebrews (also called the Palestinian, Royal Persian, and Chaldean cubit). This unit (0.637-0.644 m.) when multiplied by 10^7 yields a good approximation of the polar radius of the earth.[32]

Closely related to terrestrial units are gravitational units derived from lengths of pendulums whose swing has a certain duration, or half-period, when the pendulum is at a spot on earth with a given gravitational intensity. The gravitational intensity of the earth is a composite of several factors, the most important of which are latitude, distribution of the many materials making up the earth, and altitude. In ancient Babylonia there was in use a "double cubit" of 990-996 mm. length. Lehmann[33], a 19th century archeologist working on the site of ancient Babylonia discovered that the double cubit was almost exactly the length of the second's pendulum for the latitude of Babylon, 31° north. At this latitude and approximately sea level, the theoretical length of a second's pendulum is 992.35 mm., assuming no aberration in the earth's density in this area. [34]

Lehmann concluded that the theory of the pendulum must have been known to the early Babylonians, who derived it from the plumb-line employed in their many building operations. Most reviewers, however, consider that the double-cubit having a length almost identical to the second's pendulum to be a sheer coincidence and that, clever as the ancient Babylonians were, they had not yet discovered the important physical principle of the pendulum.

Our system of measuring angles is sexagesimal-decimal in character and is generally believed to be Babylonian in origin. It is purported to have come about this way. In the course of their astronomical observations, it was ascertained that on the days of the equinocti, the apparent diameter of the sun as it crossed the meridian was 1/360 of the visible heavens (30 minutes). Furthermore, by using a water clock, where water flows through a small orifice from one jar to another, it was noted that the amount received in the 12 hours from sunrise to sunset (on the days of the equinocti) was 360 times as much as the the amount received while the sun was traversing an arc equal to its own diameter, i.e., it took the sun two minutes to traverse its own diameter. This afforded an accurate method of measuring time and became the foundation of the sexagesimal system, the underlying principle of Babylonian metrology. [35]

Still, the Babylonians did not divide their circle into 360 or 720 parts, but 60. (It was Hipparchus, a couple of millenia later, who founded trigonometry and the system of the 360 degree circle.) The Babylonians, aware that the radius of a circle is equal to the length of the chord of an arc one-sixth the circumference, first divided the full circle into six parts and then further subdivided each of the six parts by 10. Thus, up to the time of Hipparchus, a circle was divided into 60 parts.

EVIDENCE OF ANCIENT METROLOGIC SYSTEMS

Thus far in this chapter, we have been describing the most popular ancient (and, in some cases, medieval) units of linear measure and their relationship, where existing. We now examine the evidence that formal linear measurement systems existed in ancient times.

Mesopotamian Systems

Several archeological finds, taken together, have permitted scholars of ancient metrology to propose likely systems in use during the third millenium B.C: a) The Senkereh Tablet, dating from 2500 B.C., was discovered on the site of the ancient Sumerian town of LARSA in 1850 and now resides in the British Museum; b) 8 headless statues of a seated person identified as Gudea, the ruler of LAGASH during the Neo-Sumerian (or Ur-III) period, c. 2120-2006 B.C., were unearthed at Telloh, an Arab village near the LARSA site, in 1881. Two of the statues show Gudea with a tablet on his lap, one tablet being blank except for a graduated rule near the edge and a stylus ready for right-handed use, the other tablet shows an architectural plan of a temple. These statues are now in the Louvre, in Paris; c) Cuneiform texts, clay tablets inscribed by stylus with a type of picture-writing antedating Egyptian hieroglyphic, have been found at several sites of ancient Sumeria; and d) Archeological work at the ruins of Babylon, in ancient Assyria (about 100 miles northwest of Sumer) provided evidence of the use of Assyrian units of measurement very similar to the older Sumerian.

Based on the foregoing and the interpretive work of such men as the Rev. W. Shaw-Caldecott, M. E. de Sarzec, and M. J. Oppert, two separate Early Babylonian systems of linear measurement have been proposed by Hallock and Wade, and Berriman. For purposes of comparison, these are shown in Table 2 as

Unit	Equivalency		Origin			Modern Reference	
	In Other Babylonian Units	In Metric	S.T.	G.R.	C.T.	HsW	Be.
Se (Line)	1/180th Gin = 1/30th Shusi	0.56mm		✓	✓		✓
Line	1/180th Palm (Handbreadth)	0.55mm	✓	✓	✓	✓✓	
Sossus	3 Lines = 1/60 Palm	1.66mm	✓	✓✓		✓✓	
Shusi (Digit)	1/30th Kus (cubit) ≈ 1/6 Palm ≈ 10 Sossus	16.8mm		✓	✓	✓	✓
Digit (Thumb?)	1/3 Palm	33.2mm	✓	✓		✓✓	
Palm (Handbreadth)	60 Sossus = 180 Lines	99.6mm	✓	✓		✓✓	✓
Gin	1/5 Kus (Cubit) = 1/60 Gar ≈ 1 Palm	100.8mm		✓	✓	✓	
Small Ell (Cubit)	3 Palms	29.88cm	✓	✓	✓	✓	✓
Foot	2/3 Kus (Cubit) = 20 Shusi	33.6 cm		✓✓	✓	✓✓	
Medium Ell (Cubit)	4 Palms	39.8 cm	✓	✓✓		✓✓	✓
Large Ell (Cubit)	5 Palms	49.8 cm	✓	✓		✓	
Kus (Cubit)	30 Shusi = 3/2 Foot ≈ Large Ell (Cubit)	50.4 cm		✓	✓	✓✓	✓
Double Cubit	10 Palms	99.6 cm	✓	✓✓	✓	✓	
Small Reed	4 small Ells = 12 Palms	1.95 m.	✓	✓✓		✓✓	
Medium Reed	6 medium Ells = 24 Palms	2.39 m.	✓	✓✓		✓✓✓	✓
Large Reed	6 large Ells = 30 Palms	2.99 m.	✓	✓	✓	✓✓	
Gar	2 large Reeds = 12 large Ells (Cubits) = 60 Palms	5.98 m		✓✓	✓	✓	
Gar	12 Kus (Cubits) ≈ 2 large Reeds	6.05 m		✓	✓	✓	✓
Ush (Stadion)	60 Gar = 720 large Ells (Cubits)	358.8 m		✓✓	✓	✓	
Small Kasbu (Para-sang)	15 Ush (Stadions) = 900 Gar = 10,800 large Ells (Cubits)	5382 m.-		✓	✓	✓	
Kasbu (Parasang)	2 small Kasbus = 30 Ush (Sta-dions) = 1800 Gar = 21,600 large Ells (Cubits)	Sta-10-10.164 m.		✓	✓	✓	
Beru	1800 Gar = 21,600 Kus (Cubits) ≈ Kasbu (Parasang)	10,886 m.		✓	✓	✓	✓

Abbreviations: S.T.= Senkerah Tablet (2500 B.C.) Larsa ruins, 1850; G.R.= Gudea's Rule (2175 B.C.) Lagash ruins, 1881; C.T.= ancient cuneiform mathematical texts; HsW= Hallock's Wade, Evolution of Weights and Measures," pp.13-18, MacMillan N.Y., 1906; Be.= A.E.Berriman, "Historical Metrology," pp.51-67, incl., E.P.Dutton, New York, first published by J.M.Dent & Sons, London, 1953.

Table 2 : Composite Listing of Early Babylonian (3rd Millenium, B.C.) Units of Linear Measurement with Other Babylonian and Metric Equivalents

Table 2

Table 3 : Rationalized Sumerian (Early Babylonian) System of Linear Measurements (c. 2175 B.C.)

Unit	Equivalency in Other Sumerian Units	Metric Length
Line (Se) Sassus	1/180th Gin(palm) = 1/30th Shusi(digit)	0.55-0.56 mm.
	3 Lines = 1/60th Gin (palm)	1.66 mm.
Shusi (digit)	1/30th Kus(cubit) = 1/20th foot = 1/6th Gin(palm) = 10 Sassus	16.8 mm.
Thumb or Knuckle (?)	1/3rd Gin(palm) = 2 Shusi(digits)	33.2-33.6 mm.
Gin (palm)	1/5th Kus (cubit) = 6 Shusi (digits)	99.6-100.8mm.
Foot	2/3rds Kus (cubit) = 20 Shusi (digits)	336mm = 0.336m.
Kus (cubit)	1/12th Gar = 1/6th Reed = 3/2 Foot = 5 Gin(palms) = 30 Shusi (digits)	0.498-0.504 m.
Double cubit	2 Kus (cubits) = 3 Feet = 10 Gin = 60 Shusi	Approx. 1 m.
Reed	6 Kus (cubits) = 9 Feet = 30 Gin = 180 Shusi	Approx. 3 m.
Gar	2 Reeds = 12 Kus = 18 Feet = 60 Gin	Approx. 6 m.
Ush (stadion)	60 Gar = 120 Reeds = 720 Kus = 1080 Feet	Approx. 360 m.
Small Kasbu (Small Parasang)	15 Ush (Stadions) = 900 Gar = 1800 Reeds = 10,800 Kus	Approx. 5,400 m. = 5.4 km.
Kasbu (Large Parasang)	2 Small Kasbu = 30 Ush = 1800 Gar = 3600 Reeds = 21,600 Kus	Approx. 10.8 km.

Table 3

a "Composite Listing . . . ". Then, by re-titling Hallock and Wade's digit of 1/3 palm (as thumb or knuckle)[36] and ignoring the small and medium cubits and reeds (which are not found in Berriman's system), one obtains remarkable agreement between the two sources. A consolidated, rationalized Sumerian (Early Babylonian) linear system based on these sources is shown in Table 3.

Egyptian Systems

As in the case for Mesopotamia, artifacts supplemented by texts give testimony to the existence of formal measurement systems in Ancient Egypt. The cubit rod, the Nilometer, the remains of ancient monuments and structures, and the Rhind Mathematical Papyrus are, collectively, the bases for the construction of linear (and other) measurement systems by Egyptologists and metrologists.

The Cubit Rod

The most important, by far, of the artifacts was the cubit rod (already mentioned) because of its portability, the great number which have been found, and its adaptability for inclusion on its faces of a veritable mine of information. This is not to say that all cubit rods were similarly inscribed nor even of the same length. Still, lengths were generally restricted to a) the 24 digit cubit (the short cubit) of approximately 0.45 metre; b) the 28 digit cubit (the Royal cubit) of approximately 0.525 metre; and the reed, a 6 Royal cubit rod approximately 3.15 metres in length.

Some of the cubit rods were of a specific length but otherwise blank. Most, however, were inscribed, some regularly and some with a mix of regularly and irregularly spaced inscriptions. The regularly spaced inscriptions were generally intended to denote the lengths of digits, the palm, the small span, the large span, and the foot. In the case of the Royal cubit rod, the length of the short cubit was marked at the 24th digit point.

Where cubit rods were inscribed with a mix of regularly and irregularly spaced markings, metrologists have speculated that the intent was to set forth the length of non-Egyptian as well as Egyptian units. The HARMHAB (HOREMHAB) cubit rod found at Memphis early in the 19th century (dating from Harmhab's reign, 1330-1303 B.C., and now resting in the Turin Museum) is a good example. Whether this was intended to be so or not, all of the cubits described earlier in the chapter - the Greek, Roman, short Assyrian, Talmudist or large Assyrian, and the Sacred Hebrew, Palestinian, Royal Persian, and Chaldean - can be found on the rod directly, or by adding certain of the divisions or by taking multiples thereof. Likewise, it is possible to find markings equivalent in length to the Greek finger and palm as well as other non-Egyptian basic units.

The Nilometer

In early pharaonic times (perhaps as early as 3000 B.C.), the government established a series of measuring stations along the Nile to gauge the water level at these points throughout the year, but especially during the flood season. These Nilometers had as their primary purpose the establishment of a basis for the setting of taxes - low flood levels meant a lesser crop, higher levels a greater yield. Reference to records kept over the years made it possible to predict, with some accuracy, the crop yield which could be expected after subsidence of the flood levels. Taxes were based upon the expected yields per unit area of farmland and, of course, the area of the farm.

Another purpose of the Nilometer was to gauge the rapidity with which the level of the river rose with

24 digit cubit
Nilometer

Figure 5: Sketch of a 24 digit cubit Nilometer at Elephantine Island.
Set in a masonry wall bordering a lagoon of the Nile, with stone steps leading to it for reading the water level at high water and servicing of the stone gauge at low water, the Nilometer was a durable and reliable instrument. While it is likely that this Nilometer represented the short cubit of approximately 0.45 metre, it is possible that the cubit was the large Assyrian cubit of approximately 0.54 metre, composed of 24 fingers of 22.5 mm. (See text, this chapter, section on cubits, and chapter 4, Table 4, "Medieval Islamic Units of Length.")

the advent of the flood season. The more rapidly it rose, the higher the maximum level which would be reached. Here, again, reference to records of prior years permitted quite accurate predictions. Indications that the flood was going to be higher then the previous year gave the government time to warn farmers along the river to raise their dike levels and reinforce them with the sun-dried mud bricks which were readily available (and were used for a variety of purposes).

Early 19th century measurements conducted in Lower Egypt by LePere [37] indicated the Nilometers there to be marked off in short cubits of 24 digits. Later 19th century measurements of Nilometers at Elephantine by Petrie,[38] the noted Egyptologist and metrologist, showed these to be marked off in Royal Cubits of 0.525 m, not further sub-divided.

Land Measurement--The Rhind Mathematical Papyrus[39]

Our knowledge of land measurement in Ancient Egypt is somewhat better based than that of the Nilometer, thanks to a written document called the Rhind Mathematical Papyrus.

The oldest mathematical documents in existence are two Egyptian papyrus rolls which contain information traceable to the 12th Dynasty kingdom during which Amenemhet III reigned, c. 1900 B.C. One of the rolls, the Golenischev (named after its former owner) reposes in Moscow. The other, the Rhind papyrus, is in the British Museum, except for important fragments which are in the possession of the New York Historical Society.

In the winter of 1858, a young Scottish antiquary named A. Henry Rhind, sojourning in Egypt for his health, purchased at Luxor a rather large papyrus roll said to have been found in the ruins of a small ancient building at Thebes. The document, originally a roll 18 feet long and 13 inches high, had important parts - necessary for an understanding of the whole - missing. Fortuitously, these parts turned up fifty years later at the New York Historical Society to whom they had been given by the collector Edwin Smith.

The Rhind Papyrus, according to its writer, the scribe Ahmose,[40] was copied by him during the 33rd year of the reign of the 16th Dynasty Hyksos king Aauserre Apopi II, c. 1580 B.C. The original, he says, was written when Amenemhet III of the 12th Dynasty was on the throne. Amenemhet I was the founder of the 12th Dynasty although some give this distinction to Senusret I, the Greek Sestoris. In any event, Amenemhet III and Senusret III are considered by many to be among the greatest rulers of Ancient Egypt.

It was during the reign of Sestoris (Senusret III), says Herodotus, that the Egyptians first learned geometry and it was from Egypt that this knowledge passed to Greece. The generally accepted account of the origin and early development of geometry is that the ancient Egyptians were obliged to invent it in order to restore the landmarks and boundaries of property which had been destroyed by the inundations stemming from the annual flooding of the Nile. In addition, where the course of the river had been permanently altered, it was necessary to survey, establish, and record the new boundaries. Taxes assessed on landowners was based on the size of their plot (and the degree of inundation as predicted by Nilometer readings). Plot shapes were not always rectangular, hence the requirement for methods to quickly determine the areas of non-rectangular shapes.

The Rhind scroll, it turns out, was a practical handbook of Egyptian mathematics and remains the principal source of knowledge as to how the ancient Egyptians counted, reckoned and measured. Less a treatise on mathematics than a collection of mathematical exercises and practical examples, the Rhind illustrates the use of fractions, the solution of simple equations and progressions, and the mensuration of areas and volumes.

One of the translations of the Rhind Papyrus utilized by students of ancient Egyptian metrology is that by T. E. Peet, done in 1923. In this it is disclosed that there was a linear unit called khet, measuring 100 (royal) cubits and a unit of area called setat that was I khet square. The square khet, therefore, was equal to 10,000 (1 myriad) square (royal) cubits. However, from the context, it is clear that the setat should be visualized as 100 parallel strips 1 khet long and 1 cubit wide. The name for such a strip was 'a cubit of land.' While there is no evidence that the setat was called 'a hundred of land,' 10 setats was called 'a thousand of land.' Herodotus mentions another important unit in connection with the practice (by grateful pharaohs) of awarding 12 arurae of land to Egyptian warriors. [41] This is to be interpreted as 12 setats and is approximately equal to 13 Roman jugera.

But, how big is a linear cubit? The Rhind Papyrus tells us little about that. The Papyrus is valuable in revealing the mathematical capabilities (and weaknesses) of ancient Egyptians and in, at least, mentioning the units of land measurement employed and their relationship to one another. For an indication of the absolute value of the length units mentioned, we must look to the cubit rod and, to a lesser degree, the Nilometer.

It is simply impossible to construct an "ancient system of Egyptian metrology" because elements of that system are defined differently in the literature and the relationships between different units varied with time. The best that can be done is to present a "rationalized" system such as that which formed the basis for the Ancient Greek, the Roman, and the Greek Olympic systems of metrology during the last five centuries of the pre-Christian era. In so doing, however, we are in essential agreement with Berriman, Hallock and Wade, Gillings,[42] and Barnard[43]. Such a system, consistent also with our own prior treatment of anatomical, multiple, botanical, and arbitrary units is presented in Table 4.

Ancient Greek and Roman Systems

Pelasgic ruins (preceding 500 B.C.) found in Greece, Asia Minor, and Southern Italy give evidence that the Greek counterpart of the Egyptian digit and t'ser (foot) were employed almost exclusively in the dimensioning of the original structures. Two separate decimal systems, one based on the digit and one based on the foot evolved, related by the binary equivalency: 1 foot = 16 digits.

The digit based system was about as follows:

> 1 Pelasgic digit = approximately 1/54th metre = 18.52 mm.
> 10 digits = 1 handlength = 185.2 mm
> 100 digits = 1 orguia = 1.852 m
> 1000 digits = 10 orguias = 1 amma = 18.5 m
> 10,000 digits = 10 ammas = 1 stadion = 185.2 m

The foot based system appears to have looked like this:

> 1 Pelasgic foot = 16 Pelasgic digits = 296.3 mm
> 10 feet = 1 pole (arkana) = 2.963 m
> 100 feet = 10 poles = 1 plethron = 29.63 m

Augmented by other anatomical units of measurement long in use in Egypt and Mesopotamia which could be related binarily, sexagesimally, or in combination to the foregoing, an Ancient Greek (Pelasgic) system emerged which became the prototype for the Roman system and which, at one time or another, included the following units:

> 1 Pelasgic digit = approx. 18.52 mm
> 1 Knuckle = 2 digits = 37.04 mm
> 1 Handbreadth = 6 digits = 111.12 mm
> 1 Lick (Little Span) = 8 digits = 148.16 mm

1 Handlength = 10 digits = 185.2 mm
1 Span = 12 digits = 222.24 mm
1 Foot = 16 digits = 296.3 mm
1 Pygme = 18 digits, 3 handbreadths, or 1 lick + 1 handlength = 333.36 mm
1 Pygon = 20 digits, 5 palms, 2 handlengths, or 1 foot + 1 palm =
      1 Egyptian Remen = 370.4 mm
1 Cubit = 24 digits, 6 palms, 3 licks, 2 spans, or 3/2 feet = 444.5 mm
1 Step = 40 digits, 10 palms, 5 licks, 4 handlengths, or 1 foot + 1 cubit = 740.8 mm
1 Xylon = 72 digits, 18 palms, 12 handbreadths, 6 spans, 4 pygmes, or 3 cubits = 1.333 m
1 Orguia = 100 digits = 1.852 m
1 Pole = 10 feet or 160 digits = 2.963 m
1 Amma = 1000 digits = 18.52 m
1 Plethron = 100 feet, 10 poles or 1600 digits = 29.63 m
1 Polybius Stadion = 600 feet, 60 poles, 6 plethrons = 177.78 m

From this welter of length units two other systems of measurement emerged, the Roman, which was essentially a streamlined version of the foregoing, and the Greek Olympic system, both of which were post 500 B.C. The origin and basis of the Greek Olympic system is shrouded in dispute. Some metrologists, among them Berriman, take the position that the Pelasgic digit and foot were replaced by the Olympic finger and foot, both of which were larger than their older counterparts in the ratio 625/600 = 25/24 = 1.04167.

The other position, the chief protagonist for which is Petrie, is that there never was a Greek digit as large as the Olympic finger of 19.29 mm and the the Pelasgic digit of 18.52 mm persisted as part of the Olympic system, 16 2/3 such digits making the Olympic foot of 308.64 mm and 25 making the Olympic cubit of 463 mm. Berriman being the later investigator and in full possession of all of Petrie's arguments, we have elected to adopt his position. Thus, Table 5, which sets forth our own understanding of the Greek Olympic and the Roman systems of linear measurement, is essentially Berriman augmented by compatible data provided by other investigators, including Petrie.

Conclusions

We have seen that there is a significant degree of similarity between the structures of the various systems of linear measurement encountered, but some rather wide variation between the specific lengths of some of the basic units which appear in most systems, i.e. the palm, foot, and cubit. Even the digit (finger) is not immune. As we progress into itinerary, geographic, and geodetic measurements in which larger units such as stades (stadions, stadiums), miles, schoeni, parasangs, leagues, and degrees of latitude and longitude are involved, differences in evaluation of the basic units can be expected to have important and, frequently, troublesome consequences.

Table 4 : Rationalized Egyptian System of Linear Measurement, the Basis for the Ancient Greek, Roman, and Greek Olympic Systems of Linear Measurement.

Unit	Equivalency in Other Egyptian Units	Metric Length
Digit	$1/4$ Palm = $1/24$ Short Cubit = $1/28$ Royal Cubit	18.52 mm.
Palm	$1/6$ Short Cubit = $1/7$ Royal Cubit = 4 Digits	74.0 mm.
Handbreadth	5 Digits	92.6 mm.
Fist	6 Digits	111.1 mm.
Small Span	12 Digits = 3 Palms = $1/2$ Short Cubit	222.2 mm.
Large Span	14 Digits = $1/2$ Royal Cubit	259.3 mm.
T'ser (Foot)	16 Digits = 4 Palms = $2/3$ Short Cubit	292.6-300 mm.
Remen	20 Digits = 5 Palms = $5/4$ T'ser = $5/6$ Short Cubit = $\sqrt{2}/2$ Royal Cubit	Approx. 370.4 mm.
Short Cubit	24 Digits = 6 Palms = 2 Small Spans = $3/2$ T'ser	Approx. 450 mm.
Royal Cubit	28 Digits = 7 Palms = 2 Large Spans = $\sqrt{2}$ Remen = Short Cubit + Palm	Approx. 525 mm
Measuring Reed	6 Royal Cubits	Approx. 3.15 m.
Khet, Hayt, or Schoenia	100 Royal Cubits	Approx. 52.5 m.
Stade (Eratosthenian)	300 Royal Cubits *	Approx. 157.5 m.
Stade (Ptolemaic or Royal Egyptian)	400 Royal Cubits *	Approx. 210 m.
Ater (called Schoenus by Greeks)	40 Eratosthenian Stades * = 12,000 Royal Cubits	Approx 6.3 Km.

* There were other stades (stadions, stadia) and schoeni and even at least one other logical definition of the Eratosthenian stade. See Chapter 4.

Table 5: Two Ancient Systems of Linear Measurement the Greek Olympic and the Roman, and Their Relationship to Each Other

(See also "General Notes" for a more realistic perspective in which this table should be viewed.)

Greek Olympic System

Unit	Metric Length	Finger	Palm	Foot	Cubit	Plethron	Stadion
Finger	0.01929	1	1/4	1/16	1/24		
Knuckle	0.03858	2					
Palm	0.07716	4	1	1/4	1/6		
Handbreadth	0.11574	6	1½		1/4		
Lick (little span.)	0.15432	8					
Handlength	0.1929	10	2½				
Span	0.23148	12					
Foot	0.30864	16	4	1	2/3		
Pygme [1]	0.34722	18					
Pygon [2]	0.3858	20					
Cubit	0.463	24	6	3/2	1		
Step [3]	0.7716	40	10	2½			
Xylon	1.3888	72	18		3		
Fathom	1.8518	96	24	6	4		
Pole	3.0864	160	40	10	6⅔		
Cable	18.5184	960	240	60	40		
Plethron	30.864	1600	400	100	66⅔	1	1/6
Stadion	185.184	9600	2400	600	400	6	1
Ride	740.736	38,400	9600	2400	1600	24	4
Roman Mile	1481.472	76,800	19,200	4800	3200	48	8

Notes:
1: One pygme = One lick + One handlength
2: " pygon = " foot + " palm
3: " step = " foot + " cubit

Roman System Based on Ancient Greek System

Unit	Metric Length	Digit	Foot	Pace	Stadium
Digit	0.0185185	1	1/16		
Uncia	0.0247		1/12		
Palm	0.074074	4	1/4		
Foot (pes)	0.2963	16	1	1/5	
Palmipes	0.37037	20			
Cubit (ulna)	0.4444	24	1½		
Step	0.74074	40	2½		
Pace (passus)	1.48148	80	5	1	
Pole	2.96296	160	10	2	
Actus	35.556	1920	120	24	
Stadium	185.185	10,000	625	125	1
Roman Mile	1481.48	80,000	5000	1000	8
Miglio Geografico	1851.85	100,000	6250	1250	10

Notes:
1: One palmipes = One palm + 1 pes
2: Miglio Geografico may date from a few centuries after ancient Roman system proper

Table 5, part 1

General Notes

1. The foregoing table is based primarily on data taken from the following sources: A.E. Berriman's "Historical Metrology", William Hallock's and Herbert T. Wade's "The Evolution of Weights and Measures and the Metric System", H. Arthur Klein's "The World of Measurements", W.M. Flinders Petrie's "Inductive Metrology", and the Encyclopaedia Brittanica.

2. The relationship between units holds, in general, for systems based on larger feet such as the Babylonian (0.3142 to 0.316/m.) and the Drusian (0.3349 to 0.3415m.) both of which were used in Egypt, Greece, and Italy as well as elsewhere. A single value is given for each unit length rather than a range of values which actually would suggest in order to comply with the relationships shown and the largest unit in the tables for the ancient systems, the Roman Mile.

3. W.M. Flinders Petrie, the noted 19th century Egyptologist and metrologist, does not concur with other metrologists on the subject of the Greek finger, shown in table to have a length of 0.01929 (1/12th of the Olympic foot). His position is that the digit used with the Olympic foot was the Egyptian, used throughout Egypt, Assyria, Persia, Syria, Asia Minor, Greece, and Italy. This varied from 0.018295 to 0.01879 m. in length.

3. (Continued) Petrie set 6 Olympic feet (or 4 cubits) equal to 100 Egyptian digits, which holds for the values shown above at right for the-digit and at the left for the Olympic foot.

4. There were cubits in use in ancient times other than those shown above. Among them the Royal Egyptian and Babylonian, used in Egypt, Assyria, Persia, Syria, and Asia Minor, having a length of 28 digits. Within Egypt, there was also a short cubit of 24 digits. Two Assyrian cubits used also in Egypt, Persia and Italy were the 0.543 m. and the 0.494m. units, based upon 24 digits of 0.0226 m. and 0.02058m, respectively. The Sacred Hebrew, Palestinian, Royal Persian, and Chaldean cubit had a length of from 0.637 to 0.6445m. Petrie maintained that this unit was used in the dimensioning of Indian mounds at several points in North America.

5. With the significant number of feet and cubits which existed in ancient times, which could be taken in various numbers to make larger itinerary units such as the stade, mile, schoenus, parasang, and league, it is not surprising that a variety of lengths of each of the itinerary units resulted.

Table 5, part 2

Chapter 3

THE ANCIENTS' VIEWS AND
ESTIMATES OF THE SIZE AND
SHAPE OF THE EARTH

INTRODUCTION

It is customary in the western world to place the earliest civilizations and the earliest development of astronomical skills in Mesopotamia and Egypt. In both regions the priesthood acquired an accurate knowledge of the periods of the sun, moon, and planets and were able to predict the positions of these bodies among the stars and the recurrence of lunar eclipses. Astronomy being closely tied to astrology, the practice of both remained in the hands of the priests among whom speculations as to the origin and construction of the world and the size and shape of the earth was not an active pursuit.[1]

Greek cosmological speculation was at first, highly mythological in nature, like that of the Eastern nations and Egypt. Sometime around the middle of the 6th century B.C., however, the era of Greek philosophy began. Increasingly, mythological trammels were cast off as scholars endeavored to find the laws which regulated the phenomena of nature.

Homer, the great epic poet whose period has been variously placed as between 1200 and 850 B.C., or later, held that the earth was a plane disc surrounded by a constantly moving ocean river, "Oceanus", forever flowing round and round the earth. The vault of heaven, he conceived as a solid concave surface, like the "firmament" of the Jews. This extended to "the ends" of the earth so as to rest on it. Evidently additional support was required for Homer speaks of additional tall pillars in the charge of Atlas "which keep the heaven and the earth asunder". The notion that Atlas himself supported the heavens developed in later times.[2]

The foregoing concept, with minor changes, was held for several centuries. It was the view of the Ionian school of philosophy founded by Thales (b. 640 B.C.) at Miletus in Asia Minor. Thales, himself, is regarded as the pioneer in the sciences of geometry and astronomy among the Greeks. Anaximander (611-547 B.C.), a student of Thales, nevertheless wandered from his mentor's knee, holding that the earth was cylindrical in shape and poised in space. The habitable part was disc-shaped and considerably larger than Homer's world. Anaximander held that the material substances of the universe derived from the element, "apeiron," which contained and governed all things, being immortal and imperishable. Yet Anaximander was an able astronomer and geographer, inventor of the celestial globe, who taught the obliquity of the ecliptic. He is credited with the introduction of the gnomon and sun-dial into Greece. Anaximenes (fl. 6th cent., B.C.), a student of Anaximander's, rejected his teacher's views, visualizing instead an earth in the shape of a rectangle, buoyed up and cushioned by compressed air. He held further, that all substances of the material universe were derived from one natural element, air, as a result of rarefaction or condensation. Hecataeus (fl. c. 500 B.C.) was the youngest of the Miletan philosophers and the first to record, in prose, a systematic description of the world as it was then known to the Greeks. Despite his extensive geographical knowledge, he was a believer of the disc-shaped world and the circumfluent ocean.

Pythagoras (b.c. 582 B.C.), Greek philosopher and founder of a school, semi-religious and semi-philosophical, at Crotona, Italy, is generally credited with first conceiving of the earth as a sphere. Diller [3] says it may, however, have been Parmenides (6th-5th cent. B.C.). Further, the idea appears first in writing in "Plato's 'Phaedo' where Socrates (d. 399) alludes to a current controversy on the shape and size of the earth . . ." Socrates, Plato's teacher, indicates that " . . . somone has convinced him that the earth is really a sphere at rest in the center of a swirl, very large, containing many inhabited regions 'like ours' in the form of valleys or depressions where mist and water collect . . . " Diller indicates the idea was developed further in Plato's *Timaeus*. However, it was not until Aristotle (384-322 B.C.) that the idea of a spherical earth is developed in a practical form with demonstrable proof and an estimate of size.

ARISTOTLE

One of the greatest of the Hellenic philosophers, Aristotle (384-322 B.C.), was born in Macedonia. At the age of 17, he went to Athens where for the next 17 years he was associated with Plato. In 343 B.C., he was recalled to Macedonia by Philip to tutor his son, Alexander. In 334 B.C., when Alexander entered Asia "to subdue the world," Aristotle returned to Athens to found his Peripatetic school at the Lyceum. For the next 12 years, he was absorbed in scientific research and in lecturing, activities made possible by gifts from Alexander.

In his treatises *Meteorologica* and *Decaelo*, Aristotle undertakes, inter alia, to prove that the earth is a sphere, arguing that when heavy particles are moved uniformly from all sides towards a centre, a body must be formed, the surface of which is everywhere equidistant from the centre; and even if the particles were not equally moved towards the centre, the greater would push the smaller on, until the whole was everywhere uniformly settled round the centre. He advances more solid arguments, however, based on observation. Referring to lunar eclipses, he notes that as this progresses, the edge of the shadow is always circular without showing any of the changes to which the line terminating the illuminated part of the moon is subject in the course of a month, so that the earth which throws the shadow must be a sphere. Then he notes that a very slight journey north or south is sufficient to change the horizon sensibly and expose different stars; some stars are visible in Egypt and the neighborhood of Cyprus which are not seen to set when we go south. This shows that the earth is not a large sphere, as a small change of position makes so much difference in the appearance of the heavens.

Aristotle finishes the astronomical part of *Decaelo* (On the Heavens) with the statement "And those among the mathematicians who attempt to calculate the extent of the circumference maintain that it is about 400,000 stadia, from which it follows that the bulk of the earth must not only be spherical, but not large in comparison with the size of the other stars." Dreyer/Stahl[4] speculate that it is likely that the estimate of 400,000 stades may have been made by Eudoxus of Cnidus (409-356 B.C.), the first scientific astronomer, who is known to have travelled in Egypt and in Greece. Further, they assume the length of the stade (stadion, stadium) to be 157.5 metres (referring to it as the Ptolemaic stade after the Macedonian dynasty which ruled Egypt from 323 to 30 B.C.). Thus, Dreyer/Stahl equate Aristotle's estimate of 400,000 stades to 63,000,000 metres, or 39,146 statute miles. However, the stade is known to have varied from time to time and place to place, thus this equivalency is highly speculative.

One of the older forms of the stade, called variously the Hebraic, Eratosthenian, and other names (as we shall see) measured 148.148 metres. It was composed of 500 Ancient Greek feet of 0.2963 metre. (This foot, also known as the Pelasgic foot, later was adopted by the Romans.) Applying this length stade to

Aristotle's estimates of 400,000 stades would have yielded a circumference for the earth of 59,259,200 metres, still almost 50% greater than the figure of 40,000 kilometres which we shall use as our standard for the circumference of a spherical earth.[5]

The literature discloses that there was a still older version of the stade in use in Egypt, much smaller than the Hebraic or Ptolemaic stades. Gossellin reports[6] that Herodotus (484? - 425? B.C.) observed a measurement in Egypt of the distance from the Mediterranean to Thebes. The result was 6360 stadia (stades). Gossellin then says " . . . the stadium made use of in Egypt at the time of Herodotus consisted of 1111 1/4 to a degree on the grand circle . . ." By Gossellin's time (fl. 1790), the length of a degree of latitude had been measured enough times and in widely varying latitudes to permit an estimate of its length at about 28.5° N (the median latitude between Thebes and the Mediterranean). Thus, Gossellin knew the degree length there had to be about 56,870 toises, or 110.85 kilometres. From this, it follows that the stadium in use in Egypt in Herodotus' time must have measured 110, 850 metres ÷ 1111 1/4, or just about 100 metres. This is the shortest stadium (stade) found in the literature.

Now, 400,000 stades of 100 metres per stade yields an earth's circumference of 40,000 kilometres, a perfect result! However, because the details of the measurement observed by Herodotus are so vague, the stade length of 100 metres so obscure, and no one has been able to prove that this was the stade length used by Aristotle or Eudoxus, geographers, astronomers, and metrologists have not greeted this estimate with wild acclaim.

ARCHIMEDES

Archimedes (d. 212 B.C.) was the next, of record, to estimate the size of the earth. The consensus among writer/historians is that he gave a circumference of 300,000 stades and that the real source for this figure was Dikaearchus of Messana (d. 285 B.C.).

The details of this estimate are attributed by Dreyer, our source[7], to Posidonius (135-50 B.C.). The following account is a rationalization of Posidonius' and, to a lesser degree, Dreyer's description of the event.

An anonymous astronomer/geographer, believed to be Dikaearchus, reported (erroneously) that Gamma Draco, the brightest and most southerly star in the head of the constellation Draconis, passes through the zenith at Lysimachia, in Thrace. He also reported that at Syene, in Upper Egypt, Delta, the central star of the constellation Cancer, passes through the zenith. The difference in declination of the two stars being taken (erroneously) as 1/15th the circumference of the celestial sphere (24°) and the distance between the two sites (erroneously) as 20,000 stadia, an earth's circumference of 15 x 20,000 = 300,000 stadia was obtained.

The estimate was a rough one for the actual difference in latitude between the two sites is only 16° 25' and the actual difference in declinations of the two stars was about 29° 20', rather than the 24° taken for both. With regard to the earth surface estimate, the distance between the parallels of Lysimachia and Syene is not 20,000 stadia, but only 12,313, if the stadia is taken as 148.15 metres, and 11,582, if taken as 157.5 metres. While the theoretical approach taken was sound, the application was less than perfect.

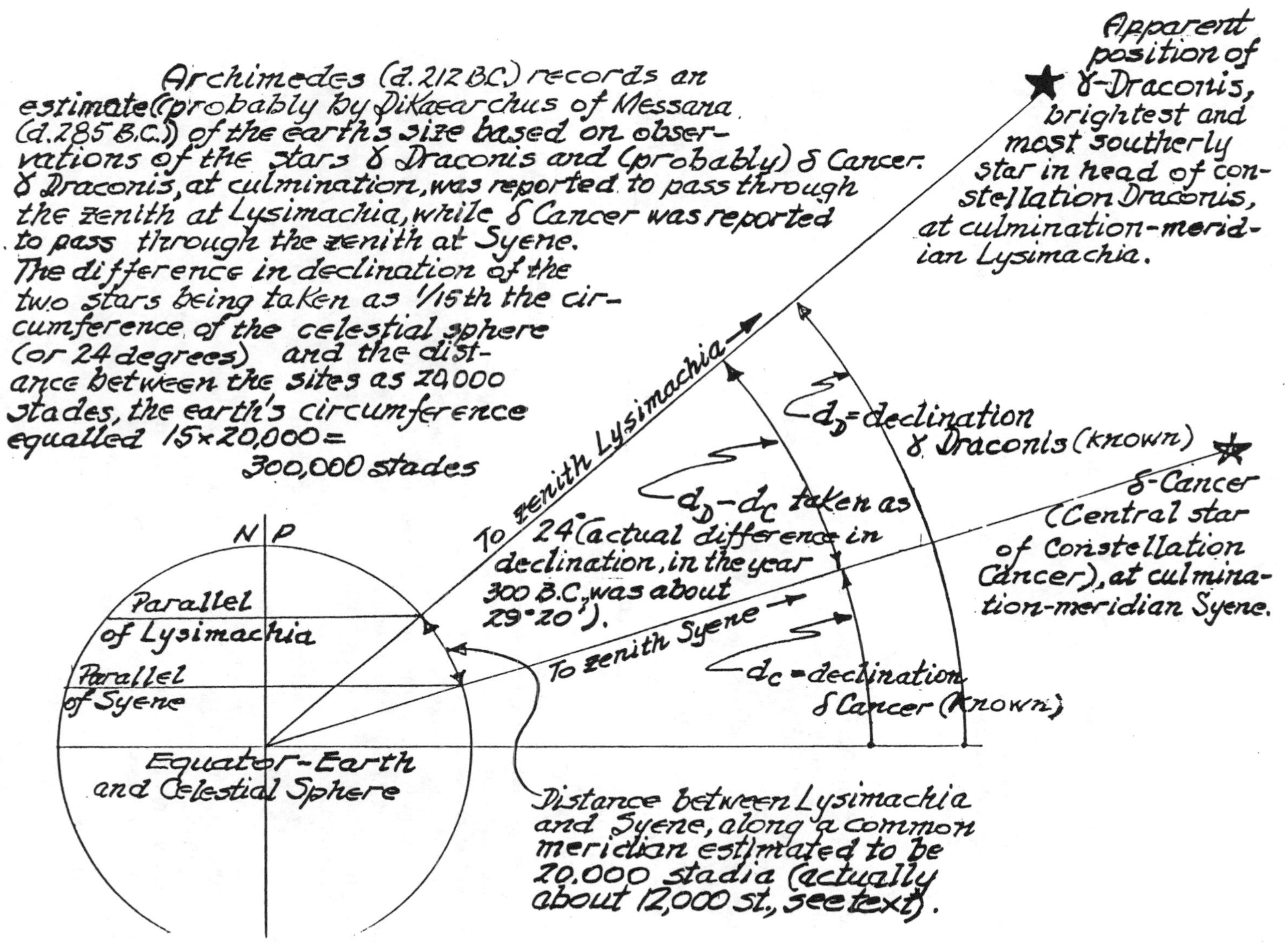

Archimedes (d.212 B.C.) records an estimate (probably by Dikaearchus of Messana (d.285 B.C.)) of the earth's size based on observations of the stars δ-Draconis and (probably) δ Cancer. δ-Draconis, at culmination, was reported to pass through the zenith at Lysimachia, while δ Cancer was reported to pass through the zenith at Syene. The difference in declination of the two stars being taken as 1/15th the circumference of the celestial sphere (or 24 degrees) and the distance between the sites as 20,000 stades, the earth's circumference equaled 15×20,000 = 300,000 stades

Apparent position of δ-Draconis

δ-Draconis, brightest and most southerly star in head of constellation Draconis, at culmination-meridian Lysimachia.

δ-Cancer (Central star of Constellation Cancer), at culmination-meridian Syene.

d_δ=declination δ Draconis (known)

$d_\delta - d_c$ taken as

$d_\delta - d_c$ (actual difference in declination, in the year 300 B.C., was about 29.20°).

To zenith Syene

d_c =declination δ Cancer (known)

zenith Lysimachia

To zenith Lysimachia

Distance between Lysimachia and Syene, along a common meridian estimated to be 20,000 stadia (actually about 12,000 st., see text).

NP

Parallel of Lysimachia

Parallel of Syene

Equator-Earth and Celestial Sphere

Figure 1: Dikaearchus determines the length of a degree of latitude with the help of δ-Draconis and δ-Cancer.

ERATOSTHENES

His Measurement of the Length of a Degree of Latitude

The next and most celebrated determination is that of Eratosthenes of Cyrene and Alexandria (276-194 B.C.), philosopher, mathematician, astronomer, and geographer. Eratosthenes noted that at Syene, just north of the first cataract of the Nile and later known as Aswan, a gnomon[8] cast no shadow at noon on the day of the summer solstice. At the same time at Alexandria (considered by Eratosthenes to lie on the same meridian as Syene) the meridian zenith distance (90° -the altitude) of the sun was 1/50 of the circumference of the heavens (7.2°). This arc, therefore represented the difference in latitude between the two sites. The meridional distance between the parallels of Alexandria and Syene was taken as 5000 stadia. Consequently, the circumference of a great circle of the assumed sphere was 50 x 5000 = 250,000 stadia. Later, Eratosthenes himself or a successor substituted 252,000 stadia, presumably to admit an even division into Babylonian sexagesimals of 4200 stadia each, equivalent to 700 stadia per degree (The degree, however, was not invented until Hipparchus (c. 190-125 B.C.) who also invented trignometry and the method of fixing terrestrial positions by circles of latitude and longitude).

Some versions of Eratosthenes' measurement indicate that at noon on the day of the summer solstice the sun was clearly reflected from the bottom of a deep open well at Syene, the gnomon being employed only at Alexandria where the shadow cast at noon indicated the sun had a zenith distance of 1/50th of the zodiac. Either way, the zenith distance of the sun at Alexandria, and therefore the difference in latitude of the two sites equalled 7.2° , or 7° 12'. This is remarkably close to the correct figure of 7° 6' 42" for the difference in latitude of the Museum at Alexandria and the small cataract at Syene.

Strabo[9], quoting Cleomedes (Kleomedes), also reports that the distance from the small cataract to the sea was 5300 stadia, on the authority of Eratosthenes - the 5000 stadia figure (later enlarged to 5040) being, like the length of a degree, round numbers. The figure of 5300 stadia for the distance between Alexandria and Syene had been obtained as an "itinerary" measurement in which professional pacers, called bematistes, were employed. However, the 5300 stades probably represented the sum of the distances which had been paced off between the many communities lining the Nile between Alexandria and Syene. The Nile is not a particularly straight river as a reference to Figure F-1, Chapter 2, will show. The reduction from 5300 to 5000 stades was probably an attempt to approximate the "airline" distance between the two sites which Eratosthenes believed (mistakenly) to lie on the same meridian. The meridian of Alexandria lies about 3° west of the meridian of Syene. A true airline course from Syene to Alexandria would be about 20° west of north. By considering the two sites to lie on the same meridan Eratosthenes was unknowingly inflating the true meridional distance between the parallels of the two sites by about 6%.

There were other flaws in Eratosthenes' measurements, as has been pointed out by Delambre.[10] Alexandria and Syene are not at the same elevation (an almost negligible consideration) and the gnomon yields the zenith distance not to the sun's centre, but to its upper edge, a difference of 15 minutes, or 3 1/2% in the case of the Alexandria observation. Still Delambre recognizes the sheer brilliance of the accomplishment which has earned for Eratosthenes the distinction "Father of Geodesy".

In any event, the values of 700 stadia for the degree and 252,000 stadia for the earth's circumference were adopted by Hipparchus (b. 190 B.C.), the greatest of the Greek astronomers, by Strabo (63? B.C.-A.D. 21?), famed geographer and historian and Pliny (A.D. 23-79), Roman naturalist, encyclopedist, and writer. However, Strabo's position on Eratosthenes' measurement is apocryphal, as we shall see.

Cleomedes (Kleomedes) to whom we are indebted for this account written about 50 B.C. [11], goes on to say that Eratosthenes also made observations at the same two sites at noon on the day of the winter solstice. In this case, the shadows cast by gnomons at each site were measured and the same end result, 1/50, was obtained. Cleomedes gave no details. However, the principle employed by Eratosthenes in his initial determination on the day of the summer solstice holds for any day of the year and any heavenly body. The difference betwen the meridian zenith distances of the sun measured at the two sites, or the difference between the altitudes, represents the difference in latitude of the two sites. The ratio of the difference in latitude to the entire circumference of the celestial sphere (i.e. 360°) is equal to the ratio of the distance between the two sites (measured along their common meridian) to the circumference of a meridional great circle[12]. Figures 2 a,b, and c, illustrate Eratosthenes' measurements.

Interpreting Eratosthenes' Result

But what was the length of the stadium adopted by Eratosthenes? Unfortunately, there is significant disagreement on this.

Strabo and St. Isidore (fl. A.D. 600-636), bishop and writer on natural history and geography, equated Eratosthenes' stade to 1/8th Roman mile, or 185.2 metres[13]. Thus, this version of Eratosthenes' measurement yielded an 87.5 Roman mile degree and an earth's cicumference of 31,500 miles, equivalent to 129.6 and 46,667 kilometres respectively - and 16-2/3% too large.

Pliny (A.D. 23-79), Roman writer on natural history, agreed with Strabo's interpretation of Eratosthenes but also reported that Eratosthenes had rated his stade as 1/40th the Greek schoenus[14]. Some claim[15] the Greek schoenus measured 6,300 metres, so that the stade would have had a length of 157.5 metres, the degree 110.25 kilometres (or 74.4 Roman miles), and a great circle 39,690 kms (or 26,790 Roman miles) - less than 1% smaller than the correct figures.

Another interpretation of the schoenus, taken from the Heronian Tables, is that it equals four Roman miles[16]. This is the position of Macrobius (fl. c. A.D. 395-423), famed Roman grammarian, philosopher, geographer and prolific writer in Latin. Macrobius said[17] Eratosthenes' degree of 700 stades was equal to 70 Roman miles equivalent to 103.7 kilometres. The earth's circumference, then, was 252,000 stades, 25,200 Roman miles, or about 37,334 kilometres - figures about 6-2/3% too small. This stade, at 1/10th the Roman mile, measured 148.15 metres.

Before leaving the subject of the schoenus, it is important to note that Strabo, citing Artemidorus, indicated[18] that the schoenus (a Greek name for the Egyptian "ater") varied in different parts of Egypt from 30 to as much as 120 stadia. It is clear, however, that this was not simply a case of the schoenus varying in length but the stade, as well.

The three differently sized stades, 148.2, 157.5, and 185.2 metres, are the units which have been associated with Eratosthenes' measurement. These were not, as we have seen in the case of Gosselin's 100 metre stade and as we shall see as we progress to other measurements and interpretatons, the only stades in use in the Ancient World.

By the time of Eratosthenes, Egypt had been subject to many powerful foreign influences. As was pointed out in Chapter 2, Egypt was conquered and ruled by a succession of foreign powers from about 900 B.C. Perhaps the most important impacts on Egypt's metrological systems, pre-Eratosthenes, occurred as a result of the conquests by Assyria (671-652 B.C.), Persia (525-404 and 342-332 B.C.), and Macedonia (332 B.C. up to Eratosthenes' day).

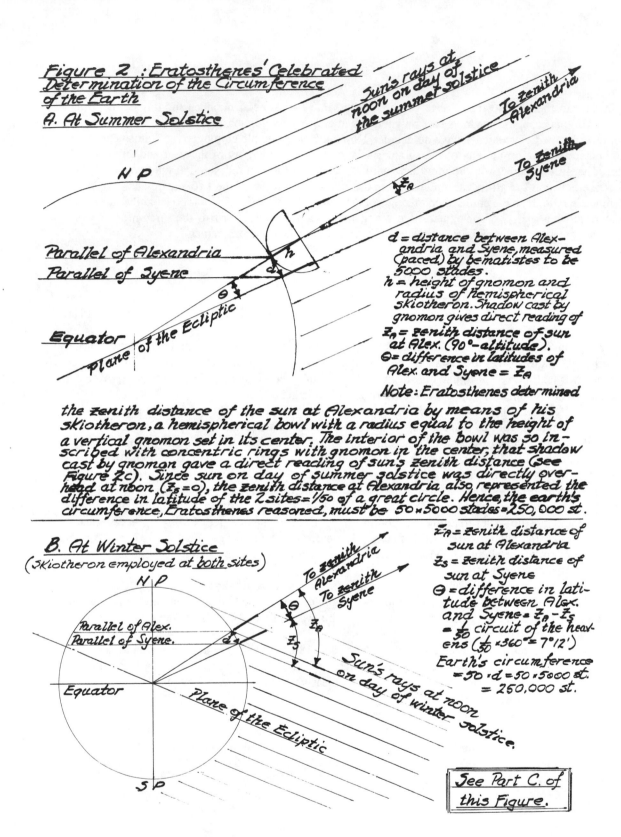

Figure 2 : Eratosthenes' Celebrated Determination of the Circumference of the Earth

A. At Summer Solstice

Sun's rays at noon on day of the summer solstice

To Zenith Alexandria

To Zenith Syene

N P

z_A

Parallel of Alexandria

Parallel of Syene

h

d

Θ

Equator

Plane of the Ecliptic

d = distance between Alexandria and Syene, measured (paced) by bematistes to be 5000 stades.

h = height of gnomon and radius of hemispherical skiotheron. Shadow cast by gnomon gives direct reading of

z_A = zenith distance of sun at Alex. (90° – altitude).

$Θ$ = difference in latitudes of Alex. and Syene = z_A.

Note: Eratosthenes determined the zenith distance of the sun at Alexandria by means of his skiotheron, a hemispherical bowl with a radius equal to the height of a vertical gnomon set in its center. The interior of the bowl was so inscribed with concentric rings with gnomon in the center, that shadow cast by gnomon gave a direct reading of sun's zenith distance (see Figure 2c). Since sun on day of summer solstice was directly overhead at noon (z_s = 0), the zenith distance at Alexandria also represented the difference in latitude of the 2 sites = 1/50 of a great circle. Hence, the earth's circumference, Eratosthenes reasoned, must be 50 × 5000 stades = 250,000 st.

B. At Winter Solstice
(skiotheron employed at both sites)

N P

To zenith Alexandria

To zenith Syene

Θ

z_A

z_s

Parallel of Alex.

Parallel of Syene.

d_s

Equator

Plane of the Ecliptic

Sun's rays at noon on day of winter solstice.

S P

z_A = zenith distance of sun at Alexandria

z_s = zenith distance of sun at Syene

$Θ$ = difference in latitude between Alex. and Syene = $z_A - z_s$ = $\frac{1}{50}$ circuit of the heavens ($\frac{1}{50} \times 360° = 7°12'$)

Earth's circumference = 50 · d = 50 × 5000 st. = 250,000 st.

See Part C. of this Figure.

Figure 2a, 2b

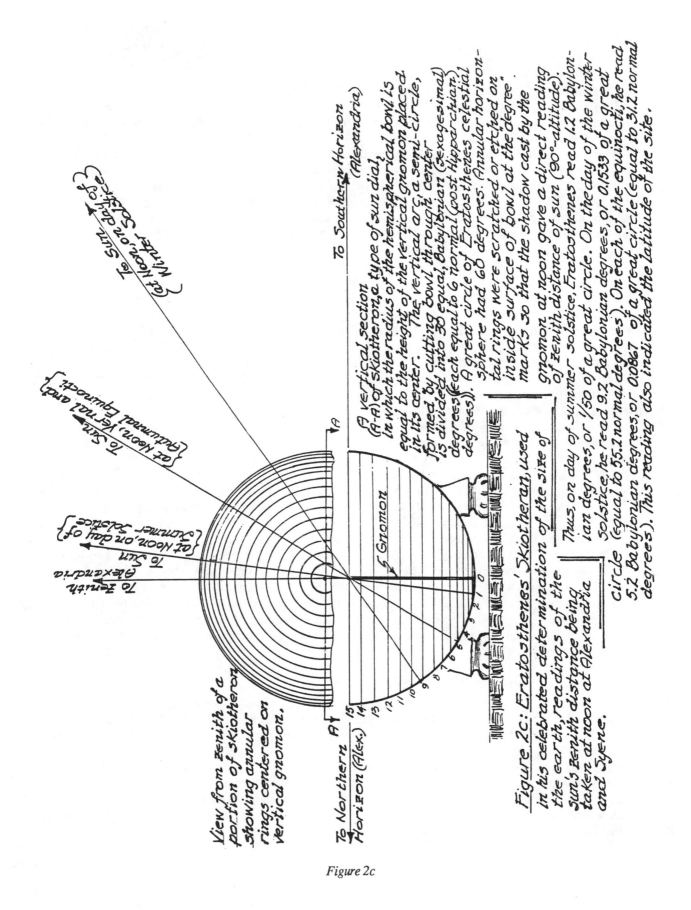

Figure 2c

Less apparent, but important nevertheless, were the changes brought about locally in the wake of waves of Jewish immigration which occurred after the fall of the Temple (in Jerusalem) in 587 B.C., when freed from bondage in Assyria and Babylonia by Cyrus II of Persia, in 538 B.C., and during the entire period of Persian dominance in Egypt.

Many foreign units of measure had been introduced and were being employed, particularly at the seat of governmental power, in border areas, and where immigrants had established thriving communities such as that at Elephantine. One such unit was the Assyrian cubit of 0.494 metre, 300 of which yielded the 148.2 metre stade, 3000 made the Roman mile of 1482 metres, 9000 made the Persian parasang of 4446 metres, used for itinerary measures, and 12,000 made the 5928 metre Egyptian schoenus, used (by the Greeks and Romans, also) for itinerary measures. Thus, there is an historical basis for the use by Eratosthenes of the 148.15 metre stade.

Before leaving the metrological aspect of Eratosthenes' determination of the meridional circumference of the earth, it must be pointed out that there are those[19] who hold that there was but one stade utilized in ancient times and it was the "Olympic" of 185.185 metres, equal to 600 Greek feet of 0.3086 metre or 625 Roman feet of 0.2963 metres. An aspect of this unit which convinces them, if not others, is that 8 such stadia make the Roman mile of 1481.48 metres, 10 make the geographical mile of 1851.85 metres which is almost exactly equal to the length of a meridional minute of arc, and 600 make a meridional degree.

One of the arguments used by the "exclusionists," those who insist there was only one stade, the Olympic, is that Strabo and Pliny[20] knew no other. This is only figuratively true for Strabo, who leaned heavily on Polybius, reports[21] that Polybius (who travelled extensively throughout the Roman Empire) reckoned 8 1/3 stadia per Roman mile, whereas most people counted 8 stadia per mile. Assuming the Roman mile Polybius and Strabo were referring to was the 1481.48 metre unit, then Polybius stadium was equal to 177.78 metres. To complicate matters still more, Drabkin[22] calls Polybius' unit the "Olympic stadium," concurring with Cuntz and others who call the 185.2 metre unit the "Italian" stadium.

Polybius was not being perverse when he set the Roman mile/stade relationship at 8 1/3. It had become customary by his time to set the stade, or stadion, as the Greeks sometimes called it, at 600 Greek feet or 400 Greek cubits (a cubit usually was equal to 1 1/2 feet wherever both units were employed). The problem was that all units tended to vary from place to place, and with time. Clearly, Polybius was basing his stade on the Pelasgic or Ancient Greek foot of approximately 0.296 metre.

There have, thus far, been described 5 stades varying in length from about 100 m. to 185.2 m. all of which could have been employed by the great Greek astronomer/geographer in his celebrated "double-measurement" of the circumference of the earth. Three of these units enjoy prestigious sponsorship (the 148.2, 157.5, and 185.2 metre units), but only one, if any, could have been the unit employed. As we progress to other of Eratosthenes' activities and to the double measurements of other astronomer/geographers and their interpretations, we will meet new descriptions of this ubiquitous unit.

Eratosthenes Views on the Habitable Earth[23]

Eratosthenes was the first mathematical geographer and is generally accorded the distinction of having raised geography to the rank of a science. He wrote a comprehensive work on geography, consisting of three books and a map of the earth.

The first book, a sort of Introduction, gathered together--for the first time - a review of the work of his predecessros, chiefly Homer, Anaximander (the disciple and fellow citizen of Thales, the first great Greek philosopher/astronomer), and Hecataeus, the Milesian, one of the earliest Greek prose writers. The review was critical, intended to bring out what his predecessors thought about the form and nature of the earth and what he, Eratosthenes, thought of their views.

The second book was devoted to mathematical geography -the size of the earth, the determination of latitudes, the relation between latitude and habitability, the relations between the earth and the rest of the celestial bodies which revolved about the earth daily, the definition of "horizon", "meridian", "great circle", "equinoctial or equatorial line", the "poles," the "northern and southern tropics", the "ecliptic" and its obliquity, and a description of the movements of the sun, moon, planets, and the stars.

The third book was devoted to political geography and gave descriptions of the various countries, derived from the works of earlier geographers and the reports of travelers.

Connected with his *Geography* was a map of the inhabited earth, as he conceived it to be, in which towns, mountains, rivers, lakes, and climates were marked. The map was limited to about one-third of the northern hemisphere from east to west and from the equator to the arctic circle. Eratosthenes divided the world into a Northern Division and a Southern Division, the dividing line, parallel to the equator, running through familiar places which he supposed were on the same parallel. This line stretched from the Sacred Promontory (the westernmost point of the Iberian peninsula), through the Straits of Gibraltar, touched the Strait of Sicily, the southern capes of the Peloponnesus, ran thorugh Attica, Rhodes, the Gulf of Issus at the eastern end of the Mediterranean, and then along the southern edge of the Taurus Range which, in the far east, formed the northern boundary of India. Other parallels besides the Equator, the Arctic Circle (which he believed ran through Thule,[24] Iceland), and the Tropic of Cancer (which he believed ran through Syene), were lines through Alexandria, the Hellespont, and the mouth of the Borysthenes (Dnieper).

In addition to his irregularly spaced parallels, Eratosthenes had a main meridian which ran from the Equator to the Arctic Circle, passing through Syene, Alexandria, Rhodes, the Hellespont, and the mouth of the Borysthenes, all of which he believed to have a common meridian. Additional meridians ran through the Sacred Promontory, the Pillars of Hercules, Carthage and Sicily, the Gates of Thapsacus (the busiest ford across the Euphrates River), the Strait of Hormuz and the Caspian Gates (a mountain pass), the mouth of the Indus River, the mouth of the Ganges River, and Cape Coniaci - which was then believed to be the easternmost point in India.

Figure 2d is a representation of Eratosthenes' concept of the habitable earth as described in Strabo's *Geography*. Eratosthenes' map, itself, has been lost to posterity. The basic map shown was apparently drawn for William Smith's *Atlas of Ancient Geography,* first published in London in 1874. It has appeared in virtually every atlas or history of ancient geography published since that time. We have superimposed on this map data from Strabo's *Geography* attributable to Eratosthenes which highlight his geographical views. The east-west dimensions shown are related to the parallel of Athens. While his equatorial circumference of the earth was the same as the meridional, which he had determined to be 252,000 stades, the circumference at the parallel of Athens (which he took to be 36° 55' 43") was 201,445 stades. And, since the length of his habitable world was 77,800 stades at this parallel, it represented 139°2' of longitude. Thus, the distance (roughly) from the offshore islands of West Africa (not firmly located in Eratosthenes time) westward to the islands off the coast of India (still more poorly envisioned) was put at 220° 58'. Eratosthenes' put the length of the Mediterranean at 47° 54' and the length of Asia, eastward from the eastern end of the Mediterranean, at 78° 38'. No matter how unauthoritative Eratosthenes sources of information were, these order of magnitude east-west dimensions and distances would not be improved on for over a millenium.

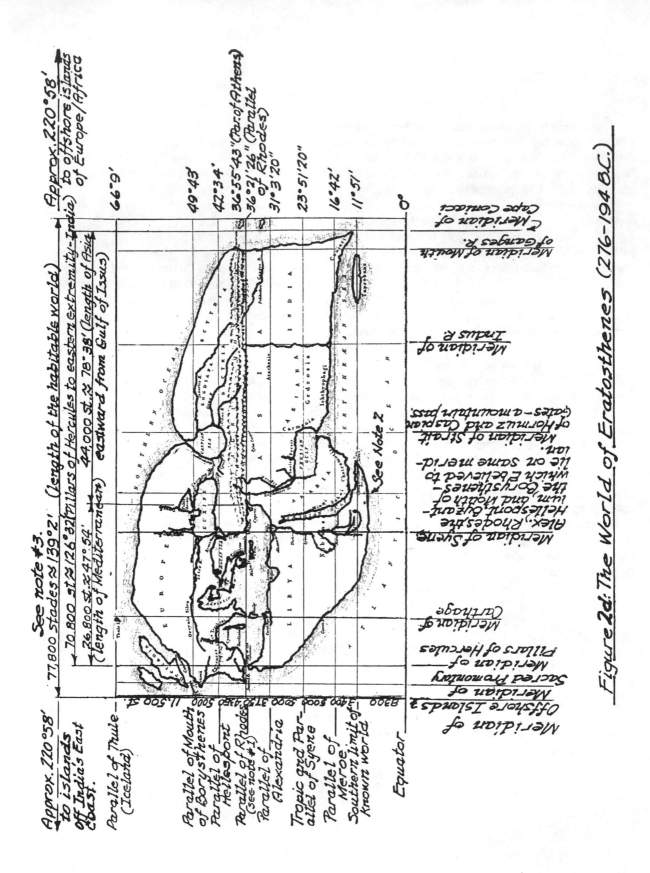

Figure 2d. See notes and credits on next page.

Credit and notes for Figure 1, page 84: The World of Eratosthenes (276 - 194 B.C.)

Eratosthenes was the first mathematical geographer, achieving immortality for his celebrated and surprisingly accurate "double measurement" of an arc of a meridian between Syrene (Aswan) and Alexandria, in Egypt. From his measurements he determined that a degree of latitude on the Babylonian scale (60° to a circle) equalled 4200 stades, equivalent in the Hipparchian scale (360° to a circle) to 700 stades. In his map of the habitable earth, he provided meridians and parallels only as required to locate important places. As director of the famed Alexandria Museum (and Library), he was privy to an enormous store of geographical information provided by travelers, military commanders, professional bematists (pacers), mariners, and earlier geographers - much of it inaccurate.

Notes:

1. *Eratosthenes believed that Rhodes, Strait of Messina, Pillars of Hercules, the southern extremities of the Peloponnesus and Attica, the Gulf of Issus, and the southern edge of the Taurus Mountains lay on the same parallel.*

2. *The ford across the Euphrates at Thapsacus was an important point of caravan routes. The meridian of this ford is shown on map with legend. (See reference note 2)*

3. *All east-west linear distances refer to the parallel of Athens which Eratosthenes beleived to be the longest in the habitable world and 400 stades (34' 17") north of the parallel of Rhodes. Erathostenes figured the circumference of the earth along the parallel of Athens to be :*
252,000 stades x cos 36° 55' 43" = 201, 445 stades.

Credit:

The basic map used here is from William Smith's Atlas of Ancient Geography, London, 1874, and has appeared in many works on geography, cartography, and history of science texts. This map and the data added by this author are based upon Eratosthenes' own geographical publications as presented in several translations in French, German, and English of The Geography of Strabo.

HIPPARCHUS

One of the greatest of classical Greek astronomers, Hipparchus, was born at Nicea in Bithynia around 190 B.C., but carried out most of his celestial observations at Rhodes and Alexandria, probably during the period 160-125 B.C. While his chief title to fame was his catalogue of 1080 stars, he also is recognized as the inventor of trigonometry and the originator of the method of fixing terrestrial positions by means of circles of latitude and longitude.

Hipparchus never wrote any geographical work of his own, but devoted an entire treatise to the criticism of that of Eratosthenes. While this treatise, like Eratosthenes' work, are no longer extant, we know of both through the less than perfect *Geography* of Strabo. The thrust of Hipparchus' criticism is that Eratosthenes had failed to base his location of geographical features on astronomical observations. Eratosthenes had, indeed, recognized the desirability of doing so. While the means of determining latitude were everywhere at hand, the reports received by Eratosthenes were frequently grossly in error. Longitude was another matter. If Eratosthenes was aware of any method other than itinerary measurements for fixing longitudinal positions, it was not mentioned by Strabo.

Hipparchus was the first to indicate the method of fixing comparative longitudes by the different local times that eclipses were observed in different areas, one hour of time difference equalling 15° . Hipparchus introduced the 360° system, Eratosthenes having used the older Babylonian system of dividing the circle into 60 sesagesimals. Despite Hipparchus' knowledge of a method for fixing longitudes, he appears to have been no more successful than Eratosthenes in siting known geographical features accurately.

Hipparchus divided the habitable world into zones of latitude, or "Klimata," bounded by parallels selected on the basis of the number of daylight hours on the longest day of the year, i.e., the summer solstice[25].

On most of the fundamental points in Eratosthenes' work, Hipparchus was content to adopt the conclusions of the earlier geographer. This included Eratosthenes' determination of 252,000 stades for the earth's circumference and 4200 stades to a Babylonian sexagesimal (700 stades to the Hipparchian degree). Pliny later wrote that Hipparchus had added somewhat less than 25,000 stades to Eratosthenes' estimate of the earth's circumference. However, authorities on the history of ancient geography, such as E. H. Bunbury, discount Pliny's statement, considering Strabo's testimony pertaining to Hipparchus' acceptance of Eratosthenes' estimates too explicit to be set aside[26].

POSIDONIUS

His Measurement of the Length of a Degree of Latitude

The next recorded attempt to determine the size of the earth was that of Posidonius (135-50 B.C.), a philosopher and astronomer, born in Apameia (Syria) but later a member of the famed Stoic school of philosophy at Rhodes. Prior to settling at Rhodes, Posidonius travelled extensively, even as far as Spain. We are indebted to Cleomedes for the following account[11].

Posidonius, during his residency at Rhodes, became aware of the fact that the bright star Canopus culminated just on the southern horizon there, while its meridian altitude at Alexandria was observed as "a quarter of a sign." A "sign" being 1/12th part of the zodiac, or 30° , the altitude of Canopus at Alexandria was 1/48th zodiac = 7.5°. This then, represented the difference in latitude between the two sites. The distance between Alexandria and Rhodes being taken as 5000 stadia, the predominant figure reported by mariners plying the route between the two places, the earth's meridional circumference figured out to be 48 x 5000 = 240,000 stadia, pretty close to Eratosthenes' 252,000 stadia.

This was not a particularly good determination for the following reasons. First, the difference in latitude between Rhodes and Alexandria is not 7° 30', but only 5° 15'.

Posidonius reported Canopus culminating just at the horizon at Rhodes. If, indeed, such an observation were possible (not likely because of the instantaneous nature of the occurance) a correction for refraction of 36' 36" would have had to be applied, i.e. the real altitude would have been -0° 36'36" (the star would actually have been below the horizon). Dreyer/Stahl, in reporting this measurement[27] indicate that the declination of Canopus in 100 B.C. was -52° 40.2' and the latitude of the port of Rhodes was 36° 26.6' so that the real altitude of Canopus at culmination would have been 90° - (52° 40.2' + 36° 26.6') or 0° 53.2'. Because of refraction, the apparent altitude would have been about 1° 16'.

At the Museum of Alexandria, which Dreyer/Stahl report has a latitude of 31° 11..7', the true altitude of Canopus at culmination was 90° - (52° 40.2' + 31° 11.7') = 6° 8.1'. This, however, because of refraction, would have appeared to be about 6° 16'.

Thus, the true difference in latitude was about 5° 15', the apparent difference in latitudes, because of refraction, should have been just about 5° (which is what Ptolemy later reported), but Posidonius' figure for the difference was 7.5°.

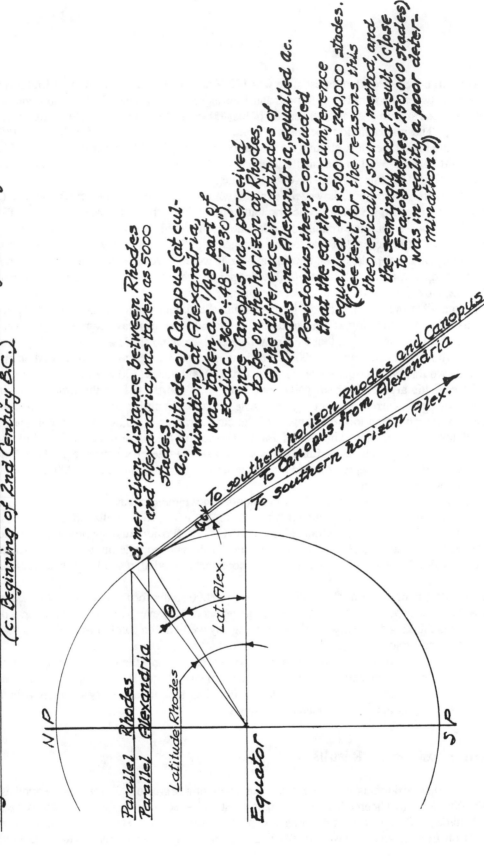

Figure 3 : Posidonius' Determination of the Circumference of the Earth (c. Beginning of 2nd Century B.C.)

d, meridian distance between Rhodes and Alexandria, was taken as 5000 stades.

a, altitude of Canopus (at culmination) at Alexandria, was taken as 1/48 part of zodiac (360°÷48 = 7°30').

Since Canopus was perceived to be on the horizon at Rhodes, θ, the difference in latitudes of Rhodes and Alexandria, equalled a.

Posidonius, then, concluded that the earth's circumference equalled 48 × 5000 = 240,000 stades. (See text for the reasons this theoretically sound method, and the seemingly good result (close to Eratosthenes' 250,000 stades) was in reality a poor determination.)

To southern horizon Rhodes and Canopus

To Canopus from Alexandria

To southern horizon Alex.

☆ Canopus

N P

Parallel Rhodes
Parallel Alexandria

Latitude Rhodes

θ

Lat. Alex.

Equator

S P

87

However, Strabo says Posidonius' estimate was 180,000 stades[28]. He also says that Eratosthenes had put the distance between the parallels of Rhodes and Alexandria at 3750 stades[29] (from observations of the shadows of gnomons at both sites and application of his 700 stadia per degree of latitude equivalency). As Diller puts it[30], "The contradiction (between Cleomedes 240,000 stades and Strabo's 180,000 stades posited for Posidonius) is the more striking for the simple ratio (4 : 3). Moreover the arc from Alexandria to Rhodes would become 3750 in the smaller estimate (3/4 x 5000 = 3750) and this is the very number Eratosthenes calculated for this arc from his own estimate . . ."

This observation does not, however, explain what change Posidonius made in his Canopus observations. If he retained his incorrect determination of 7° 30' as the difference in altitude of the star's culmination at the two sites, his circumference for the earth would be 48 x 3750, or 180,000 stades as Diller proposes. But, if he accepted Eratosthenes' distance of 3750 stades, knowing it had been determined by sundial readings at the two sites, then it would appear he would have to accept the difference in latitudes that this distance represented, 3750 stades ÷ 700 stades per degree = 5.357° = 5° 21' 26". This would bring him back to 252,000 stades for the earth's circumference.

Drabkin, in a wide ranging discussion[31] of Posidonius' estimate which includes all of the foregoing considerations, adds the following as a suggestion: " . . . Posidonius may have combined (1) a proposition of Eratosthenes well known to him, that the sun at the zenith casts no shadow over a circle 300 stades in diameter . . . , with (2) an estimate of 1/600 of the circumference, i.e. 36', as the apparent diameter of the sun. These data would give a terrestrial circumference of 180,000 stades (300 x 600)."

Elsewhere in his article, Drabkin points out that " . . . shortly before the time of Marinus and Ptolemy[32] there were current, apparently, estimates of the circumference of the earth that varied considerably. Plutarch refers to the estimates of the earth's radius at 40,000 stades as a sort of mean estimate." Plutarch's rough approximation of 240,000 stades for the circumference should of course, be 251,327 stades, almost the same as Eratosthenes' estimate. Plutarch's estimate is, however, of no scientific value since it was simply obtained by averaging two other estimates, 180,000 and 300,000 and dividing by 6 to obtain the radius.

Now, Strabo's troublesome remark about Posidonius' measurement was [33] " . . . of the more recent measurements, we prefer those which diminish the size of the earth, such as that adopted by Posidonius, which is about 180,000 stadia . . ." Strabo also reported[34] "He (Posidonius) supposes that the length of the inhabited earth is about 70,000 stadia, being the half of the whole circle on which it is taken; so that, says he, starting from the west, one might, aided by a continual east wind, reach India in so many thousand stadia."

The second statement confirms the first, for "about 140,000 stadia for the earth's circumference at the parallel of Athens corresponds to 180,000 stadia at the equator. However, it represents a radical departure from Eratosthenes thinking as regards the distance along any well-chosen parallel from Spain, westward, to India, almost halving Eratosthenes' estimate.

Strabo's coverage of Posidonius' cosmographical views, particularly the two foregoing quotations had extraordinary results. It reappeared in the writings of Ptolemy, Roger Bacon, Pierre d'Ailly, and other philosopher/geographer/astronomers. Christopher Columbus saw it in d'Ailly's *Imago Mundi* and it became the centerpiece of his own cosmographical views.

Interpreting Posidonius' Results

We have seen that Posidonius' measurement of the length of a degree of latitude was reported as 666 2/3 stades (240,000 ÷ 360) by Cleomedes and 500 stades (180,000 ÷ 360) by Strabo. Cleomedes' version and the 10 to 1 stade to Roman mile ratio was accepted by some in the Hellenic cultural sphere as an improvement on Eratosthenes' measurement. Thus, to this group, Posidonius' degree measured 66 2/3

Roman miles (about 98.8 kms.). The great circle had a circumference of 24,000 Roman miles (35,568 kms.) The 66 2/3 mile degree and 24,000 mile great circle were adopted widely in western Asia, in the areas which were to become Islamic, although the character of the mile is questionable, i.e., it probably varied somewhat from the Roman mile of 1480 to 1482 m. - in certain places and at certain times. (More on this later.) The interpretation employing the 10 to 1 stade to Roman mile ratio resulted in figures about 11% too small.

Since Strabo maintained, erroneously, that there was but one stade, that which went 8 to the Roman mile, there was born the 62.5 Roman mile degree and 22,500 mile earth's circumference, equivalent to 92.6 and 33,333 kilometres, respectively - both figures about 16-2/3% too small.

STRABO

Despite the foregoing, it is clear that Strabo did not finally desert Eratosthenes. In Chapter V of Book II, Strabo finally gets around to stating what he thinks the earth is like, as against what others think. Here, we find Eratosthenes, with numerous modest "corrections." Thus, the length of his habitable world is 70,000 stadia, his length of the Mediterranean is just about what the older geographer had estimated - 26,500 stades and his equatorial and meridional earth's circumference - 252,000 stades. But, was the stadium of Strabo's day the same as the stade employed by Eratosthenes 200 years earlier? We think not!

Before leaving Strabo, it is important that another of his geographic views which was echoed by succeeding cosmographers for almost 1500 years be aired. In Chapter I of Book I, while on the subject of Homer, the poet (whom Strabo accorded the distinction of being the first geographer), he philosophizes:

"Perception and experience alike inform us that the earth we inhabit is an island: since wherever men have approached the termination of the land, the sea, which we designate ocean, has been met with: and reason assures us of the similarity of those places which our senses have not been permitted to survey. For in the east the land occupied by the Indians, and in the west by the Iberians and Maurusians is wholly encompassed (by water), and so is the greater part on the south and north. And as to what remains as yet unexplored by us, because navigators, sailing from opposite points, have not hitherto fallen in with each other, it is not much as anyone may see who will compare the distances between those places with which we are already acquainted. Nor is it likely that the Atlantic Ocean is divided into two seas by narrow isthumuses so placed as to prevent circumnavigation: how much more probable that it is confluent and uninterrupted! Those who have returned from an attempt to circumnavigate the earth do not say they have been prevented from continuing their voyage by any opposing continent, for the sea remained perfectly open, but through want of resolution and the scarcity of provision . . . "

The foregoing was mulled over by every philosopher, geographer, and educated mariner. Those who agreed with this philosophy repeated it in their own writings. Columbus read about it in Cardinal d'Ailly's *Imago Mundi* and it strengthened his conviction that a westward voyage from Spain to India was a practical undertaking.

PLINY

Caius Plinus Secundus, commonly known as Pliny the Elder, to distinguish him from his nephew of the same name, was born in A.D. 23 at Verona or Comum, in northern Italy. During his lifetime he held various public offices including that of procurator in Spain. His last official appointment was admiral of the fleet at Misenum where he was killed by the 79 A.D. eruption of Vesuvius (which also buried Herculaneum and Pompeii). He wrote, among other works, a *Natural History* in 37 books, reading about 2,000 volumes and collecting 20,000 facts in preparation. It was during his years as a public official that he read this enormous array of documentation, all in his spare time and frequently having segments read to him while he was dining. Thus, it was not until a comparatively advanced period of his life that he devoted himself in earnest to the writing of his *Natural History*.

Unfortunately, this enormous essay towards a physical description of the Universe was deficient in many ways. There is no attempt to distinguish the valuable material from the worthless. Perhaps this function was beyond his capability for Pliny did not possess either an orderly or highly scientific turn of mind. The mass of "facts" (true or false) is presented without any clear and lucid arrangement. As Humboldt has put it ". . the elements of a general knowledge of nature lying scattered almost without order in his great work . . .". The great naturalist Cuvier considered that part of Pliny's work which might now be considered natural history as far inferior to Aristotle's works in the same disciplines almost 400 years earlier. Bunbury considers the geographical portions as perhaps the most defective parts of the whole work and far inferior to the writings of Eratosthenes and Strabo, the appearance of any scientific comprehension of his subject being almost totally absent.

The great value of Pliny's work, according to Bunbury, " . . . lies in its important contribution to the political or statistical geography of the countries that were in his time organized as provinces under the Roman Empire"

With regard to Eratosthenes' estimate for the circumference of the earth, Pliny says " . . . his (Eratosthenes') process of reasoning was so ingenious, that it was impossible not to believe it. It was, indeed, generally adopted although Hipparchus had corrected it by the addition of about 26,000 stadia". As explained previously (under our coverage of Hipparchus), most geographers and history of science writers discount the 26,000 stadia addition claimed for Hipparchus. Strabo having been explicit in indicating Hipparchus' acceptance of the 252,000 stadia circumference. General acceptance of Eratosthenes' is confirmed by Vitruvius ("de Architectura" vol. i., chapter 6, para. 9), the different estimate of Posidonius having been either overlooked or discredited.

What then was Pliny's intepretation of Eratosthenes' 252,000 stades in terms of the Roman mile? Earlier, we quoted Pliny as saying Eratosthenes had set the schoenus equal to 40 stades. Despite this, Pliny went along with Strabo. He converted from stades to Roman miles at 8 to 1 ratio, thus, 252,000 stades became 31,500 Roman miles. And this is the figure which crops up so frequently in medieval geography. The schoenus, which according to Hultsch had a length of 6,300 metres, and according to d'Anville and Lehmann-Haupt (via Diller) equalled 5,926 metres, balloons to 40 x 185.2 metres = 7,407.5 metres.

MARINUS OF TYRE

Marinus of Tyre, about whom very little is known except for two circumstances, was a great geographer of the end of the first century A.D., or the first half of the second century. What we know of Marinus and his work is chiefly due to what Ptolemy, who followed Marinus, said about him in his own *Geographiae*. However, much of the fame of this ancient, diligent, and imaginative geographer stems from the fact that his views as to the great extent of the Asian continent, from west to east, enjoyed a resurgence of confidence some thirteen hundred years after his time. By this time, the reports of travelers (emissaries, missioners, and traders) who had made it all the way to the court of the great Mongol Khans in Cambaluc (Peking) had been published.

The most famous of these travelers was Marco Polo and the most avid and impressed of his readers was Christopher Columbus. But it would be a mistake to think that *The Book of Ser Marco Polo* impressed only laymen and those with limited formal education. Equally struck by the geographical implications of Polo's narrative were such philosophers/cosmographers as the great Florentine physician and astronomer Paolo Toscanelli and the German geographer Martin Behaim, both of whom flourished in the 15th century.

Up to the time of Marinus of Tyre, concepts of the habitable world were essentially those of Eratosthenes, essentially endorsed (although encased within varying blankets of criticism) by Hipparchus, Posidonius, and Strabo. Marinus departed boldly from this mold. First, he adopted Posidonius' measurement/estimate of the earth's circumference, as publicized by Strabo - 180,000 stadia. Next, he established a zero meridian at the Fortunate Islands which he conceived to lie 2 1/2° west of the Sacred Promontory of Spain. Then, taking note of the great number of travelers' reports which had accumulated since Strabo's day, he undertook to correct the conceptions of earlier geographers, particularly as regarded the extensions of the two great continents - Africa and Asia - to the south and east.

Based upon the reports of military expeditions and travelers' itineraries, he concluded that the breadth of the habitable world was 87°, stretching from the parallel of Thule at 63° north latitude to that of the promontory of Prasum, in Africa along the Indian Ocean coast; this he estimated to be at 24° south latitude.

However, it was Marinus' extension of the habitable world's length which was to have such enormous consequences. His estimate of the length of the Mediterranean seemed little changed from that of Eratosthenes', being 24,800 stadia along the parallel of Rhodes, taken as 36° N. Eratosthenes had estimated 26,800 stades along the parallel of Athens, approximately 37° N. Eratosthenes' estimate was equivalent to about 48° of longitude since his degree had a length of 700 stades at the Equator and 559 stades at the latitude of 37° N. Marinus' estimate, however, based as it was on the 500 stade degree which measured just about 400 stades along the parallel of Rhodes (36° N), was a whopping 62° of longitude. Evidently, sea distances were reported to both geographers in stades which went 700 to the degree (the "sea" stade). Whether all land distances were reported in stades which went 500 to the degree (sometimes called the "land" stade) is questionable. It is likely that land distances were reported in a variety of units - Roman miles, schoeni, parasangs, and a veritable spectrum of stades.

Relying upon the reports of merchants in the silk trade, Marinus placed Sera, the capital of the Seres and the most easterly inland city of which he had knowledge, on the same meridian as Cattigara, the farthest point known to him from mariners' itineraries. Both of these he placed no less than 225° east of the Fortunate Islands.

Marinus was vague about the location of the sea to the east of Asia, but attributed to that continent an unknown extension to the east. Thus, later geographers felt justified in extending the eastern limit of Asia still further when information of the kind provided by Marco Polo and others became available.

How Marinus came to make such colossal distortions in the earth as it is and as it had been conceived by previous geographers is not easy to explain. The information on which he (and all) geographers relied was probably no worse than that previous geographers had contended with. It would seem that he not only failed to discount exaggerated estimates in itineraries, but also misinterpreted bearings, considering legs of journeys most of which had significant north-south components to be essentially east-west. Then, finally, he converted from stades to degrees using a highly questionable ratio.

Marinus drew his parallels and meridians as equidistant straight lines, forming right angles to each other. At the same time, the parallel of Rhodes, 36° N, was the map's main parallel. Only on this parallel were the degrees of longitude set off according to their due proportion to those of latitude (since Marinus spaced his meridians 15°, or 1 hour, apart, the space between his meridians equalled cosine 36° x the space allotted to 15° of latitude). Thus, both north and south of latitude 36° , the map was distorted. However, the error introduced in mapping Mediterranean countries was small. While areas further north and south were affected more significantly, the geographical data available to Marinus about distant countries was so unreliable as to make map distortion not an important consideration. When the equator is selected as the main parallel, Marinus' grid becomes quadratic.

Despite the many failings of Marinus' work, it became the basis of Ptolemy's *Geographiae* and unless the later geographer had some particular reason for departing from Marinus' conclusions, he adopted them. The greatest criticism Ptolemy had about Marinus' work, aside from the awkward grid system, was the inordinately greater length which Marinus gave to the habitable world, a length which Ptolemy reduced from 225° to 180°.

PTOLEMY (CLAUDIUS PTOLEMAEUS)

His Geographic and Geodetic Views

Claudius Ptolemaeus, more commonly known as Ptolemy, was a native of Egypt who lived, worked, and wrote at Alexandria about the middle of the second century, A.D. A philosopher, astronomer, and geographer, his scientific writings exerted an influence over succeeding centuries second only to that of Aristotle. He is best known for his great astronomical work which the Arabs in the 8th century re-named *The Almagest*, meaning "the greatest" or "best." His doctrines were incorporated in 13 books written about 140 A.D. The "Ptolemaic system" is his theory for explaining the movements of the heavenly bodies which he supposed revolved about the earth in circles.

Ptolemy also wrote the *Geographik Syntaxis,* or *Atlas of the World* in 8 books, 6 of which are essentially tables listing the known places of the world and giving their latitudes and longitudes - many grossly in error. However, his was the first effort to put geography on a scientific basis and many of his rules and directions for the construction of maps are still followed. As Nordenskiold has, so aptly, put it, " . . . the principles of geography may be said still to be published with Ptolemy's alphabet."

Ptolemy has freely admitted that his *Geography* was based on Marinus' prior efforts. This is apparent from the exaggerated length given to the Mediterranean and to the Eurasian continent (although he did reduce Marinus' length of 225° to 180°), to his erroneous notion that the Indian Ocean was landlocked, and his placement of the zero meridian at the Fortunate Islands, vaguely understood to be 2 1/2° west of Cape St. Vincent. The Island of Ferro, westernmost of the Canaries, which continued up until the latter part of the 19th century to be taken as the prime meridian, is actually 9° 20' west of the Sacred Promontory.

One of the severest criticisms registered against Ptolemy is that while he deplored the failure of prior geographers to base the geographical location of places (i.e. latitude and longitude) on astronomical observations, with very few exceptions, he did no better. He credits Hipparchus (the inventor of trigonometry and one of the most brilliant of ancient astronomers) with having pointed out how to do this. Reading the altitude of the North Star (which was considered, erroneously, to be fixed on the celestial sphere) gave latitude directly. This method could be employed during the hours of darkness on any day of the year. Hipparchus' method for determining longitude was more complicated, had fewer opportunities for its employment, and was dependent on accurate timekeeping. It depended upon the simultaneous observation of the earth's eclipse of the moon at a number of important places and the recording of the

Text continues on page 98

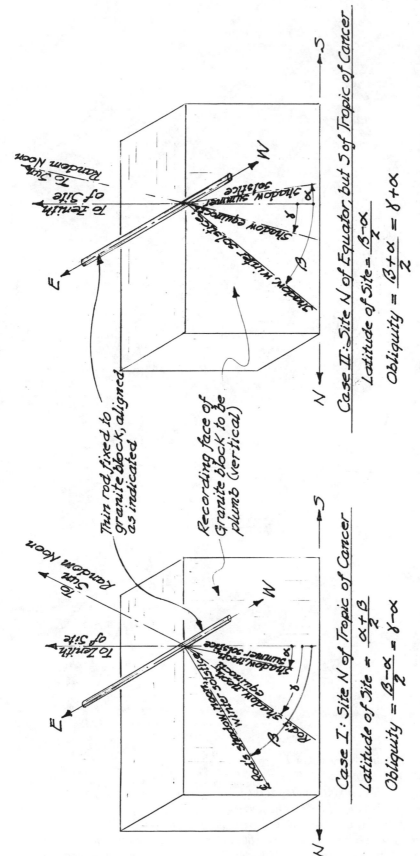

Case I: Site N of Tropic of Cancer

Latitude of Site = $\dfrac{\alpha + \beta}{2}$

Obliquity = $\dfrac{\beta - \alpha}{2} = \gamma - \alpha$

Case II: Site N of Equator, but S of Tropic of Cancer

Latitude of Site = $\dfrac{\beta - \alpha}{2}$

Obliquity = $\dfrac{\beta + \alpha}{2} = \gamma + \alpha$

Figure 4: Ptolemy's Solar Method for the Determination of the Latitude of a Site and the Obliquity of the Ecliptic (Requiring a Year's Observations).

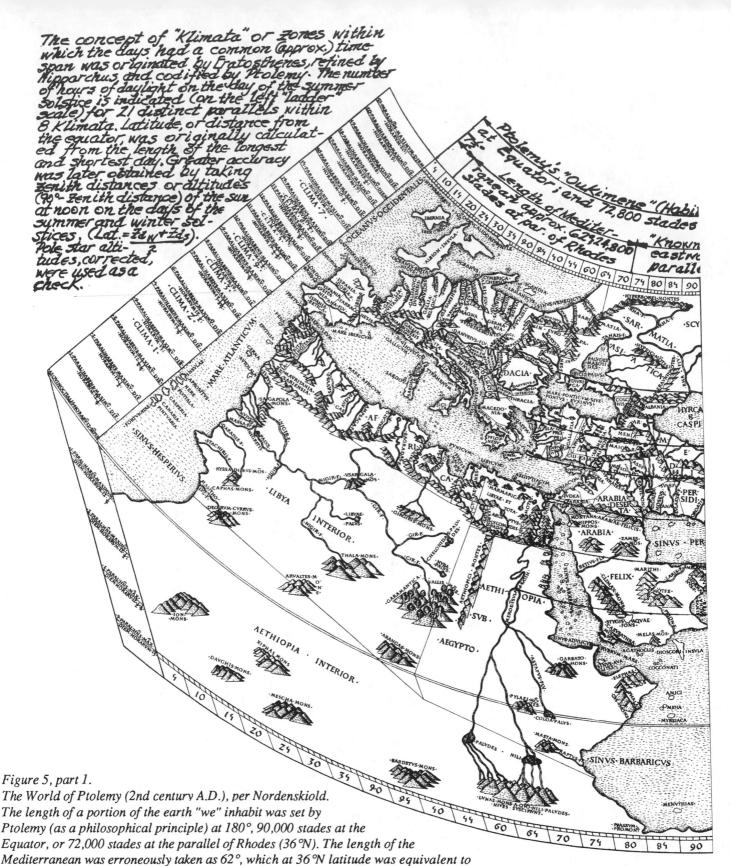

The concept of "Klimata" or zones within which the days had a common (approx.) time span was originated by Eratosthenes, refined by Hipparchus, and codified by Ptolemy. The number of hours of daylight on the day of the summer solstice is indicated (on the left "ladder" scale) for 21 distinct parallels within 8 klimata. Latitude, or distance from the equator, was originally calculated from the length of the longest and shortest day. Greater accuracy was later obtained by taking zenith distances or altitudes (90° - zenith distance) of the sun at noon on the days of the summer and winter solstices, (Lat. = $\frac{z_{d_W} + z_{d_S}}{2}$). Pole star altitudes, corrected, were used as a check.

Figure 5, part 1.
The World of Ptolemy (2nd century A.D.), per Nordenskiold.
The length of a portion of the earth "we" inhabit was set by
Ptolemy (as a philosophical principle) at 180°, 90,000 stades at the
Equator, or 72,000 stades at the parallel of Rhodes (36°N). The length of the
Mediterranean was erroneously taken as 62°, which at 36°N latitude was equivalent to
24,800 stades. The length of Asia, from the Gulf of Issus to the 180th meridian, the eastern end of the known world,
was set at 110 1/2°, equivalent to 44,200 stades at parallel of Rhodes. The zero meridian was placed at the Fortunate
Islands (Canaries), 7 1/2° west of the Pillars of Hercules (Gibraltar). Ptolemy's degree of latitude, and his degree of
longitude at the Equator, equalled 500 stades. His circumference of a spherical earth was therefore 180,000 stades.
Erastosthenes had said a degree was equal to 700 stades and the circumference - 252,000 stades. Who was in error? See
text for a discussion of this interesting and highly controversial subject. See page 96, 97 for sources and credits and for
further details of the eastern and western "ends" of the "Habitable World".

94

Figure 5, part 2.

The original Ptolemy general ("World") map is believed not to have had a grid of parallels and meridians, although many of the editions published in the 15th and 16th centuries did. On the conical projection shown, the careful scanner will note some "fudging" on the partial parallels, which are really leaders superimposed on the map connecting the scale at the right with the places critical for our comparison of Eratosthenes' and Prolemy's linear units. Not all of the critical places are precisely at the latitudes indicated in the text of Ptolemy's "Geographia".

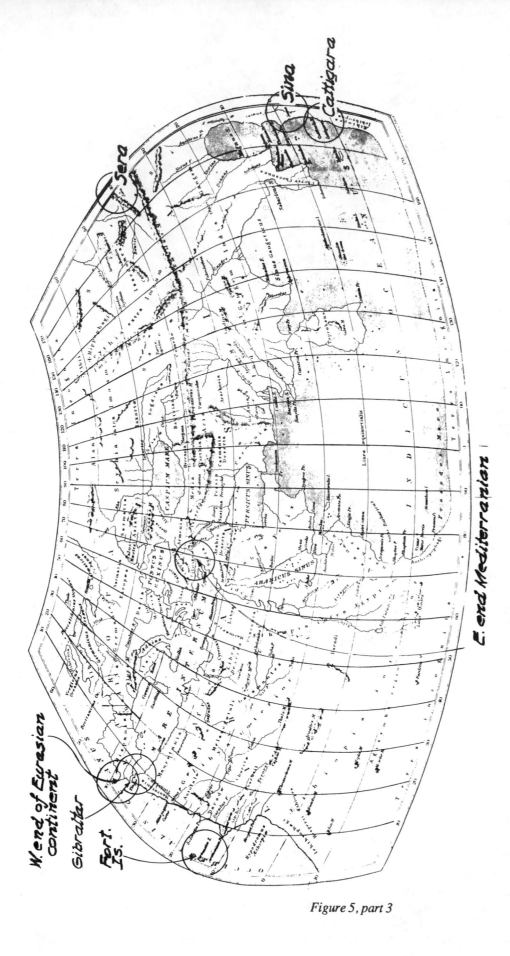

Sera

Sina

Cattigara

W. end of Eurasian continent

Gibraltar

Fort. Is.

E. end Mediterranian

Figure 5, Part 3: Map of the World According to Ptolemy (per Bunbury)

Marinus and Ptolemy both placed the prime meridian (0°) at the Fortunate Islands (Canaries), encircled; the western end of the European-Asian continent 2 1/2° east of the Fortunate Islands; the western end of the Mediterranean at the Pillars of Hercules (Gibraltar), 7 1/2° east of the Fort. Is.; the eastern end of the Mediterranean at the Gulf of Issus (encircled), 69 1/2° east of the prime (making the Mediterranean 62° long; and the easternmost places in the known world - Sera, Sina, (Thinae), and Cattigara very close to the 180th degree, E., meridian (for Ptolemy) and about the 225th degree, E., meridian (for Marinus). Neither Marinus not Ptolemy accepted the inevitability of a coastline east of the 180th (or 225th) meridian, in Terra Incognita, but most mariners, cartographers, and at least some cosmographers of the Middle Ages did.

Figure 5, part 3

About Sources and Credits

There is some doubt as to whether Ptolemy drew any of the maps associated with his "Geography," himself. Moreover, since the original manuscript of this important work has not survived, it is difficult to say how much of the text is ascribable to Ptolemy and how much to the translators who prepared the Byzantine manuscripts (the earliest known) and the Latin, Italian, German, French, Syriac, Arabic and other translations over the course of twelve centuries. The maps which have evolved appear to have been drawn by three or four hands. Between 1475 and 1605, with the double stimuli of the printing press and the "Age of Discovery," about 40 editions of "Ptolemy" were published. The map shown on pages 94 and 95 is from the 1973 Dover Publications (N.Y.) edition of A.E. Nordenskiold's famous "Facsimile-Atlas", the English edition of which was first published in Stockholm in 1889. It is a photocopy of the general ("world") map, of a folio of 27, published in Rome in 1490 by "Domitus" Calderinus, printer Conrad Scheinheim, and is based on Donis' (Nicolaus Germanus) copy of Ptolemy's "originals". The numerical latitudinal data superimposed is taken from Muller's translation of Ptolemy's "Geography", I. 21-24, and Nobbe's translation of Ptolemy, "Geography", VII. 5, as reproduced in Cohen and Drabkin's "A Source Book in Greek Science", McGraw-Hill Book Co., New York, 1948.

The Longitudinal data superimposed on the map is a consensus of information taken from Nordenskiold, p. 4b; from E.H. Bunbury's "History of Ancient Geography", Vol. II, "Map of the World According to Ptolemy", facing p. 578, Dover Pubs., N.Y., 1959 and shown here on page 96; and from J.O. Thomson's "History of Ancient Geography", p. 338, Biblo & Tannen, N.Y., 1965.

Permission to reproduce the maps from Nordensköld's "Facsimilie-Atlas" and Bunbury's "History of Ancient Geography" has been granted by Dover Publications, New York.

Figure 5, part 4

precise local time that this occurred at each place. The difference in the time of eclipse at any two sites represented the difference in longitude, one hour being equivalent to 15° (24 hours equals 360°). While Ptolemy endorsed this method, it is evident he did not employ it himself.

Ptolemy also told how to determine latitude by observing the sun's angle of elevation above the horizon over a period of at least one year. (See Figure 4.) This was a variation of the gnomon method employed by Erathosthenes.

Ptolemy is credited with being the first to describe how to construct a map depicting the earth's spherical surface on a plane. Since his habitable earth was only 180° from west to east and about 79° from north to south (the northern limit being 63° north latitude and the southern about 16° south latitude), the task was not as formidable as it would become after the "Age of Discovery." While he mentions several different methods for doing this, he uses only two projections for the 27 maps in his own atlas: the conical projection for his "world" map and the equidistant rectangular or equidistant cylindrical projection of Marinus for the 26 special maps (Figure 5 illustrates the former).

During the darkness of the Middle Ages, Ptolemy and his method of map drawing were pretty much forgotten, at least in the west. Instead of his clear and intelligible maps, based - at least for latitude - on astronomical observations, maps were produced with no pretense at proper proportioning, covered with figures of princes or other underwriters, monsters, and fantastic legends of early Christian or heathen origin.

The foregoing is simply intended to give a very rough sketch of the brilliant astronomer who did more to put geography and cartography on a scientific basis than anyone before (and probably since) his time. This chapter being primarily concerned with tracking the estimates of various writers of the ancient world as regards the size of the earth, we turn now to Ptolemy's views on this subject.

It is strange indeed that Ptolemy who followed Hipparchus so closely in his mathematics and astronomy, who respected Posidonius as a philosopher, but who knew that Posidonius' estimate for the difference in altitude of Canopus at Rhodes and Alexandria was grossly in error, should nevertheless accept Posidonius' estimate of 180,000 stades as attributed to him by Strabo. The answer may lie in his acceptance of so many of Marinus' conclusions, this being simply another - albeit very important - one. However, another plausible explanation has been advanced. Ptolemy (and Marinus, too) may have used a longer stade than that used by Eratosthenes. But which stade?

Interpreting Ptolemy's Geodetic Views

One post-Ptolemaic writer who had no doubt as to Ptolemy's stade to Roman mile relationship (although he preferred Eratosthenes' concepts to those of the later geographer) was Macrobius. He believed Ptolemy used the Strabo 8 stade to the mile ratio (ref. note 17). Thus, his version of Ptolemy's degree and earth's circumference was the same as that attributed by Strabo to Posidonius: a degree measured 500 stades or 62.5 Roman miles (92.6 km.); the earth's circumference measured 180,000 stades or 22,500 Roman miles (33,333 kms.).

Another version of Ptolemy which became popular among mariners (and some cosmographers) of Columbus' time was based on Polybius' understanding of the stade to Roman mile ratio, 8-1/3 to1. The degree of 500 stades based on his ratio measured 60 Roman miles (easily divided into minutes and seconds) or 89 kilometres. The 180,000 stade earth's circumference translated into 21,600 Roman miles, or 32,000 kilometres - about 20% less than the actual circumference.

Dreyer/Stahl say: "In his 'Geography', Ptolemy gives the length of a degree equal to 500 stadia and the earth's circumference equal to 180,000 stadia. As this is to the value of Posidonius (240,000 st.) exactly as the itinerary is to the Egyptian stade, Ptolemy evidently used the official Egyptian stade of 210 metres, which is practically 1/7 of a Roman mile (barely four feet less) and was therefore a convenient unit for a subject of the Roman Empire and an inhabitant of Egypt to use. He simply adopted the value of Posidonius and expressed it in terms of a different unit."

The Dreyer/Stahl view then translates into a 105,000 metre degree and a 37,800 Km meridional circumference for the earth, both figures only 5.5 percent low.

Diller and Drabkin agree that Ptolemy used a longer stade unit than that used by Erathosthenes; they even agree that the Ptolemaic unit was one-third larger than the Eratosthenian. However, just as these references differed with Dreyer/Stahl as to the length of the stade used by Eratosthenes, so do they differ on the length of the stade used by Ptolemy. Both of these authorities say Ptolemy used the "Philetaerian" stade of 7 1/2 to the Roman mile, i.e. 197.5 metres. This unit was composed of 400 Assyrian cubits of 0.494 metres, as against 300 of the same cubits for the stade they (Diller and Drabkin) claim Eratosthenes used.

The Diller/Drabkin view translates into a 98,750 metre degree and a 35,550 Km. meridional circumference. Taking the stade posited by Bunbury, Leake, and the rest of the "exclusionists", the 185.185 metre Olympic stade, Ptolemy's degree measures 95,592.5 m. and the meridional circumference of the earth equals 33,333 Km. These figures, some 16 2/3% to low, are the estimates usually ascribed to Ptolemy even though all he ever said was 500 stades made a degree and 180,000 stades made the earth's meridional circumference.

In an effort to assess the various claims and counterclaims, we have gone back to what Eratosthenes and Ptolemy, each, said was the meridional distance, in both degrees and stades, between known and important geographic points of interest. Strabo is the authority for Eratosthenes' data and Ptolemy's *Geographia* for his own. The map drawn for Dr. Smith's *Atlas of Ancient Geography* with some of Eratosthenes most important determinations of latitudes, and latitudinal distances, and longitudinal distances superimposed thereon, Figure 2d, will be used to illustrate the older geographer's views.

Figure 5, including the dimensional data and notes superimposed thereon, provides a counterpart for the presentation of Ptolemy's views.

Table 1 is, as titled, "A Comparison of the Employment of the Stade (Stadion, Stadium) by Eratosthenes and Ptolemy in Measurements Along a Merdian."

Without choosing sides, for both are somewhat in error throughout the range of the table, it is very clear that Ptolemy was using a unit some 40% larger than Eratosthenes. The data for Rhodes, Alexandria, Syene, and Meroe are more reliable than for the points farther north. These show an average ratio of distances from the equator to specific points, as given by Ptolemy to those of Eratosthenes almost identical to the ratios of their degree lengths (500 and 700 stades, respectively), namely 0.714. Further, comparing Eratosthenes' distances in stades to the actual distance in kilometres, it would appear that he may have been using a stade 159.8 metres in length. A similar comparison for Ptolemy would indicate his stade length may have been 225.4 metres. These are distinct possibilities and would tend to give both Eratosthenes and Ptolemy high marks for their geodetic as well as their astronomical and mathematical capabilities. These figures are closer to the Dreyer/Stahl estimates of 157.5 and 210 metres, respectively for the Eratosthenes and Ptolemy stades, than to any other estimates proposed in our review thus far.

Still, a failure on Ptolemy's part to define his stade led many astray, as we shall see.

Table 1: A Comparison of the Employment of the Stade (Stadion, Stadium) by Eratosthenes and Ptolemy in Measurements Along a Meridian

Sites	Actual		According to Eratosthenes		According to Ptolemy		Ratio Col.3b/Col.2b	Ratio Col.1b/Col.2b	Ratio Col.1b/Col.3b
Columns →	Latitude 1a	Distance to Equator 1b	Latitude 2a	Distance to Equator 2b	Latitude 3a	Distance to Equator 3b	4a	4b	4c
Thule (Iceland) (Reykjavik)	64°	7111 Km.	66.14°	46,300 stades	63°	31,500 stades	0.680	153.6 m. per stade	225.1 m. per stade
Mouth of Borysthenes (Dnieper)	46.52°	5169 Km.	49.72°	34,800 stades	48.5°	24,250 stades	0.696	148.53 m. per stade	213.15 m. per stade
The Troad (Hellespont)	40.13°	4459 Km.	42.57°	29,800 stades	40.92°	20,458 stades	0.686	149.63 m. per stade	217.96 m. per stade
The Port of Rhodes	36.45°	4049 Km.	36.36°	25,450 stades	36°	18,000 stades	0.706	159.09 m. per stade	224.94 m. per stade
Alexandria (Museum)	31.2°	3467 Km.	31°	21,700 stades	30.97°	15,483 stades	0.712	159.77 m. per stade	223.92 m. per stade
Syene (Elephantine)	24.08°	2676 Km.	23.86°	16,700 stades	23.83°	11,917 stades	0.714	160.24 m. per stade	224.55 m. per stade
Meroe	16.85°	1872 Km.	16.71°	11,700	16.42°	8208 stades	0.702	160.0 m. per stade	228.07 m. per stade

General Notes:

1) Assume a spherical earth with an equatorial or meridional circumference of 40,000 Km.
2) Eratosthenes concluded from his skiotheran/bematistes measurements that the earth had a circumference of 252,000 stades and a degree of latitude measured 700 stades.
3) Ptolemy concluded from Posidonius, Strabo, and Marinus, and from his own observations, that the earth's circumference was 180,000 stades and a degree 500 stades.
4) The ratio $\frac{180,000}{252,000} = 0.714$

Table 1, part 1.

5. Latitudes according to Eratosthenes are taken from the Hamilton/Falconer translation of "The Geography of Strabo" Bk 1, Chap. 1, P 11; Bk 2, Chap. 5, P 24; and Bk 1, Chap. 4, P 2 and 3. Distances from equator are determined from latitudes and Eratosthenes' relationship 1° = 700 stades.

6. Latitudes according to Ptolemy are taken from Cohen and Drabkin's "Source Book in Greek Science", pp. 169-172. Authors have indicated original source as Müllers Translation of Ptolemy's Geography", I.21-24 ("Principles of Cartography) - in German.

7. It should be understood that Eratosthenes used the Babylonian system of 60° to the circumference of a great circle. Thus, his degree lengths were actually 4200 stades, rather than 700. The Hipparchian system of 360° to the circumference, used by Ptolemy, is applied to Eratosthenes' estimates in order to obtain equivalent values, susceptible to comparison.

Table 1, part 2.

101

Table 2: A Comparison of Various Stades (Stadions, Stadia) Found in the Literature

Identification and Note Reference	Stade Length in Metres Note #1	Stade Composition: Unit Length (Name) × Number	No. of Stades per Roman Mile. Note #2
The Herodotus-Gossellin stade based on an Egyptian itinerary measurement between Thebes and mouth of Nile R. Note #3	99.75	0.494 (Assyrian cubit) × 200	14.85
Pliny-Macrobius Pliny-d'Anville. Note #4	148.2	0.2963m. (Pelasgic Foot) × 500 5928 m. (Schoenus) ÷ 40 0.494m. (Assyrian cubit) × 300	10
Pliny-Hultsch, Pliny-Dreyer/Stahl Note #5	157.5	6300m. (Schoenus) ÷ 40 0.525m. (Royal Cubit) × 300	9.41
Lehmann-Haupt Note #6	165	0.275m. (Plinian Foot) × 600 0.550m.(Talmudic Cubit) × 300	9
The Polybius "Palasgic" stade. Note #7	177.9	0.2963m. (Pelasgic Foot) × 600	8.33
The Strabo-Pliny Olympic stade and Roman stadium Note #8	185.2	0.3086m. (Olympic Foot) × 600 0.2963m. (Roman Foot) × 625 0.3704m. (Egyptian Remen) × 500	8
The Drabkin Philetaerian stade. Note #9	197.6	0.494m. (Assyrian cubit) × 400	7.5
The Ptolemaic or Royal Egyptian stade of Hultsch Note #10	210	0.525m. (Royal cubit) × 400	7
Lehmann-Haupt Note #6	296.4	0.988m. (Double Assyrian cubit) × 300	5
The Babylonian and Assyrian "Ush (Stadion) Note #11	355.68	0.494m. (Assyrian cubit) × 720	4.166
Klein's "Superstadion" Note #12	1896	?	0.78

Table 2, part 1.

NOTES IN SUPPORT OF TABLE 2

Note #1:

One metre = 3.28084 U.S. Customary Feet.

Note #2:

Remarkably resistant to change, the Roman mile has been reported as short as 1472.5 metres (Lithré) and as long as 1488 metres (Diller). One of the more popular lengths, 1482 metres, is used for this table.

Note #3:

Gossellin says this stade ran 1111¼ to the degree of latitude, i.e.

1 stade = 56,870 toises = 110,850 m. = 99.75 m.
$$\frac{56,870 \text{ toises}}{1111\frac{1}{4}} = \frac{110,850 \text{ m.}}{1111\frac{1}{4}} = 99.75 \text{ m.}$$

Note #4:

Pliny XII 53 says: "Schoenus patet Eratosthenis ratione Stadia XL." Macrobius put Eratosthenes' stade at 10 to the Roman mile or 1482 = 148.2 m.
$$\frac{1482}{10} = 148.2 \text{ m.}$$

d'Anville cited the Heronian Tables which put 1 schoenus = 4 Roman miles = 5928 m. Then from Pliny's 1 schoenus = 40 stades,
$$1 \text{ stade} = 5928 \div 40 = 148.2 \text{ m.}$$

It will be noted that 12,000 Assyrian cubits of 0.494 m. = 5928 m.

Note #5:

Hultsch, "Griech. und Röm. Metrologie", Berlin, 1882, p. 364, says: 1 schoenus = 6300 m. = 12,000 Royal cubits × 0.525 m. per cubit. Hultsch accepts Pliny, but not the Heronian Tables.

Note #6:

Drabkin credits Lehmann-Haupt with showing that there were at least seven different stades in use at various times and places, among them units measuring 9 and 5, respectively, to the Roman mile. (The stade compositions in the table for the 165 m. and 296.4 m. stades are conjectural.) See I.E. Drabkin's "Posidonius and the Circumference of the Earth," ISIS, 1943, p. 510.

Note #7:

Polybius said the Roman mile equalled 8 stades and 2 plethra. Since the Greek stade equalled 600 Greek feet and the plethra 100 feet, this put the Roman mile at 5000 of Polybius' feet, which was its definition in terms of Roman feet of 0.2963 m. Hence, Polybius must

have been employing Pelasgic (Ancient Greek) feet, which were the same length as the Roman. It follows that the Polybius Stade has a length of 600 × 0.2963 m. = 177.78 m.

Note #8:

This stade is the overwhelming favorite of writers who touch on ancient geographical matters. Bunbury and Leake claimed there was no other stade employed in the Greek cultural orbit.

Note #9:

This stade, whose composition as shown in table is conjectural, is the unit which both I.E. Drabkin and Aubrey Diller say may have been employed by Ptolemy in defining the length of a degree of latitude as 500 stades. It is known to have existed in later times. It will be noted that it is 4/3 as long as the 148.2 m. stade. This stade is not a credible multiple of the Philetaerian foot, reported by Petrie to have a length of 0.3529 m.

Note #10:

Dreyer/Stahl give as their authority for the 210 m. stade Hultsch (see Note #5), pp. 61 and 355. It will be noted that this stade is 4/3 as long as the 157.5 m. unit.

Note #11:

Hallock and Wade describe the Ush (stadion), a Sumerian (early Babylonian) unit of length, as being equal to 720 large elles (cubits) of 0.493-0.498 m. The Assyrian cubit of 0.494 m. was an outgrowth of the large Sumerian elle.

See chapter 2 and pp. 13-15, "Evolution of Weights and Measures," by Hallock and Wade.

Note #12:

Klein describes "...another stadion—a sort of superstadion—of 1896 meters length (which) was recognized in the ancient world. It was based upon the foot, and it was very nearly ten times the length of the digit-based Greek stadion and the later Roman stadium." This is the largest stade (stadion,stadium) found in the literature.

See p. 61, "World of Measurements," by Arthur Klein.

Table 2, part 2.

103

Table 3: Summary of Ancient Estimates of the Length of a Degree and the Circumference of a Spherical Earth, from Aristotle (384-322 B.C.) to Ptolemy (fl. middle of 2nd Century, A.D.)

Geographer	Length, in Stades		Reporter and Interpretation of Length of Stade, in Metres	Resultant Length			
	Degree	Earth's Circumf.		Degree		Earth's Circumf.	
				Kms.	R.M.	Kms.	R.M.
Aristotle (Eudoxus?)	1111.11	400,000					
			This reporter: Conjecture #1 157.5 metres	175	118	63,000	42,510
			This reporter: Conjecture #2 = 148.15 meters	164.6	111	59,259	39,984
			This reporter: Conjecture #3 = 100 meters	111.1	74.9	40,000	26,964
Archimedes (Dikaearchus?)	833.33	300,000	Dreyer/Stahl report that both the difference in latitude and the distance between the two sites (Lysimachia and Syene) as reported were grossly in error. Hence, further evaluation is not warranted.				
Eratosthenes	700	252,000					
			Macrobius, Aubrey Diller, I.E. Drabkin: 148.15 to 148.8 m.	103.7 to 104.2	70	37,334 to 37,512	25,200
			Dreyer/Stahl, George Sarton. 157.5 metres	110.3	74.4	39,690	26,790
			Strabo, Pliny, E.H. Bunbury, et al: 185.2 metres	129.6	87.5	46,667	31,500
Posidonius per Cleomedes	666⅔	240,000	Cleomedes does not specify Posidonius' stade length, but 148.15 to 148.8 m. is considered a likely possibility.	98.8 to 99.2	66⅔	35.556 to 35.712	24,000
Posidonius per Strabo	500	180,000	Strabo, Pliny, Bunbury, et al: 185.2 metres	92.6	62.5	33,333	22,500
Strabo	700	252,000	Strabo, Pliny, Bunbury, et al: 185.2 metres	129.6	87.5	46,667	31,500
Marinus	500	180,000	Bunbury and "Exclusionist" school: 185.2 metres	92.6	62.5	33,333	22,500
Ptolemy	500	180,000					
			Macrobius, Bunbury and "Exclusionist" school: 185.2 metres	92.6	62.5	33,333	22,500
			Diller and Drabkin: 197.6 - 198.4 metres	98.8 to 99.2	66⅔	35,568 to 35,712	24,000
			Dreyer/Stahl: 210 metres	105	70.9	37,800	25,515

Note: **Variations** in stade lengths result from different specifications, see Table 2. The different specifications also caused the Roman mile to vary somewhat — in this table, from 1481.5 to 1488 metres.

104

SUMMARY AND CONCLUSIONS

It is useful at this point to summarize, in tabular form, the stades (stadia) mentioned in the foregoing, and a few others found in the literature. Table 2 presents such a listing, gives the absolute length of each in metres and relates each to the Roman mile. Several conclusions may safely be made:

1. In the Egyptian-Mesopotamina-Syro/Palestinian-Graeco/Roman ancient world, there were a variety of stades (stadions, stadia, ush) in use at one time or another and these itinerary units of length varied widely, if not wildly - from about 100 m. to almost 1900 m.

2. In Egypt under Macedonian and Roman rule there were a variety of feet and cubits in use which when multiplied by 500 or 600, for feet, and 300 or 400, for cubits, yielded 7 stades of lengths varying from about 148 m. to 210 m. At least 5 of these stades have been posited by undeniable authorities as having been employed by Eratosthenes and Ptolemy in their pronouncements on the length of a degree and the circumference of the earth. Three different stades, varying from about 148 m. to 185 m. (25%) have been posited for Eratosthenes and 3, varying from 185 m. to 210 m. (13.5%) for Ptolemy.

3. Historical and empirical evidence, alike, point to a stade of about 148 m. or 157.5 metres length as the Eratosthenian unit and 197.6 or 210 metres length as the Ptolemaic unit.

4. The length of the stades employed by Posidonius and Marinus are conjectural, but were probably Eratosthenian for Posidonius and the Olympic stade of about185 m. for Marinus.

5. Both Strabo and Pliny propagated the myth that there was but one stade, the Olympic and Roman unit of 600 Olympic feet of 0.3086 m. or 625 Roman (Ancient Greek) feet of 0.2963 m., i.e., approximately 185 metres.

    Table 3 presents in tabular form a "Summary of Ancient Estimates of the Length of a Degree and the Circumference of the Earth . . . " as described in the foregoing text. In assessing this summary, it is clear that the philosophical approach taken by the ancients, from Aristotle (or Eudoxus) through to Ptolemy, to determine the size of a spherical earth was sound. The methods employed to measure the altitude, or zenith distance of heavenly bodies was relatively crude, but the way in which the distance between observational sites was determined was worse. Eratosthenes' employment of bematistes (professional pacers) to measure distances of about 500 miles was, by comparison with estimates of sea distance, relatively good.

    As we can see in Table 2 by comparing the actual distance between Alexandria and Syene, 791 kilometres, with Eratosthenes' reported 5000 stades, his bematistes' efforts could be rated fair if we accept the Diller/Drabkin version of the stade used. 148.15 metres per stade results in a distance between the observational sites of 740.75 kilometres. The Dreyer/Sarton stade of 157.5 metres results in even a better measurement, 787.5 kilometres. But the Olympic stade of 185.2 metres yields 926 kilometres, some 17% too high.

    In the case of Ptolemy's estimates, while the literature records no details on how he made his estimates, since no other methods than those employed by Eratosthenes were available to him, this must be the presumption. If we are to accept the position of the "exclusionists" that Eratosthenes and Ptolemy both employed the Olympic stade, then we have decided that bematistes, who in Eratosthenes' time measured the Alexandria-Syene and Meroe-Syene distances as 5000 stades each, were considerably more accurate than those, a few centuries later, who measured the same distances and found them to be 3566 and 3709 stades respectively. For at 185.2 metres per stade, these measurements translate to 660.4 and 687 kilometres, respectively (the actual Syene-Meroe distance is 804 kms.) The conclusions of Diller, Drabkin, Dreyer/Stahl and others that the length of the stade changed from Eratosthenes' time to Ptolemy's seems inescapable.

Arguments aside, Table 3 shows us that Eratosthenes' Earth circumference varied from 37,334 km. to 46,667 km., depending on the interpretation of the stade length employed; Posidonius' figures for this dimension varied from 33,333 km. to 35,712 km.; Strabo put the Earth's circumference at 46,667 km., as did Pliny; Marinus used Strabo's attribution to Posidonius - 33,333 km.; and Ptolemy's figures for the same dimension varied from 33,333 km. to 37,800 km., depending upon interpreter.

Assessing what the early geographers had to say about the distribution of land and water in the east west direction along the parallels of Athens or Rhodes, we have the following: Eratosthenes allotted about 139° to the length of the Eurasian continent plus offshore islands, thus making it 221° from the Fortunate Islands westward to the islands offshore Scythia and India; Posidonius made it 180° for both dimensions; Strabo was close to Eratosthenes in his views; Marinus allotted not less than 225° for the length of the Eurasian continent and because he was hazy as to the character of his eastern limit of the known world, gave the impression that this dimension might be greater than 225° and the corollary sea width westward from the Fortunate Islands to Asia less than 135°; Ptolemy reduced the Eurasian continent to 180°, but - like Marinus - was hazy about the eastern terminus, thus creating the impression that the sea width westward from the Fortunate Islands to Asia was less than 180°.

It was from the pronouncements of Eratosthenes, Posidonius, Strabo, Pliny, Marinus, and Ptolemy that philosophers, astronomers, and geographers of the Christian and Muslim eras took their leads. It will be interesting to see how they "picked and chose" from this store of conflicting information.

106

Chapter 4

MEDIEVAL COSMOGRAPHY
THROUGH THE 12TH CENTURY

INTRODUCTION

The period from the middle of the 4th century A.D. through the 12th century was one of great turmoil in Europe and Western Asia. It saw the decline and fall of Rome as the seat of great temporal power, the rise of Christianity, the birth and spread of Islam, the migrations and invasions of the Barbarians (chiefly Germanic tribes plus the Huns of Asia), and the sea-borne raids and invasions of the Vikings. This period has been dubbed the "Dark Ages," characterized by decaying towns, isolated manors, scattered monasteries, squabbling robber barons, and a general decline in knowledge and culture. However, as more than one writer has observed, these ages were probably never as dark as our ignorance of them, and - in any event - it could only be applied to Western Europe. In the East, Byzantium, studded with cosmopolitan cities such as Constantinople, Thessalonica, Antioch, and Alexandria, carried on the classical traditions under Christian auspices with nary a break in its stride.

As early as the third century A.D., Rome's trade with India, the Persian Gulf, Arabia, and the eastern coast of Africa, a trade which had been carried on by Roman merchants in Roman bottoms, began to pass into the hands of intermediaries, Persians, Abyssinians, and Yemenites. Roman traffic with trans-Alpine Europe suffered a similar decline. The gradual cessation of direct intercourse with the pagan world, particularly the Far East, resulted in a shrinking of Rome's territorial frontiers and a consequent loss of earth knowledge.

The convert Christian Emperor Constantine dedicated his capital, Constantinople, on the site of Greek Byzantium in A.D. 330, continuing the shift of the Empire's center of gravity to the east. The Eastern Romans, or Byzantines, thought themselves to be the true heirs of both Christianity and the classical tradition.

In A.D. 378, Goths defeated Romans at the Battle of Adrianople establishing the superiority of cavalry over infantry and spurring further Barbarian raids and invasions. By A.D. 400 Rome was in sharp decline, its once invincible empire shared by two capitals with Constantinople dominant, and its frontiers breached by the first major waves of invaders. With the capture of Rome itself by Visigoths in 410, the later sacking of the city by Vandals, and the deposing of the last emperor of the West, Romulus Augustus, in 476 by Odoacer the Ostrogoth chieftan, the focus of Western European development moved north of the Alps.

The date of A.D. 476 also marks the great schism when Eastern and (what was left of the) Western Roman Empires became separate entities and the contacts of the Latin world with Greek civilization became more and more tenuous, to the incalculable loss of the West.

The Germanic invasions of the 5th and 6th centuries undermined Roman life and culture, caused a collapse of the imperial administrative system, and a breakdown of political and commercial life within the Empire. Ideas and information, thenceforth spread only slowly and against great resistance. Culture became regional and stagnant.

But even as the Western Roman Empire, then officially Christian, was dissolving as a political and military force and its institutions collapsing, one among these held firm - the church. The main network of bishoprics held, with Rome's first among them, and, following St. Benedict in the 6th century, monasteries became centers of learning and piety. Reacting against temporal chaos and bolstered by Augustinian philosophy, the church was the light of the Dark Ages and the first truly European institution.

However, just as the Latin mind with its utilitarian bias had been a factor in the decline of ancient science and the silencing of the scientific spirit during the hey-day of Roman prosperity and power, the ecclesiastical expediency which followed the fall of Rome bid fair to destroy all hope of a scientific revival. The mind of Pope Gregory I (Pope 590-604), for instance, was closed to all that was irrelevant to the knowledge of God and the soul. Secular study and the whole round of man's mortal interests lay outside the field of his intellectual endeavor.

Yet the early fathers of the church, avowed opponents of pagan culture, were the very ones to keep it alive. They wrote in Latin, taught it in their schools, and the rules and examples were essentially those of classical practice. The training of students was based on anthologies and textbooks, the material for which was culled from the standard works of antiquity. And, once it was seen that Christianity was the heir of a past which had strong cultural ties with the present and that to disregard this heritage altogether would be to do the Faith lasting disservice, the doctors of the Church set about re-cultivating the classical soil.

In the 6th century, Justinian, the Byzantine Roman emperor, succeeded in pushing back the Persians (perennial adversaries of the Hellenes), in recovering the mastery of the North African provinces by destroying the Vandal Kingdom, and for a time established imperial dominion in Italy by overthrowing the Ostrogoths.

After the death of the prophet Muhammad in 632, Islamic forces swept across North Africa, invaded the Iberian peninsula, and crossed the Pyrenees into Gaul. Their penetration northward was finally stopped at Poitiers by Charles Martel, in 732. After that, Arabic political influence in Europe was largely confined to Spain and Sicily. Arabic influence on the sciences, however, and in particular astronomy and geography, was more profound and will be treated in greater detail in this chapter.

Of the many agrarian Germanic kingdoms which emerged after the fall of Rome, that of the Franks, under Charlemagne after 768, rose to prominence. Charlemagne was crowned Emperor of the Romans by the Pope in 800. The Carolingian Empire endured for nearly a century, until its disruption by Magyar and Viking invasions.

In 814, Magyars, a people of Finnish-Ugric stock, swept out of the area of present day Hungary into many elements of the Carolingian Empire and toward the heart of the Byzantine Empire at Constantinople. While the Carolingian Empire toppled as a result, it was soon succeeded by a number of kingdoms set up by the Treaty of Verdun of 843. The Byzantine Empire survived the Magyar onslaughts.

Between the 8th and 11th centuries, the Scandinavians, known to posterity as the Vikings, raided and traded over much of Europe. Early in this period, they invaded the British Isles and northwestern mainland of Europe. Later their descendants established kingdoms and states that ranged from Britain to Kiev. Viking trade routes laced Europe and its adjacent sealanes and even reached to North America.

The Magyars and Vikings appear to have been incorporated into Latin Christendom about A.D. 1000. In effect, this marked the birth of Western European society and the high Middle Ages.

We now turn to the review of evidence of geographic and geodetic knowledge during the period described.

CAPELLA

Minneo Felice Marziano Capella, also know as Martianus Capella, was a transplanted Carthaginian who achieved some fame in 3rd century Rome as an astronomer and geographer. His geography appears to have become somewhat of a standard in the Medieval period[1].

Delambre, who clearly faults Capella's astronomy and arithmetic, reports[2] that this Roman geographer put the earth's circumference at 31,500,000 paces. At the same time he indicates that the stade in Ptolemy's 500 stade degree has 125 paces, making the degree 62,500 paces and the earth's circumference 22,500,000 paces, or 22,500 Roman miles (33,300 kilometres). Since 31,500,000 is to 22,500,000 as 700 is to 500, it would appear that Capella has himself opted for the 700 stade, 87,500 pace, 87.5 Roman mile degree which Strabo and Pliny said was Eratosthenes' measure.

What does Capella say about Eratosthenes? He attributes to Eratosthenes and Archimedes, both, great circles of 406,010 stades, essentially Aristotle's round number 400,000[3]. This makes Eratosthenes degree about 1128 stades, quite close to the degree of 1111 1/4 which Gosselin said was in use in Egypt in Herodotus day (see Chapter 3.)

MACROBIUS (THEODOSIUS)

Ambrosius Theodosius Macrobius (fl. c. A.D. 395-423), Roman grammarian, philosopher, geographer, and prolific writer in Latin, was highly regarded and widely read by men of learning during the Middle Ages. *His Interpretatio in Somnium Scipionis* contains a series of essays on metaphysical and cosmographical topics and some curious speculations on the shape and extension of the inhabited world. Nordenskiöld conjectures [4] that "Macrobius considered the earth to be divided by the ocean currents into four large islands, of which two were in the northern and two in the southern hemisphere. The equatorial zone, which separates the northern and southern islands was supposed to be a sea impassable from heat . ." The sea between the two islands in the northern hemisphere evidently posed no obstacle to navigation.

Now the latter speculation was just the sort of idea Columbus could be expected to latch onto, but Humboldt remarks in *Kritische Untersuchungen,* p. 166, that Columbus never referred to it in any of his letters. Of course, it must be remarked that Columbus almost to his dying day believed that there was but one land mass in three sections, the Eurasian-African, with the African attached to the Asian part by the umbilical land passage at the southeast end of the Mediterranean just north of the Red Sea. In any event, unless Cardinal d'Ailly treated Macrobius well, it is not likely that Columbus would be impressed.

An edition of Macrobius was published at Brescia (Italy) in 1483 under the title *In Somnium Scipionis Expositio* . This edition and several later ones published in Brescia and Venice contain a map of the world which, speculatively, is traceable to a sketch in an old manuscript. It is shown in Figure 1. Only two of Macrobius' four islands are shown in the map which represents a hemisphere, the other two, being on the underside of the map, are not shown since nothing was known about these islands.

The map shown is certainly a far cry from Ptolemy's sophisticated map [5] of three centuries earlier or Eratosthenes' rectangular projection of over six centuries earlier[6] which put mapmaking on a mathmatical basis. Nor can it be assumed that Macrobius was unaware of his illustrious predecessors, for he had very well-developed views about each. We see, here, the beginning of the trend during the "darkness" of the Middle Ages in which maps were produced with no pretense at proper proportioning and which later would become adorned with monsters, figures of princes (or other underwriters), and illustrations for fantastic legends.

(5) MACROBII IN SOMNIUM SCIPIONIS EXPOSITIO,
BRIXIÆ 1483.

Figure 1 : A Map of the World from the 1483 edition of Macrobius published at Brescia (Brixiae) some 1060 years after that Roman philosopher/geographer's death. Nordenskiöld, from whose "Facsimile-Atlas" this is taken, indicates it was "probably copied from a sketch in some old manuscript". The map shown "is the first printed map on which the currents of the sea are denoted", according to Nordenskiöld. Macrobius supposed the earth to have been composed of four large islands, of which only one, the Eurasian-African, was inhabited. The sea between the northern and southern islands in each hemisphere, he considered to be impassable because of the equatorial heat, but those between the islands on the same side of the equator posed no barriers.

On the source

The Map of the World shown above is map number XXXI (5) of A.E. Nordenskiöld's Facsimilie Atlas, first published in Sweden in 1889, re-published in English by Dover Publications, New York, in 1973. It is reproduced by courtesy of Dover Publications, NY.

With regard to the size of the earth, Macrobius preferred Eratosthenes' figure of 252,000 stades for the earth's circumference to Ptolemy's 180,000 stadia. Macrobius[7] considered Eratosthenes' stade to be equal to 500 (presumably Roman) feet[8]. Ten of these stades made the Roman mile and 700 made a degree. Thus:

1 stade	=	500 x 0.2963 metre/Roman foot
	=	148.15 metres;
1 Roman mile	=	10 stades (or 5000 ft)
	=	1481.5 metres;
1 degree		= 700 stades = 70 Roman miles = 103.705 km; and the earth's circumference
		= 360 x 103.705 km = 37,334 km = 25,200 Roman miles.

Macrobius' degree and earth's circumference were only 6.67% too low. Moreover the 148.14 (148.2) metre stade has been posited for Eratosthenes by modern writers on the history of science such as Aubrey Diller and I. E. Drabkin. The more usual make-up of this stade (sometimes called the Hebraic stade) is 300 Assyrian cubits of 0.494 metre/cubit.

Macrobius' interpretation of Ptolemy's position (essentially the same as Capella's) was based on the assumption that Ptolemy used a stadium equal to 625 Roman feet. Eight of these stadia made the Roman mile and 500 made the degree. Thus:

1 stade	=	625 x 0.2963 metre/Roman foot = 185.19 metres.
1 Roman mile	=	8 stades (or 5000 ft.)
	=	1481.5 metres;
1 degree	=	500 stades = 62.5 Roman miles = 92,595 metres; and the
earth's circumference		= 360 x 92,595 metres = 33,333.3 kilometre = 22,500 Roman miles.

While these results are 16.7% less than the actual earth's circumference and degree length, they represent the views of a majority of those who have attempted, through the ages, to interpret Ptolemy's geodetic dicta. To what extent Macrobius and Capella are responsible for the acceptance of the interpretation of Ptolemy by succeeding generations of philosophers/astronomers/geographers - despite their belief that Ptolemy was in error - is conjectural. Still, these are the earliest authoritative assessments of Ptolemy in the literature.

Kimble[9], in describing Macrobius' views on Eratosthenes and Ptolemy, refers to the former geographer's stade as the "sea" stade and the latter's as the "land" stadium. Nordenskiöld[10] also refers to the stade, 700 to the degree, as that used by "sailors" in ancient times.

Macrobius played an important role in making Eratosthenes and Ptolemy, and their geographic views, as he interpreted them, known in the Middle Ages.

SIMPLICIUS

This 5th century writer, born in Cilicia (in southeast Asia Minor), studied under Ammonius of Alexandria, one of the founders of Neo-Platonism. He is best known for the commentary he wrote on Aristotle's *Decaelo* , for it is one of the most highly quoted sources on this great work.

Taking issue with Aristotle's statement that some mathematicians had estimated the circumference of the earth to be 400,000 stades, he sets forth a method which, he says, he used to determine the correct length. Searching with his diopter (a Greek instrument useful in determining altitudes and zenith distances), he found two (un-named) stars whose declinations differed by exactly one degree. Then, using the same instrument, he found two sites on a common meridian such that the stars passed through the zeniths of the sites. Following this, he measured the distance between the two sites and found it to be 500 stades. Thus, he concluded that a great circle of the earth had a circumference of 180,000 stades as Ptolemy had reported in his Geographie.

Aside from the fact that Simplicius gives us no clue to the length of his stade, Delambre[11] doubts that Simplicius ever made such a measurement. Dreyer/Stahl[12] doubt that anyone ever did. The method is an overly simplified version of that of Archimedes/Dikaerchus described and illustrated in the last chapter.

ARYABHATA

Aryabhata of Kusumapura or Pataliputra (c. A.D. 476), an Indian astronomer of some repute, is another of the ancients who anticipated Copernicus. Dreyer/Stahl[13] quote him as saying "The sphere of the stars is stationary and the earth, making a revolution, produces the daily rising and setting of stars and planets." Further, they indicated Aryabhata set the diameter of the earth equal to 1050 yôjans of 7.6 miles (presumably statute) each. Thus, the circumference would be 3298.67 yôjans, or 25,069.91 miles and a degree would be 69.64 miles (statute) or 112,072 metres. This is a very respectable estimate.

However, Ahmad indicates[14] Aryabhata's figure for the earth's circumference was equivalent to 33,177 English (statute) miles, some 32.3% greater than the figure given by Dreyer/Stahl and their source, Colebrook. A circumference of 33,177 miles makes the diameter 10,560.56 miles, which, when equated to 1050 yôjans, gives the yôjan a length of about 10 miles. On the assumption that Aryabhata's figure for the earth's diameter was indeed 1050 yôjans and that our sources have attributed different lengths to the yôjan, let us probe that unit further.

Sarton tells us[15] " . . . (a) yôjana is a measure of length (in India) difficult to define exactly: there were a long yôjana and a short one (about 9 miles and 4 1/2 miles). The word was also used to denote a day's march, about 12 miles, but variable."

Petrie reports[16] use of the Indian hasta (cubit) in ancient Asia Minor and Greece, indicating this unit had an approximate length of 0.454 metre. Since the usual relationship between the hasta and the yôjana was 16,000 hastas = 1 yôjana, this would give the yôjana a length of 7264 metres or just about 4 1/2 miles, the same as Sarton's lower limit.

Two popular elles (cubits) in countries which became Islamic were the "canonical," equal to 0.49875 m., and the "black", equal to 0.5404 metres. These would yield yôjanas of 7980 and 8646.4 metres, respectively, equal to 4.96 and 5.37 miles, both significantly smaller than even Colebrook's 7.6 miles. Hinz, however, reports[17] there was an elle in use in India (similar in length to the Persian unit of the Middle Ages) measuring 0.68 metres and a larger unit with a length of 0.91 metres. The first yields a yojana of 10,880 metres, or 6.76 miles. The larger elle yields a yôjana of 14,560 metres, or 9.05 miles. And Berriman tends to confirm this, indicating[18] that there is an ancient gaz still in use in North India, measuring 33 English inches (0.8382 m.) and one in use in Bombay measuring 27 English inches (0.6858 m.). These yield yôjanas of 13,411 metres (8.33 statute miles) and 10,973 metres (6.82 stat. miles) respectively, which bracket Colebrook's 7.6 miles, but do not necessarily prove Ahmad's version of Aryabhata's estimate wrong.

The gaz was a Persian unit of length and has been reported as varying between 0.68 and 1.04 metres in the various areas where it was employed. The upper limit would accommodate Ahmad's figure. As for the yôjana, as was shown in Chapter 2, in the analysis of Hiuen Tsiang's story about the Chinese traveler in India, a yôjana (yôjan) can be evaluated as any of a variety of lengths between 7200 m. and 26,000 m., depending on which specification we elect: 384,000 fingers, 16,000 cubits, or 16.30 or 40 li; and the size of the fingers, cubits or li utilized. Thus, Aryabhata's 1050 yôjans for the earth's diameter could have indicated a circumference as small as 14,759 statute miles (23,752 km) or as large as 53,292 miles (85,765 km).

Between 539 B.C. and Aryabhata's time, India had been subject to Persian, Greek, and Roman influences (as well as Chinese). Unquestionably, Aryabhata was trying to express in yôjanas, in use in his area at that time, the geodetic conclusions of earlier geographers such as Eratosthenes, Strabo, Pliny, and Ptolemy as he understood them from the entirely different units in which they were expressed.

BRAHMAGUPTA

Brahmagupta, an Indian astronomer and mathematician from Ujjain in present Madhya Pradesh (fl. ca. 628 A.D.), was the author of the Brahma-sphuta-Sidhanta' (known in the Arab world as Sind-Hind) and Khanda-Khadyaka (known as Arkand). Both of these works, on astronomy, were brought to Baghdad in 771 A.D. and translated into Arabic. Ahmad [19] credits Brahmagupta with an estimate for the earth's circumference of 50,936 English (statute) miles, more than double the actual value and close to the maximum estimate shown above for Aryabhata.

Ahmad also reports that Brahmagupta believed that Ujjain, just south of the Tropic of Cancer, had a latitude of 16 1/4°. Such a gross error, not uncommon during ancient and medieval times, can easily explain the distorted estimate Brahmagupta made of the earth's circumference, if he made a measurement of a degree himself. It is, however, at least as likely that his error was due to a mis-understanding of the true length of the unit in which an earlier measurement was expressed.

HOCHING-TIEN

In the mid-5th century, A.D., this Chinese astronomer undertook to determine the length of a degree of latitude by measuring the length of shadows cast by 8 pied gnomons at the villages of Tong-Feng and Tong-King at noon on the day of the summer solstice. Gaubil (Père Antoine, S. J., 1689-1759), the original source for this account, was a French missionary who went to China in 1722 and evidently spent the rest of his life there. Gaubil reports that the length of the shadow at Tong-Feng was 1 pied, 5 pouces while at Tong-King, farther south, the shadow was only 3.2 pouces. Since 12 pouces make the French pied (foot), it is evident that the zenith distance of the sun at noon at Tong-Feng was $\tan^{-1} 17/96$, or 10.04°. At Tong-King it was $\tan^{-1} 3.2/96$, or 1.91°. The difference in zenith distances (which is equal to the difference in latitudes of the two sites) thus was 8.13°.

The distance between the two villages is given by our source, Delambre [20], as 1000 Lys, but he does not say that the two villages were on the same meridian. On the assumption that they were, the degree would have measured 123 lys (lis). Chapter 2 (herein) indicates that the Chinese li varied between 536 and 650 metres in different places and at different times. Using these figures, it is evident that Hoching-Tien's degree had a length between 65,928 and 79,950 metres (some 41% and 28% too low, respectively).

Y-HANG[21]

In 721 A.D., in the wake of the furor of an announced eclipse which had failed to materialize, a Buddhist priest, one Y-Hang, was called to the court of the reigning Chinese emperor and assigned a number of astronomical, geodetic, and geographic tasks. Among these was the astronomic determination of the latitude of a number of sites within China proper and Cochin China. To accomplish this he had constructed a number of gnomons, globes, astrolabes and quadrants. Two bands of observers were dispatched with instructions to measure and record the altitude of the sun and of the polar star over a significant period of time - one band concentrating on northern sites and the other on southern. In the course of their operations, the observers were to determine the surface distance between at least two sites on the same meridian for which they had determined zenith distances of stars with known declinations.

The astronomical work of the group was reasonably accurate for they found that the polar star revolved about the celestial north pole with a radius of 3°. Nine hundred years later, Jesuit astronomers found latitudes at four sites which differed from the 8th century Chinese values by only 6', 34', 1' and 17'.

The Chinese had found a degree of latitude to have a length of 351 lys and 80 paces. The Jesuits found only 200 lys to the degree. Clearly the ly (li) varied in length rather wildly over the indicated time periods, the surface measurements of the Chinese were poorly executed, or the reporting of these events is faulty. It is not likely that the measurements of the Jesuits was significantly in error because by the 17th century measurements had been conducted in France and Holland yielding lengths for the degree betweeen 101 and 110 km, so that the Jesuits had this range of values to guide them.

ISIDORE OF SEVILLE

A bishop of the seventh century, Isidore of Seville (fl. c. A.D. 600-636), achieved considerable, lasting fame for his work on natural history and geography titled "Origins or Etymologiae." Two of the twenty books in this work, the thirteenth and fourteenth, are on geographical topics. Thomson indicates[22] the work was compiled mainly from Pliny. Worse " . . . he quotes ancient writers on the globe . . . but he fails to understand that they are talking of a globe, and not a circle, and he misconceives the zones as circles on a flat earth." Bunbury indicates Isidore derived his information not directly from Pliny, but from Solinus (fl. 3rd century, A.D.), who re-wrote Pliny and borrowed from Mela.

On the subject of the size of the earth, Isidore accepts[23] the Eratosthenes estimate of 252,000 stadia for its meridional circumference. Unlike Macrobius, however, he takes the Strabo/Pliny reckoning of 625 Roman feet to the stadium and 8 stadia to the mile. Isidore's linear system (like that of Strabo and Pliny) was as follows:

$$
\begin{aligned}
1 \text{ stadium} &= 625 \times 0.2963 \text{ metres/ft.} = 185.19 \text{ m.} \\
1 \text{ Roman mile} &= 8 \times 185.19 \text{ metres/stadium} = 1481.5 \text{ m.} \\
1 \text{ degree} &= 700 \times 185.19 \text{ metres/stadium} = 129{,}633 \text{ m.} \\
&= 129{,}633 \div 1481.5 \text{ metres/mile} = 87.5 \text{ Roman miles}
\end{aligned}
$$

$$
\begin{aligned}
\text{The earth's circumference} &= 252{,}000 \times 185.19 \text{ metres per stadium} \\
&= 46{,}667.88 \text{ Kilometres} \\
&= 31{,}500 \text{ Roman miles}
\end{aligned}
$$

Thus, Isidore's figures for the length of the degree and the earth's circumference were 16.7% too high.

ANANIAS OF SHIRAK (SIRACKI)

An Armenian writer of the 7th century by the name of Ananias of Shirak has been credited with a table of measures which has been interpreted somewhat differently by Hans von Mzik, Aubrey Diller, and Eva Taylor - all formidable metrologists.[24]

Von Mzik's interpretation of Ananias' relationship (according to Diller) is :

1 stade "by air (Asparez)" = 107 paces;
1 pace   =   6 feet; a foot = 16 fingers;
1 mile   =   7 stades,

1 stade "according to the Persians" = 143 paces;
1 mile   =   1000 paces;
1 parasang   =   3 miles.
1 degree   =   500 stades "by air"

Diller, balking at von Mzik's interpretation of Ananias' table, says " . . . the original equations were: a mile = 750 paces = 4500 feet 'by air'; and a mile = 1000 paces = 6000 feet 'according to the Persians.' Then the stade (600 feet) 'by air' is the familiar one of 7 1/2 to a mile, which we posit for Posidonius and Ptolemy . . . Ananias is not mistaken when he states that this stade goes 500 to a degree. The stade 'according to the Persians' is the one of 10 to a mile, which we posit for Eratosthenes . . .[25]" (Eratosthenes, it will be recalled, had set the degree of latitude [or of longitude at the equator] at 700 stades.)

Elsewhere, Diller identifies the mile as the Roman mile of 1488 metres (which others have called the "miglio Romano" or medieval Roman mile, a derivative of the "milliarum" or "mille passus," the ancient Roman mile of 1481.5 metres[26]). From this, the stade, 7 1/2 to the Roman mile, is 198.4 metres.

Diller also makes reference[27] to a Syriac text of the 5th century which " . . . gives 4000 cubits to a mile, which would be 6000 feet or 10 stades . . . (This) standard seems to have become common in the East, for it is the basis of the parasang of 3 miles that is widely attested in Syriac, Armenian, and Arabic sources . . . Greek sources equate the parasang and the schoenus (4 miles, 30 stades)."

Thus, it would appear - following Diller's lead - that Ananias' "by air" degree, which Diller identifies as Ptolemaic, or Posidonian, had a length of 66-2/3 Roman miles (500 stades per degree ÷ 7 1/2 stades per mile). Ananias' degree "according to the Persians," which Diller indicates is Eratosthenian, must have had a length of 70 Roman miles (700 stades per degree ÷ 10 stades per mile). The earth's circumference in the two systems was 24,000 and 25,200 Roman miles, respectively. These figures (using Diller's 1488 metres to the mile) are equivalent to 35,712 and 37,947 kilometres, respectively.

Using Diller's mile of 1488 metres to evaluate von Mzik's "by air" system, the degree of 71.43 Roman miles translates to 106,288 metres and an earth's circumference of 38,264 kilometres - both figures about 4.34% too low. Using the 1488 metre mile to evaluate the Persian system leads to inordinately small fingers, feet, and paces, which is our basic problem with Diller's Persian system[28]. Instead, we have assumed that the finger, foot, and pace in Mzik's Persian system are the same as in his "by air" system. Further, because von Mzik is silent as to Ananias' stade/degree relationship, we have assumed it to be the same as in his "by air" system, 500 stades per degree. This results in an inordinately large stade, mile, degree, and earth's circumference. Still, such lengths of the finger, cubit (or elle) of 24 fingers, pace (fathom or cord) of 4 cubits, mile of 1000 paces (or cords), and parasang of 3 miles were in use in Eastern countries both before, and after Ananias' time (see Table 4, Items 1a, 3, 12, 16, and 17, and "Ancient Sumerian Units of Length" on reverse side of Table 4).

Professor Taylor[29], in reporting Mzik's treatment of Ananias, ignores the "by air" and " . . Persians" classifications and comes up with a single system. She attributes to Ananias a stade measuring 571-3/7 palmipes or 143 paces. She also indicates Ananias used a stade to mile relationship of 7 to 1 and (by indirection) that his mile was the Roman - which, for purposes of comparison with Diller, will be assigned his stated length of 1,488 metres. Finally, she says Ananias set the degree at 71-3/7 miles. Thus, Ananias, according to Taylor, said the following:

Text continues on page 120

Table 1: Some Interpretations of a Table of Measures Traceable to Ananias of Shirak (Armenia, 7th Century)

"By Air (asparēz)" system — Hans von Mžik Interpretation

Unit	Metric Length	Relationship to Other Units in System				
		Finger	Foot	Pace	Stade	Mile
Finger	(?) 0.0207 metre	1	1/16	1/96		
Foot	(?) 0.3311 "	16	1	1/6	1/642	1/4494
Pace	(?) 1.987 metres	96	6	1	1/107	1/749
Stade	(?) 212.6 "		642	107	1	1/7
Mile	(?) 1488 "		4494	749	7	1
Degree	(?) 106,288 "				500	71.43
Earth's Circumference	38,264 Kilometres				180,000	25,714.8

System "according to the Persians" — von Mžik Interpretation

Finger	(?) 0.0207 metre	1	1/16	1/96		
Foot	(?) 0.3311 "	16	1	1/6	1/858	1/6000
Pace	(?) 1.987 metres	96	6	1	1/143	1/1000
Stade	(?) 283.9 "		858	143	1	1/7
Mile	(?) 1987 "		6000	1000	7	1
Parasang	(?) 5961 "		18,000	3000	21	3
Degree	(?) 141,931 "				500	71.43
Earth's Circumference	(?) 51,095.3 Kilometres				180,000	25,714.8

"By Air" system — Aubrey Diller Interpretation

Finger	0.02067 metre	1	1/16	1/96		
Foot	0.3307 "	16	1	1/6	1/600	1/4500
Pace	1.984 metres	96	6	1	1/100	1/750
Stade	198.4 "		600	100	1	2/15
Mile	1488 "		4500	750	7.5	1
Degree	99,200 "				500	66 2/3
Earth's Circumference	35,712 kilometres				180,000	24,000

System "according to the Persians" — Diller Interpretation

Finger	0.0155 metre	1	1/16	1/96		
Foot	0.248 "	16	1	1/6	1/600	1/6000
Pace	1.488 metres	96	6	1	1/100	1/1000
Stade	148.8 "		600	100	1	1/10
Mile	1488 "		6000	1000	10	1
Parasang	4464 "		18,000	3000	30	3
Degree	104,160 "				700	70
Earth's Circumference	37,497.6 kilometres				252,000	25,200

Taylor's Interpretation of Ananias – via von Mžik

Finger	0.0186 m.	1	1/16	1/80		
Foot	0.298 "	16	1	1/5	1/715	1/5000
Pace	1.488 "	80	5	1	1/143	1/1000
Stade	212.6 "		715	143	1	1/7
Mile	1488 "		5000	1000	7	1
Degree	106,286 "				500	71.43
Earth's Circumference kms.	38,262.9				180,000	25,714

Table 1, part 1.

<u>Notes</u>

#1: Absolute values (metric lengths) shown for von Mžik are conjectural, unlike those shown for Diller which are based on Diller's Roman mile of 1488 metres. The values shown for Mžik's "By Air" system are based upon an assumption that the mile had a length of 1488 metres. Since the same assumption for the Persian system leads to inordinately small fingers and feet (my only quarrel with Diller's Persian system), the assumption was made that the finger, foot, and pace in both of Mžik's systems were the same, thus causing all of the larger units, from the stade through the earth's circumference, to balloon. There were, however, units of the approximate lengths shown for von Mžik's Persian system (finger through parasang) in use in Mesopotamia, Persia, and other western Asia civilizations (see Table 4, Items 1a, 3, 12, 16, and 17).

#2: The meaning of "by air" is not explained by Diller, von Mžik, or, presumably, Ananias. Diller identifies the stade/mile/degree relationship as Ptolemaic or Posidonian in the "by air" system and Eratosthenian in the Persian system.

#3: Diller also cites a 5th century Syriac text in support of his Persian system relationships (if not absolute values). The Syriac text specifies 1 mile = 4000 cubits (elles). Diller's extrapolation to 6000 feet causes no problems, but his further extrapolation to 10 stades, while acceptable to this reporter, is not the only reasonable assessment of the mile/stade relationship possible. The five matrices shown at left have three different relationships: 7, 7½, and 10 stades to the mile. Relationships of 8, 8⅓, and 9.41 also enjoy respectable credentials. The literature, however, contains references to stades as large as 0.78 stade/mile and as small as 14.82 stades/mile.

#4: The reader's particular attention is invited to the 66⅔ Roman mile/degree ratio shown in Aubrey Diller's "by air" interpretation. This ratio will be seen again!

#5: Taylor ignores the "By Air" and "According to the Persians" distinctions and credits Ananias with a single system (even though her source is von Mžik). Since she gives no absolute values for any of her units, but implies that her mile is Roman, we have given it the same length as Diller's mile, 1488 metres, and sized all other units based on Taylor's stated ratios - with one exception. The finger to foot ratio, while likely, is conjectural.

Table 1, part 2.

117

Unit of Measurement → Authority ↓	Dimensions of Units in Metres (or Km.) and in Multiples of Smaller Units							
	Digit or Finger I	Foot II	Cubit or Elle III	Pace or Fathom IV	Stade, Stadion, Stadium V	Mile[4] VI	Degree VII	Earth's Circumference VIII
Capella[1]								
Eratosthenian	?	?	?	?	?	?	1128×V	406,010×V
Ptolemaic	18.52mm	16×I 0.2963m.	24×I 0.4445m.	5×II 1.4815m.	625×II 125×IV 185.2m.	5000×II 1000×IV 8×V 1481.5m.	500×V 62.5×VI 92,600m.	180,000×V 22,500×VI 33,333 Km.
Capella's Own System	18.52mm	16×I 0.2963m.	24×I 0.4445m.	5×II 1.4815m	625×II 125×IV 185.2m.	5000×II 1000×IV 8×V 1481.5m.	700×V 87.5×VI 129,630 m.	252,000×V 31,500×VI 46,667 Km.
Macrobius[2]								
Eratosthenian System (favored)	0.01852m.	16×I 0.2963m.	24×I 1.5×II 0.4445m.	80×I 5×II 1.4815m.	500×II 100×IV 148.15m.	5000×II 1000×IV 10×V 1481.5m.	700×V 70×VI 103,705m.	252,000×V 25,200×VI 37,333 Km.
Ptolemaic System (not favored)	Essentially the same as Capella's Ptolemaic System.							
Simplicius[3]	Essentially the same as Capella's and Macrobius' Ptolemaic System.							
Aryabhata[4]								
per Colebrook and Dreyer/Stahl	?	?	?	?	?	?	110,465m.	39,767 Km.
per Ahmad	?	?	?	?	?	?	148,314m.	53,393 Km.
Other Possibilities	?	?	?	?	?	?	From 65,980m. to 238,236m.	From 23,752 Km. to 85,765 Km.
Brahmagupta[5]	?	?	?	?	?	?	227,700 m.	81,980 Km.
Hoching-Tien[6]	?	?	?	?	?	?	From 65,928m to 79,950m.	From 23,734 Km. to 28,782 Km.
Y-Hang[7]	?	?	?	?	?	?	From 188,254m to 228,268m.	From 67,771 Km. to 82,177 Km.
Isidore of Seville[8]	Essentially the same as Capella's Own System.							
Ananias of Shirak[9]								
Persian System (based on Eratosthenian Stade), per Diller	15.5mm	16×I 0.248m.	24×I 1.5×II 0.372m.	96×I 6×II 4×III 1.488m.	600×II 400×III 100×IV 148.8m.	6000×II 4000×III 1000×IV 10×V 1488m.	700×V 70×VI 104,160m.	252,000×V 25,200×VI 37,498×Km.
"By Air" System (based on Philetaerian stade). Diller's Posidonian or Ptolemaic System	20.7mm	16×I 0.3307m.	24×I 1.5×II 0.496m.	96×I 6×II 4×III 1.984m.	600×II 400×III 100×IV 198.4m.	4500×II 3000×III 750×IV 7.5×V 1488m.	500×V 66⅔×VI 99,200m.	180,000×V 24,000×VI 35,712 Km.
Ananias, per Taylor	18.6mm	16×I 0.298m.	20×I 5/4×II 0.372m. (palmipes)	80×I 5×II 4×III 1.488m.	715×II 572×III 143×IV 212.6m.	5000×II 4000×III 1000×IV 7×V 1488m.	500×V 71.43×VI 106,286m.	180,000×V 25,714×VI 38,263 Km.

Table 2: A Comparison of Linear Measurement Systems, Post-Ptolemaic and Pre-Islamic.

Table 2, part 1.

118

Notes

1. 3rd century, A.D., Roman astronomer/geographer of Carthaginian origin. Sources: De-lambre's "Hist. de l'Astron. Ancienne", Paris, 1817, p. 304; "Tooley's Dictionary of Mapmakers", A.R. Liss, N.Y., 1979.

2. Roman geographer and writer, fl. c. A.D. 395-423. Sources: Kimble's "Geography in the Middle Ages", Methuen, London, 1938, pp. 8, 9; Berriman's "Historical Metrology", Dutton, N.Y. 1953, p. 17.

3. 5th century Cilician Neo-Platonist writer/astronomer, frequently quoted by later Medieval writers. Sources: Delambre's "Hist. de l'Astron. Ancienne", p. 304; Woodcock's "Dictionary of Ancient History", Philosophical Library, N.Y., 1981.

4. 5th century Indian astronomer of high repute. Sources: Dreyer's "A History of Astronomy from Thales to Kepler", Dover, N.Y., pp. 242, 243. Original source is Colebrooke's "Notes and Illustrations to the Algebra of Brahmagupta", p. xxxviii and "Essays", p. 467. Ahmad's version is from his "Muslim Contribution to Geography", Ashraf Press, Lahore, Pak., p. 116. "Other possibilities" is based upon the range of values reported for the "yojan" (yojana), in which unit Indian geodetic values were given. (See text.)

5. 7th century Indian astronomer/mathematician, author of "Sind-Hind" and "Arkand". Source: Ahmad's "Muslim Contribution ...", p. 116.

6. 5th century Chinese astronomer. Source: Delambre's "Histoire de l'Astron. Anc.", p. 372.

7. 8th century Chinese Buddhist priest/astronomer. Source: Delambre, p. 374. The range of values given for the degree and great circle of both Hoching-Tien and Y-Hang is due to the range of values reported for the Chinese ly (li). (See text, chapter 2.)

8. Bishop, natural historian, geographer, fl. c. A.D. 600-636. Sources: Thomson's "History of Ancient Geography", Biblo & Tannen, N.Y., 1965, p. 389; Kimble, pp. 23, 24.

9. Armenian writer, fl. 7th. cent. Sources: Diller's "Ancient Measurements of the Earth", ISIS, Feb. 1949, footnote 15, p. 8. Diller disputes Von Mzik's "Erdmessung, Grad, Meil, und Stadion nach den altermenischen Quellen" (Studien zur arm. Gesch. VI, 1933, pp. 90-113. Taylor's "Some Notes on Early Ideas of the Form and Size of the Earth", Geographical Journal, London, Jan, 1935, though based on Von Mzik's work, makes no reference to the "By Air" and "Persian" systems in her interpretation of Ananias.

Table 2, part 2.

a) A degree = 500 stades, 71-3/7 miles, or 106.3 kilometres: and

b) The earth's circumference = 180,000 stades, 25,714 miles, or 38,263 kilometres.

The foregoing estimates are only approximately 4.34% too small.

A better insight and appreciation of the three interpretations of Ananias is obtained by referring to Table 1, which presents all numerical data in matrix form.

The reader's attention is invited, particularly, to the 66-2/3 Roman mile degree shown in the matrix illustrating Diller's "by air" system. This relationship came to be widely held throughout the East until it was challenged by the astronomers of the Abbasid Caliph Al-Ma'mun. The challenge, whether valid or not, was to have far-reaching consequences when Christopher Columbus read about it in Cardinal Pierre d'Ailly's "Imago Mundi".

Before passing on to the Arab measurement of a degree and related interpretations, it is interesting to note that the stade attributed to Ananias by Taylor measures 1,488 ÷ 7 = 212.6 metres, a length which several history of science writers (of the late 19th and early and mid 20th century) have espoused as the true length of Ptolemy's degree. Table 2 compares the linear measurement systems presented thus far in this chapter (less those of von Mzik, about which there is excessive conjecture).

ISLAMIC ESTIMATES

The Arab Measurement of a Degree in A. D. 830

The prophet Mohammed died in A.D. 632. Almost immediately following his death a series of military campaigns were initiated emanating from Arabia. These were brilliantly executed. Damascus fell in 635, Jerusalem in 637, Mesopotamia was reduced in 638, and by the next year, 639, the Islamic era was established. Egypt fell in 640, Tripoli submitted in 643, Carthage in 698, and by 711, the conquest of Spain had begun. As Kimble puts it . . . "the Arabian Moslem who brought with him from the desert a keen sense of intellectual curiosity, a voracious appetite for learning and many latent faculties, found himself heir to older and superior cultures. These he was not slow to adopt. Within a century of the establishment of Baghdad, he was in possession of the chief philosophical works of Aristotle, the leading Neo-Platonic commentaries, and most of the medical writings of Galen, as well as Persian and Indian scientific works."

Among the earliest works to be translated into Arabic were *The Elements of Euclid* and Ptolemy's great astronomical work which the Arabs named *Almagest,* followed by Ptolemy's *Geography.* While the Almagest and Geography were treated with great respect and considered quite authoritative, the Arabs did not hesitate to attempt to improve on both these works. The Almagest was the basis for such treatises as Alfragan's (Al-Farghani) *On the Elements of Astronomy,* Al-Battani's *On the Movements of the Stars,* and Ibn Junis' *Hakimite Tables . The Book of the Description of the Earth* by Al-Khwarizmi, and a similar work (with the same title) by Al-Battani, were modeled after Ptolemy's *Geography.*

Among the early Caliphs of the Abbasid dynasty, Al-Ma'mun (fl. 813-833 A.D.) was perhaps the greatest patron of the sciences. Around A.D. 830, he founded at Baghdad a scientific academy, a library and an observatory. The first outstanding scientific work carried out by order of Caliph Al-Ma'mun was the measurement of a degree of arc of the earth's surface on the plains of Sinjar (Syria) in 830 A.D. (See Figure 2).

Reports of this operation vary, thus the following is a rationalization of what probably occurred. Under the direct supervision of the sons of Musa b. Shakir, Muhammad, Ahmad, and Hassan, two parties set out from a point between Tudmur (Palmyra) and Ar Raqqah (Nicephorium), at about 35 degrees north latitude

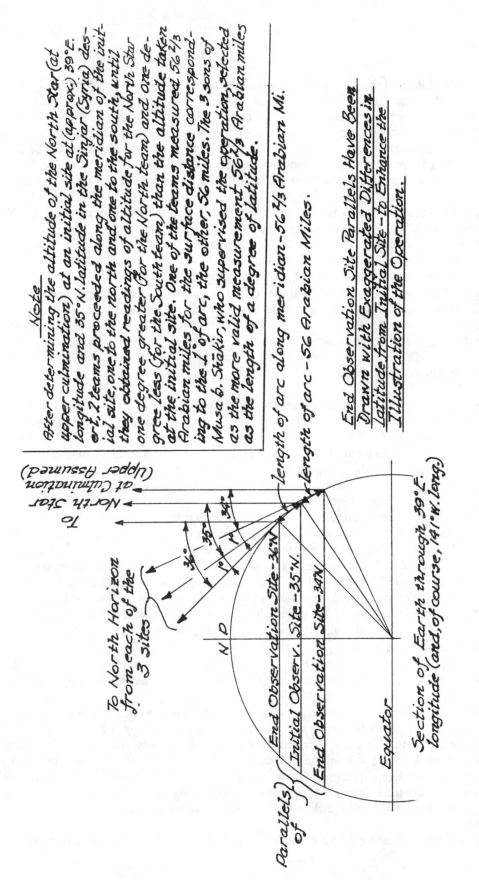

Note

After determining the altitude of the North Star (at upper culmination) at an initial site at (approx.) 39°E. longitude and 35°N. latitude in the Sinjar (Syria) desert, 2 teams proceeded along the meridian of the initial site, one to the north and one to the south, until they obtained readings of altitude for the North Star one degree greater (for the North team) and one degree less (for the South team) than the altitude taken at the initial site. One of the teams measured 56 2/3 Arabian miles for the surface distance corresponding to the 1° of arc, the other, 56 miles. The 3 sons of Musa b. Shākir, who supervised the operation, selected as the more valid measurement 56 2/3 Arabian miles as the length of a degree of latitude.

length of arc along meridian – 56 2/3 Arabian Mi.

length of arc – 56 Arabian Miles

To North Star at Culmination (Upper Assumed)

To North Horizon from each of the 3 sites

36° 35° 34°

N

End Observation Site – 36°N.

Initial Observ. Site – 35°N.

End Observation Site – 34N

Parallels of {

Section of Earth through 39°E. longitude (and, of course, 141° W. long.)

Equator

End Observation Site Parallels Have Been Drawn with Exaggerated Differences in Latitude from Initial Site – to Enhance the Illustration of the Operation.

Figure 2: Probable Method Employed by Caliph Al-Ma'mun's Astronomers, c. 830 A.D. to Determine the Length of a Degree of Latitude. (See Note Above)

after having measured the altitude of the north star at upper culmination. One party proceeded directly north and the other directly south, measuring the distance travelled from the initial point with a rope of known length with pegs at each end to facilitate and speed the measurement. When each party was able to discern a change in the altitude of the north star (at upper culmination) of one degree, the northern party an increase in altitude and the southern party a decrease, they stopped and recorded their results. One result was 56 2/3 Arabian miles and the other 56 miles. The larger result was chosen as the official length of a degree of latitude. Thus, the meridional circumference of the earth, based on this measurement, was 20,400 Arabian miles.

Now, the foregoing results, 56 2/3 Arabian miles for the degree, 20,400 miles for the earth's circumference, and therefore 3250 miles for the earth's radius (rounded from 3246.76 miles) were the results recorded in the writings of Alfragan (ninth century A.D.), Al Battani (fl. 877-918 A.D.), Ibn Yunus (d. 1009 A.D.), Al-Biruni (d. 1050 A.D.), and Abulfeda (Abul-Fida, b. 1273 A.D.)--all celebrated Islamic astronomers and geographers. However, Ibn Rustah, a Persian geographer who worked at Isfahan c. 903 A.D., puts the results of the measurements on the plains of Sinjar differently. He says the degree was found to measure 66 2/3 miles, the earth's circumference 24,000 miles, and the earth's radius 3818 miles [30]. Furthermore, it is now known that the Arabian mile, based as it usually was on 4000 cubits, tends to vary in length with the length of the cubit, which undoubtedly is the reason why Abulfeda's understanding of the length of an Arabian mile is different from other 56 2/3 Arabian mile per degree proponents. Let us see what the literature can offer on this subject.

a) Kimble says that 18 inch and 15 inch cubits, both, have been attributed to this measurement[31]. On the assumption that he is referring to Roman unciae as his inches, the metric equivalents are 0.4445 and 0.3705 metres[32]. 4000 such cubits make an Arabian mile of 1778.4 metres (for the larger cubit) and one of 1483 metres (for the smaller cubit). The smaller result is, of course, the Roman mile, and indeed a number of authorities,among them Abulfeda, Roger Bacon, Cardinal d"Ailly, Reinaud, and Vignaud, [33] opt for the equivalence of the Arabian mile and the Roman mile. 56 2/3 such miles make a degree of 83,980 metres and 20,400 yield an earth's circumference of 30,223 kilometres (and, of course, 20,400 Roman miles) - results about 25% too low!

The 15 unciae or 0.3705 metre cubit is essentially the same as the cubit in the Persian system attribtued to Ananias by Diller which employed the 10 stade/Roman mile relationship and the 1488 metre Roman mile. In that system, 6000 feet = 4000 cubits = 1 mile. Thus, the cubit (equal to 1 1/2 feet) = 0.372 metres. The difference in the lengths of 0.3705 and 0.372 metres results from the use of slightly different Roman miles - 1482 and 1488 metres - by the different writers. The important point here is that the tiny foot of 0.248 metres, which Petrie has identified as the ancient Pythic foot, and which this writer has found difficult to accept, has indirectly attained some additional respectability.

The 18 uncia cubit leads to a degree of 100,776 metres (56 2/3 x 1778.4 metres) and a circumference for the earth of 36,280 kilometres or 24,480 Roman miles - dimensions still about 9.3% too small.

Whether or not the 18 uncia cubit was actually employed in the Arab measurements, this cubit did exist, being 1 1/2 times the well-known Pelasgic foot (adopted by the Romans as their standard) which was used in Syria, Asia Minor, Greece, Africa, Italy and its colonies, Great Britain, and by the Arabs. The Roman cubit had 24 digits of 0.0185 (1/54th) metre (essentially the same length as the Egyptian digit). The inch as we know it (0.0254 meter) derived from the Roman uncia of 1/12th the Roman (Pelasgic) foot, or 1/18th of the Roman cubit, and had a length of 0.0247 metre. Thus, the English inch is about 2.834 percent longer than the Roman uncia.

b) Hinz[34] describes a canonical cubit ("Kanonische elle") specifically related to the parasang (farsah) of 3 Arabian miles, i.e., 12,000 canonical cubits of 0.49875 metres = 1 parasang of 5985 metres, from which an Arabian mile of 1995 metres (sometimes taken as 2000 metres) is derived. 56 2/3 such miles would yield a degree of 113,050 metres and an earth's circumference of 20,400 Arabian miles would be equivalent to 40,698 kilometres or 27,471 Roman miles - results only about 1.75% too large. This cubit dates back to ancient times having been known as the Assyrian cubit and the large Sumerian elle.

c) Ahmad (see ref. note 30) puts the Arabian mile of the Sinjar measurements at 6472.4 English feet, or 1972.8 metres. Dividing the latter by 4000 yields a cubit of 0.4932 metres, only 5.5 millimetres smaller than Hinz' canonical cubit. This is enough smaller, however, to give an improved result for the length of a degree of latitude, 111,792.07 m.

d) Hinz also describes a black cubit ("schwartzen elle") which was introduced by the Caliph Al-Ma'mun. This unit, he says, measures 0.5404 metres and became the common ("gewohnliche") cubit. Now, if we take 4000 such cubits, we obtain an Arabian mile of 2161.6 metres. 56 2/3 such miles yields a degree of 122,491 metres. The earth's circumference based on this unit is 44,096.7 kilometres, or 29,765 Roman miles. This is the unit which Morison/Raisz[35] attribute to the measurement carried out at Al-Ma'mun's direction.

e) Hallock and Wade[36] give Jomard and Boeckh as authorities for the employment of a black cubit of 0.51916 metres on the specific measurement of a degree we have been attempting to evaluate. This reference adds " . . . according to tradition . . . (the black cubit) . . . was the length of the arm[37] of a favorite black slave of the Caliph . . . " 4000 such cubits gave an Arabian mile of 2076.6 metres; 56 2/3 such Arabian miles gave the degree of 117,676 metres; and 360 such degrees gave an earth's circumference of 42,363.5 kilometres, or 28,595 Roman miles.

f) Next, we turn to J. L. E. Dreyer[38], the former astronomer-in-charge of the Armagh Observatory and author of the classic work, "A History of the Planetary Systems from Thales to Kepler." This work[39] contains much information on earth measurements made during classical and medieval times. In describing the measurement conducted at Caliph Al-Ma'mun's direction, he says "If the ' black cubit' is the Egyptian and Babylonian cubit of 525 mm (referring to Hultsch, "Griech. u. Rom. Metrologie, p. 309), the (Arabian) mile would be = 2100 m. and 56 2/3 miles = 119,000 metres . . . " The earth's circumference of 20,400 Arabian miles would then equate to 42,840 kilometres, or 28,917 Roman miles, figures about 7.1% too large.

g) Finally, we review two interesting references to the size of the coudée (French for cubit) used in the Arab measurement-from the works of J. P. J. Delambre, French astronomer, geodesist, and geographer of the late 18th and early 19th century. Thus:

Christman quotes Alfragan (on whom more later) in putting the mile at 4000 coudées and the coudée at 6 palms. Then, he adds, the palm has 4 digits and the coudée has 24. The width of the digit has a value of 6 grains of barley. (In Chapter 2, herein, it was concluded that when the digit is assigned 6 grains of barley, reference is being made to the width, rather than the length, of the barley grain. The width has a value of about 3.1 mm., so that the digit would have a width of about 18.6 mm.) Christman's coudée would measure about 0.446 metres which is essentially the same as Kimble's 18 inch cubit of 0.4445 m., so that the mile, the degree, and the earth's circumference would be about as stated under a) for the 18" cubit.

Christman then quotes *Abraham, son of Chaia, (who) in his Sphere*, Chapter IX says the mathematicians of the Caliph Al-Ma'moun and other officials used the unit relationships 6 grains of barley equalled the "polle", 4 polles a palm, and 4 palms made the coudée. This would make the coudée or cubit only 2/3 that of the 18 inch (uncia) cubit, or 12 uncia-about the size of the Roman foot of 0.296 m. The mile would have a length of about 1184 metres, the degree only 67.1 kilometres, and the circumference of a great circle 24,154 kms. or 16,303 Roman miles-all this if the polle is indeed a digit and Christman, Delambre and their publishers (as well as Abraham's sources and transcribers) have been faithful in their work. If so, this would be the smallest estimate of the earth's size in the literature[40].

It will be noted that the eight cubits of different length yield degrees varying from 67,100 to 122,491 metres, and earth's circumferences from 24,154 kilometres to 44,097, an enormous disparity.

Alfragan

An Arab astronomer and writer of the 9th century, born Ahmed Ben Kebir, but to whom was given (from the place of his birth) the surname Al-Farghani, which Cardinal d'Ailly corrupted to Alfragan, was responsible for the important work, *Chronologica et Astronomica Elementa*. In this work, he described the measurement of a degree ordered by Caliph Al Ma'mun (and first reported by Ibn Yunus). However, he went a step further, saying that Ptolemy had made the degree too large, 66 2/3 miles instead of the figure the Arabs had determined, 56 2/3 miles. This statement was to have profound repercussions.

When *Elementa* was translated into Latin in the 12th century, it created quite an impact on the scientific community of Europe. Roger Bacon (?1214-?1294) carried a digest of "Elementa" in his "Opus Majus", including the statement that the Arabs held Ptolemy's degree to be too large in the ratio of 66 2/3 to 56 2/3. When Cardinal d'Ailly published his *Imago Mundi* around 1419, he included in it an almost verbatim account of Alfragan's views, as taken from Bacon's *Opus Majus*. Christopher Columbus obtained his cosmographical views almost entirely from d'Ailly's work and central to these views was the notion that the world was smaller than generally accepted, i.e., Ptolemy's 500 stade degree and 180,000 stade earth's circumference.

In Europe, the interpretation of Ptolemy that had become dominant by Columbus' time was based upon the 8 stade per Roman mile relationship, so that a degree measured 62 1/2 Roman miles and the earth's circumference 22,500 such miles. This was Macrobius' interpretation of Ptolemy. However, in Alfragan's time the interpretation of Ptolemy that had become predominant in the Middle East was the 66 2/3 mile degree and 24,000 mile earth's circumference. This is very likely the system, based - at least initially - on the 7 1/2 stade per Roman mile relationship, which was described under the Diller interpretation of Ananias' "by air" relationships. Hence when the Arabs made their determination that a degree measured 56 2/3 miles, it indicated one or more of the following:

a) the Arab determination was a poor one (the view of some writers);
b) Ptolemy was wrong;
c) The Arab interpretation of Ptolemy was in error; d) the Arabian mile of 4000 black cubits was not equal to the Roman mile.

Alfragan, however, entertained no doubts on the subject. The Arab measurement of a degree, 56 2/3 Arabian miles, reduced the size of the world from Ptolemy's estimates in the ratio of 56 2/3 to 66 2/3, the implication being, also, that the Arabian mile and Roman mile were identical.

Henry Vignaud, refers to M. Reinaud's translation of Abulfeda's *Geography* [41] in supporting the conclusions of the preceding paragraph. Abulfeda is the authority for the assertion that " . . . the ancient

Text contines on page 129

All Unit Dimensions in Metres or Kilometres

Row →	Column No. → Source/Reference	I Cubit(Elle)	II = 4000×I Arabian Mile	III = 56⅔×II Alfragan's Degree	IV = 360×III + 1000 Earth's Circumfce	V = 66⅔ + 36⅔×III "Ptolemy's Degree"	VI = V ÷ 500 "Ptolemy's Stade"
A	Abulfeda, Bacon, Sacrobosco, Dante, d'Ailly, Reinaud, Vignaud; Kimble's 15" (uncial) cubit.	0.3705	1482 = Roman Mile	83,980	30,233 Km.	98,800	197.6 = Philetaerian Stade
B	Kimble's 18"inch cubit, same as Christman's 6 grain digit, 24 digit coudée (cubit).	0.4445	1778	100,753	36,271 Km.	118,533	237
C	Hinz's Canonical Elle, same as Arabian "post", Egyptian "Hand", and Persian "Ell55" (cubit).	0.49875	1995	113,050	40,698 Km.	133,000	266
D	Ahmad's Arabian Mile of 6472.6 English feet. (Essentially the same as item C).	0.4932	1972.8	111,792	40,245 Km.	131,520	263
E	Hinz's Black Elle, specifically posited for Al Mamun; Morison/Raisz, Parie.	0.5404	2161.6	122,491	44,097 Km.	144,107	288.2
F	Black Elle of Jomard, Boeckh and Hallock & Wade (length of forearm of Al Mamun's favorite Slave).	0.5191 6	2076.64	117,676	42,363 Km.	138,442	276.9
G	Black Elle of Hultsch and Dreyer/Stahl.	0.525	2100	119,000	42,840 Km.	140,000	280
H	Abraham's 4 palm coudée (cubit) as relayed to Christman, the Delambre.	0.296	1184	67,093	24,154 Km.	78,933	157.9

Table 3: Alfragan's Degree and Great Circle Circumference, as interpreted by various authorities, or simply based upon cubit (elle, coudee) lengths likely to have been used in the measurements of a degree that Alfragan reported on. Columbus, a cosmographical disciple of Cardinal d'Ailly, became convinced — as did all the sources/references of Row A — that Alfragan's Arabian mile was the same as the Roman mile, which in Columbus' time was, with the league of 4 Roman miles, the unit of distance at sea. (See note at right.)

In Columns V and VI are shown Alfragan's concept of Ptolemy's degree and stade. Alfragan, unabashedly declared Ptolemy's degree to be too long in the ratio of 66⅔ to 56⅔.

Table 3

Table 4: Medieval Islamic Units of Length

Item No.	Unit and Description	Metric Length	finger width canon. 1a	black 1b	fist width canon. 2a	black 2b	span black 3	black 4	7	10	cord 12	rod 13	chain 15	mile 16	parasang 17
1	aṣbaʿ : finger width a) based on canonical elle b) " " "black" elle	0.02078 0.02252	1	1	¼	¼	1/24	1/24							
2	qabḍa: fist width a) based on canon. elle b) " " "black" elle	0.08312 0.0901	4	4	1	1									
3	aḏ-ḏirāʿ aš-šarʿiyya: canonical elle; ḏirāʿ al-yad: Egyptian hand elle; ḏirāʿ al-barīd: Post elle; aḏ-ḏirāʿ al-Yūsufiyya: Kadi Abū Yūsuf (c.798 A.D) elle; and ẕarʿ-e šarʿī (gāz, zirāʿ): Persian elle (small)	0.49875	24		6		1				¼	⅛	1/80	1/4000	1/12,000
4	aḏ-ḏirāʿ as-saudā: black elle; aḏ-ḏirāʿ al-ʿāmma: common elle; ḏirāʿ al-kirbās: Egyptian white sackcloth elle; and aḏ-ḏirāʿ ar-raṣṣāṣiyya: Raṣṣaṣi elle used in Spain and Majorca.	0.5404	26	24	6½	6	1½	1							
5	ḏirāʿ ad-dūr: Home or Family elle.	0.503		22⅓				1-1¾ qaṣaba (black)							
6	aḏ-ḏirāʿ al-Bilāliyya: elle originated during reign of Bilāl ibn Abī Burda (739 A.D.), Kadi at Basra; aḏ-ḏirāʿ al-Hāšimiyya: small Hāšimi elle originated during reign of Abbasid Caliph al-Manṣūr (754-775 A.D.). See item following	0.60055		26⅔				1+2/9 qaṣaba (black)							

7	ad-dirā' al-Hāšimiyya: large Hasimi elle (see item immediately above); dirā' al-malik: King's (or Royal) elle; dirā' al-misāha: Surveyors' elle; ad-dirā' az-Ziyādiyya: Early Islamic (Ziyad ibn Sumayya, c. 673 A.D.); dirā' al-'amal: "Practical" Egyptian elle.	0.665	32	29 2/3	8		1 1/3	1+5 2/3 asba (black)							
8	dirā' al-bazz: Cloth elle; a) In Cairo: b) " Damascus: 1 1/12 Cairo elle: c) " Aleppo: 1 1/6 Cairo elle: d) " Tripoli: 1 1/10 Cairo elle:	0.58187 0.63036 0.67884 0.640	28		7		1 1/6								
9	ad-dirā' al-Omariyya: Caliph Omar's elle = 1/2 Measuring elle	0.72815		32 1/3		8 1/12		1 1/3 + 1/3 asba (black)							
10	ad-dirā' al-mi'māriyya: Building elle; ad-dirā' bi'n naǧǧārī: Egyptian Carpenter's elle; and Isfahan (Persian) elle.	0.798	38 2/5		9 3/5		1 3/5			1	2/5	2/10	1/50	1/2500	1/7500
11	ad-dirā' al-mīzāniyya: Measuring elle used in canal works.	1.4563		64 2/3		16 1/6		2 2/3 + 2/3 asba (black)							
12	bā' (qama): cord	1.995 (2)	96		24		4			2 1/2	1	1/2	1/20	1/1000	1/3000
13	bāb: rod or pole qasaba: Hākimī rod or pole	3.99 (4)	192		48		8	7 1/7 (approx.)		5	2	1	1/10	1/500	1/1500

#	Unit	Values (ascending columns, left → right)
14	habl: unit used in land surveying in western Andalusia (Spain).	21.616 — 960 — 240 — 40 — 1 — 1/50 — 1/50
15	qšl: chain or cable (Arabic) and tanāb (Persian).	39.9 (40) — 1920 — 480 — 80 — 50 — 20 — 10 — 1 — 1/50 — 1/50
16	mīl. Arabian mile	1995 (2000) — 96,000 — 24,000 — 4,000 — 2,500 — 1000 — 500 — 50 — 1 — 1/3
17	farsak: Persian parasang (itinerary meas.)	5985 (6000) — 72,000 — 12,000 — 7,500 — 3000 — 1500 — 150 — 3 — 1
18	barid: Persian itinerary measure, equal to Latin veredus.	23,940 (24,000) — 48,000 — 30,000 — 12,000 — 6000 — 600 — 12 — 4

Sources, Credits and Comments

The date in Table 4 is drawn entirely from Chapter III, "Langenmasse", of Walter Hinz' "Islamische Masse und Gewichte, Umgerechnet ins Metrische System", E.J. Brill, Leiden, 1955.

Hinz presents data in Chapter III indicating how many of the units of length changed, with time (usually becoming longer), from the 7th to 9th centuries, A.D. onward to the Renaissance and more modern times. Such information is not included in Table 4 in keeping with its title.

Hinz' sources are essentially 19th century German, French, and English articles and handbooks on a variety of subjects, length measurement—in many cases—being only incidental to the general theme of the work. He also has drawn on some Dutch, Iranian, and a very few Arabic sources.

The data in Table 4 is presented in matrix form and in ascending size, top to bottom and left to right, in order to heighten relationships between units, where they exist. It will be noticed that this table contains no reference to the foot (although the large Hasimi and practical Egyptian elles - Item 7 - are equal to 32 fingers, exactly 2 feet, and the cord - Item 12 - is equal to 96 fingers, exactly 6 feet or, 1 pace). Another notable omission is the stade. The reasons for Hinz' failure to even mention either the foot or stade are not stated.

mile--i.e., the Roman mile - and the Arab mile - have an identical value (Geog. t.I., p. 18) . . . and . . . the Arabs diminish the extent of the circumference of the globe . . " In the Introduction to his translation of Abulfeda's *Geography*, M. Reinaud confirms this with the remark, "The circumference of a (great) circle, according to the ancients was 8000 parasangs; according to the Arabs, it measured only 6800." The ratio of 6800 to 8000 is precisely the same as that of 56 2/3 to 66 2/3.

According to Abulfeda, then, "the ancients," i.e., Ptolemy, held the circumference of the earth to be 8000 parasangs, or 24,000 Roman miles. This is 66 2/3 miles to the degree of 98,000 metres, employing Reinaud's and Vignaud's 1482 metre Roman mile. Ptolemy, himself, had said (only) 500 stades made a degree, but if, indeed, 500 stades equalled 66 2/3 Roman miles, then Ptolemy's stade must have been the Philetaerian unit measuring 197.6 metres. Since the "moderns", the Mámun era Arabs, held the earth's circumference to be 6800 parasangs, or 20,400 Arabian or Roman miles, their degree was 56 2/3 miles and the cubit, 4000 of which made the Arabian mile, must have been Kimble's "short", 15 inch cubit measuring 0.3705 metres. This is essentially the same as Diller's cubit of 0.372 metres, 1/4000th of Ananias' Persian mile of 1488 metres (see Table 2). Here we have the basis for Christopher Columbus' 56 2/3 mile degree. He read about it in d'Ailly's "Imago Mundi" and he confirmed it - at least to his own satisfaction - in "measurements" at sea. (See chapter 7.)

Table 3 lists the various evaluations of Alfragan's 56 2/3 mile degree and 20,400 mile earth's circumference discussed in the foregoing. It also shows, for each evaluation, what length the Arabs ascribed to Ptolemy's degree and earth's circumference, as they understood it.

Table 4 provides a matrix of "Medieval Islamic Units of Length", according to Hinz. On the reverse side of the table, along with credits and sources, is shown a table of "Ancient Sumerian Units of Length," the most ancient origins of Islamic length measurement.

Ibn Khurdadhbih

As though to confound Alfragan, we have the case of Ibn Khurdadhbih (fl. c. A.D. 850) an official in the Central postal service at Samarra, near Baghdad, who was commissioned by Caliph Al-Mo'Tamid to compile a *Book of Roads and Provinces* (presumably to be of help to Moslem pilgrims enroute to Mecca). The book, of which only an abridged version remains, was evidently highly regarded and widely utilized in its day, for it provided a summary of all of the main trade routes of the Arab world and provided descriptions of such distant areas as China, Korea, and Japan. While a certain amount of fiction and unreliable hearsay were mixed with the facts presented, the book became a major source of information about the Arab world for later geographers.

Quoting from an Arabic source, probably pre-Ma'mun, he assures readers, in the preface of his book, that the earth's circumference measures 9000 parasangs. Kimble[42] tells us in a footnote that "1 parasang = 12,000 cubits = 3 miles, 1 cubit equalling 18 inches." Again, depending upon whether Kimble meant English inches or Roman unciae, the parasang is either 5486.4 or 5334 metres and the earth's circumference either 49,377,600 or 48,006,000 metres (both alternatives 20% or more too large). However, if Khurdadhbih were using the smaller cubit of 0.3705 metres, 4000 of which yield the Roman mile of 1482 metres and 12,000 the parasang of 4446 metres, the earth's circumference becomes 40,014,000 metres, almost a perfect estimate.

Al Biruni

Al Biruni (b. 972, d. ca. 1050) occupies a unique position among Muslim scholars. He was a scientist, historian, naturalist, geologist, astronomer, geographer, encyclopaedist, and mathematician[43].

On the fall of his native Khawarizm (Khiva, in modern Turkmen S.S.R.), he was captured by Sultan Mahmud of Ghazni (in modern east-central Afghanistan). Subsequently, having gained the Sultan's favor he was invited to accompany that chieftan on several of his 17 campaigns in northwestern India. There, he availed himself of the opportunity to learn Sanskrit which in turn, unlocked the door to his study of Brahmagupta, Aryabhata, and other Indian masters. During lulls in the fighting Al-Biruni was able to travel widely throughout India, thus obtaining an intimate picture of the land - its topography, climate, and people. Around 1030 A.D., he published his *Kitab al Hind,* popularly known as *Al-Biruni's India,* highly regarded in the Arab world.

Being well-versed in Greek, Syriac, Persian and Arabic, as well as Sanskrit, he studied all the works of the great scholars who had preceded him. Like most Arab scholars he respected Ptolemy for his enormous contributions to astronomy and geography, but was critical of the earlier geographer's seemingly blind acceptance of erroneous geographical information about Asia and, of course, his odd concept of a landlocked Indian Ocean.

Al-Biruni was the author of several dozen books, many of which dealt with geographical matters. No less than fifteen books and essays treated on topics such as measurement, determination of latitudes and longitudes, and finding distances between places. Five booklets described instruments and their uses, the astrolabe and its construction being a favorite topic.

An example of Al-Biruni's versatility and catholic interests is his authorship, with Ib'n Sina, of "Ikhwanassafa" which dealt with denudation, earthquakes, orogeny, tectonics and even continental drift. His astronomical and mathematical ideas were gathered together in a monumental work written in 1038 A.D., while at Ghazni, titled *Canon Masudicus.* While Al-Biruni accepted certain vague Hindu conceptions that had the earth rotating on its own axis, he was not prepared to accept the suggestion of Abu Said Sinjari that perhaps the earth might be revolving about the sun.

He is credited with the production of a round map of the world which was used in his "Kitab al Tafhim" to illustrate the disposition of the seas. In another work, *Chronology of Ancient Nations,* he derived a method for the projection of maps of the sky and the earth.

In geodetics, he did not like the method of determining longitude by comparing the local time at which a lunar eclipse occurs with that of a place whose longitude is known and converting the difference in times to degrees of longitude (one hour equals 15 degrees). His problem, a valid one, being that applications of this method resulted frequently in errors of several degrees. Instead, he preferred to measure as precisely as possible the linear distance between two points, determine their latitudes, and then apply - properly - the yardstick, 56 2/3 Arabian miles per degree of longitude (at the equator), modified for mean latitude. Of course, Al-Biruni's method could not be used where measurements involved large bodies of water. In addition, the determination of longitude is hinged, unnecessarily, to an accurate yardstick relating degrees of longitude at the equator to linear measure, further complicated by the requirement for accurate determination of latitude at each site where the longitude was required. Still, the method was readily available, which cannot be said about lunar eclipses, which never occur more than 3 times a year and occasionally skip a year.

Al-Biruni was aware of the method of determining latitude by circumpolar stars and realized its inherent advantage in accuracy it enjoyed over the method utilizing the pole star itself for that purpose. Ahmad reports that Al-Biruni applied the method of circumpolar stars to the sun to obtain latitude. Presumably this means he took the mean of the sun's zenith distance at noon on the days of the summer and winter solstices (equivalent to, but a more accurate method than, observing the sun's zenith distance at noon on the days of the vernal and autumnal equinoxes.)

In Chapter VII of *Qanun al-Mas' udi,* Al-Biruni takes up the question of the circumference of the earth. While commenting very favorably on the measurements and calculations of the three sons of Musa b. Shakir, he proceeds to describe his own efforts. Initial efforts undertaken in a level plain in Northern Dahistan in Jurjan having fallen short of success, he tried again in India. This time, he measured what

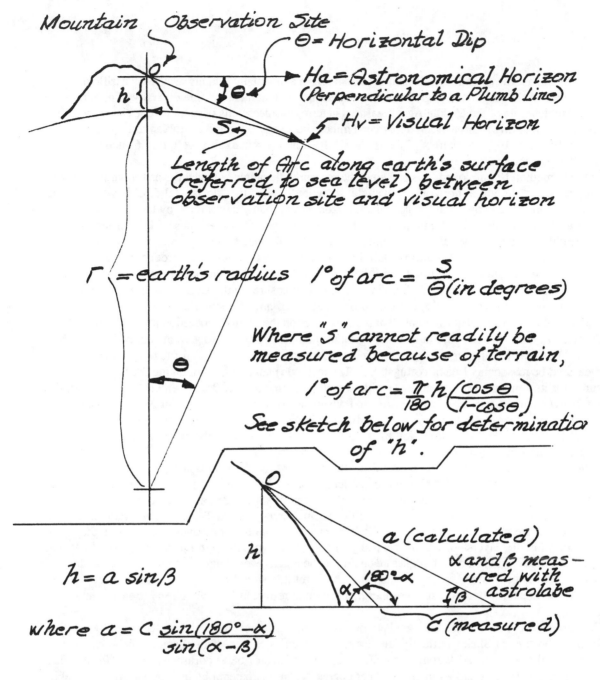

Mountain   Observation Site

Θ = Horizontal Dip

Ha = Astronomical Horizon
(Perpendicular to a Plumb Line)

Hv = Visual Horizon

Length of Arc along earth's surface
(referred to sea level) between
observation site and visual horizon

r = earth's radius $1°$ of arc = $\dfrac{S}{\Theta}$ (in degrees)

Where "S" cannot readily be
measured because of terrain,

$$1° \text{ of arc} = \frac{\pi}{180} h \left(\frac{\cos\Theta}{1-\cos\Theta} \right)$$

See sketch below for determination
of "h".

a (calculated)

α and β measured with
astrolabe

$180°-\alpha$

$h = a \sin\beta$

c (measured)

where $a = c \dfrac{\sin(180°-\alpha)}{\sin(\alpha-\beta)}$

Figure 3 : Al-Bīrūnī (972-c.1050) Measures
the Length of a Degree of Arc of the Earth's Surface in India by measuring the horizontal dip, or
depression, of the visual horizon from the astronomical horizon (upper figure). Lower figure shows
how "h", the altitude of his observation site, could
have been determined if terrain prevented an easy
measurement of "S".

Figure 3

Ahmad refers to as "the so-called horizontal depression from a mountain" and his result was a trifle over 56 Arabian miles for the length of a degree. We have noted before that there is some diversity of opinion as to what the Arabian mile of the Middle Ages measured in terms of linear units of more recent vintage which have been defined fairly rigorously. Here, Ahmad minces no words, putting the Arabian mile equal to 6472.4 English feet (1972.79 metres). On this basis, Al-Biruni did remarkably well, for 56 such miles equals 110,476.2 metres.

As to the method he employed, it is presumed he measured the angle between the astronomical horizon (see Figure 3) and the visual horizon. This angle, sometimes called the "horizontal dip," is what Ahmad refers to as "horizontal depression." The visual horizon was probably determined by mounting a sunlight reflector at the mountain site. A measuring party, utilizing rope, chains, or rods of known length, then moves steadily away from the reflector, keeping count of the distance traversed, until the reflector can no longer be seen. This point, difficult to determine with any precision, will - theoretically - be the point seen from the mountain observation site as the visual horizon. The distance traversed to the "visual horizon" divided by the "horizontal dip," in degrees, as measured with an astrolabe at the mountain observation site, will give, directly, the length of a degree of arc of a spherical earth. The method is independent of orientation, requiring only that the measurers on the ground proceed in a straight line.

Theoretically, the distance measured on the ground should be corrected to give the equivalent length at sea level. As a practical matter, all that is required is to insure that the vertical profile of the land traversed and measured be reasonably free of corrugations. The method is not precise, but taking the length of arc as measured for its equivalent length at sea level introduces an error of only 0.015 percent. If, however, the terrain between mountain observation site and visual horizon is such as to prevent a meaningful measurement of either the sea level or air line distance between the two points, the method may still be employed if a meaningful determination of the altitude of the mountain observation site can be made. Figure 3 illustrates both methods.

The major weaknesses in this method are the great difficulty in obtaining reasonably accurate measurements of either the horizontal dip or the length of arc between end points. With regard to the former, since one minute of angle represents over 1850 metres of surface arc, accuracy of small fractions of a minute are required, probably not obtainable on the instruments of Al-Biruni's day. Affecting both angular and surface measurements is atmospheric refraction of light which bends light rays as they pass through atmospheric "layers" of different density. It is not known how high up the mountain Al-Biruni estalished his observation site. The higher he could go, the greater would have been his horizontal dip and hence his chances of a good reading on his astrolabe. But this would have been offset to some degree because of the greater difference in atmospheric density and refracton between observation site and light source.

The suspicion is strong that Al-Biruni kept on taking measurements until his results were close to that obtained by the sons of Shakir in the Syrian desert two centuries earlier. This suspicion is heightened by Kazwini's attribution[44] to Al-Biruni of an earth's circumference of 6800 parasangs, or 20,400 Arabian miles. This is precisely 56 2/3 Arabian miles per degree, the measurement obtained by the sons of Shakir, whereas his own measurement, as reported by Ahmad, was essentially 56 miles.

All in all, Al-Biruni's measurement of a degree was neither as elegant, nor - in reality - as acccurate as the earlier Arab measurement on the plains of Sinjar.

Al-Idrisi

This Arab geographer of note was born in Ceuta in 1099, educated at Cordova, and travelled widely in Europe, Africa, and the Levant before settling at the court of Roger II, the Norman king of Sicily. That monarch, a man of some learning himself, saw to it that any geographic information obtainable from any part of the civilized world was made available to his royal geographer. Sicily, being a rendezvous for navigators from Mediterranean, Atlantic, and Northern waters, Al-Idrisi was in a remarkably advantageous position to stay abreast of contemporary geographic and astronomical events.

Ahmad reports[45] that in 1154, he wrote a treatise, "Amusement for Him Who Desires to Travel Round the World," also known as *The Book of Roger*. About the same time, he also made a celestial sphere and drew a map in the form of a disc representing the known world. The latter was to have a profound effect upon cartography as late as the 16th century. "One great feature of Idrisi's work" - as Ahmad puts it - "is the absence of unreserved approval of Ptolemy's ideas . . . " This,because of his own personal knowledge and experience stemming from his wide travels. That alone is enough to endear him to many geographic and other scholars over the millenia.

Kimble reports[46] that in one of his works, undertaken about 1154, he described the earth as a globe of 22,900 miles in circumference. There are no further details on this estimate. If the miles were Roman, his estimate was about 15% too small. If his miles were Arabian with a possible variation in length from 1778 to 2161.6 metres (see Table 3, this chapter), the earth's circumference could vary from 40,716,200 metres (only 1.79% too large) to 49,500,640 metres (23.8% too large).

Bhaskara Acharya

In India, from A.D. 300 for about 1000 years, astronomy was treated as a science in a number of text books, the *Siddhantas,* which were heavily influenced by Greek authors. They taught the earth was a sphere, unsupported in space. Bhaskara Acharya, about A.D. 1150 said the planets moved round the earth, all with the same linear velocity. The diameter of the earth is 1600 yojans, the distance of the moon is 51,570 yojans (or 64.5 times the radius of the earth).

Inasmuch as Aryabhata's estimate for the earth's diameter, made in the 5th century A.D. had been 1050 yôjans, it would appear that either a) the yôjan had undergone a significant change in the 700 years between Aryabhata and Acharaya; b) that Acharaya's estimate of the earth's diameter was much larger and in greater error than Aryabhata's; or c) within the broad range of possible lengths of the yôjan reported in the section covering Aryabhata, these astronomers were using units of different length.

Dreyer, our basic source for this Medieval Oriental estimate[47], is silent on the matter. However, Ahmad[48] gives Acharya's estimate of the earth's circumference as 48,714 English (statute) miles, or 78,398 kilometres. Now, Ahmad had given Aryabhata's estimate as 33,177 English miles, or 53, 393 kilometres. It is noteworthy that 1600 ÷ 1050 is quite close to 48,714 ÷ 33,177.

From the foregoing, it would appear that Acharaya's estimate was a poorer one than Aryabhata's, 700 years earlier. It would appear, further, that Dreyer (or his source, Colebrooke) is using a different yôjan to English mile relationship than Ahmad.

SUMMARY OF THE PERIOD A.D. 400 to 1150

In this chapter, the geodetic views of no less than 15 philosophers, astronomers, geographers and writers have been presented, covering a span of 750 years. All but four were Oriental (Armenian, Arabic, Indian, Chinese). Because of the uncertainty or differences of opinion as to the length of the basic units in which geodetic dimensions were expressed, more than 30 different sets of values for the length of the degree and the circumference of the earth were obtained. These ranged between approximately 66,000 and 238,000 metres for the degree and 24,000 and 86,000 kilometres for the earth's circumference.

Four measurements of a degree were carried out during this period: one in present-day Syria, one in India, two in China. The two Islamic efforts yielded 56 2/3 miles for the length of a degree and 20,400 miles for the earth's circumference. However, one segment of the Islamic world held that the measurement in Syria had yielded 66 2/3 Arabian miles for the length of a degree and 24,000 miles for the earth's circumference.

The Arabian mile of the first millenium A.D. is considered by most modern authorities on the subject to have a length of about 2000 metres. However, Alfragan and some geographers from his time to the turn of the 20th century and especially some like Roger Bacon, Sacrobosco, and Pierre d'Ailly, who influenced Columbus greatly, considered the Arabian mile identical in length to the Roman mile of 1480-82 meters. Further, Alfragan declared that Ptolemy had overstated the length of the degree and the earth's circumference in the ratio of 66 2/3 to 56 2/3.

Chapter 5

13TH CENTURY
EUROPEAN COSMOGRAPHERS
AND WRITERS

INTRODUCTION

The 13th century in Europe saw an intellectual renaissance with the production of such works as Roger Bacon's *Opus Majus,* Sacrobosco's *Tractatus de Sphaera,* Vincent of Beauvais' *Speculum* (under three headings: . . . Naturale, . . . Doctrinale, and . . . Historiale), and Albertos Magnus' *Physics, De Natura Locorum,* and others. The 13th century also produced such imaginative and widely read authors as Durante Alighieri (Dante) and the incomparable writer-travelers John de Plano Carpini, William de Rubruquis, and Marco Polo. Some of these writers were to influence cosmographic or geographic thinking for hundreds of years after their times.

The degree of coverage accorded these writers in the following is related to the influence each is known to have exerted - directly or indirectly - on that cosmographer, geographer and navigator Christopher Columbus and his contemporaries.

BACON, SACROBOSCO, AND DANTE

The Franciscan, Roger Bacon

The Franciscan, Roger Bacon (?1214-?1294), in his *Opus Majus,* undertook to assess the statements of the ancients with results he could never have anticipated. On the extent of the habitable earth, he wrote: "Ptolemy, in his book on the *Arrangement of the Sphere* , maintains that about a sixth part of the earth is habitable, due to the water, and that all the rest is covered by water..... But Aristotle maintains at the end of the second book of *De Caelo et Mundo* that more than a fourth is inhabited. And Averroes confirms

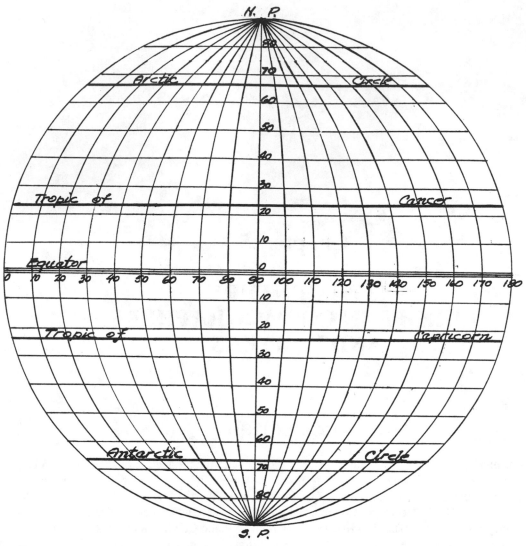

N. P.

Arctic   Circle

Tropic of   Cancer

Equator

0 10 20 30 40 50 60 70 80 90 100 110 120 130 140 150 160 170 180

Tropic of   Capricorn

Antarctic   Circle

S. P.

Figure 1 : Bacon's "Meridian Projection"

Roger Bacon, 13th century Franciscan philosopher/cosmog-
rapher, author of "Opus Majus", and source for much of
Cardinal Pierre d'Ailly's "Imago Mundi" (in the next century)
is credited by A.E. Nordenskiöld ("Facsimile-Atlas", pp. 93b,
94a, Dover Pubs, N.Y., 1973, first published in Sweden in 1889)
as the originator of the "Meridian Projection" shown above.
This projection (credited to d'Ailly by others) is characterized by equidistant, rectilinear parallels, and meridians
formed by arcs of circles passing through the poles
and dividing the equator in equidistant parts.

  Developed when the "habitable earth" was still believed
to be contained within 180° of longitude, the grid was not employed until the next century after Columbus and others
were well into the "Age of Discovery", at which time two such
grids were utilized, each representing a hemisphere.

136

this. Aristotle says that the sea is small between the end of Spain on the west and the beginning of India on the east. Seneca in the fifth book on *Natural History* says that this sea is navigable in a very few days if the wind is favorable. And Pliny teaches in his *Natural History* that it was navigated from the Arabic Gulf to Cadiz . . . therefore the width of the earth through which the Red Sea (Indian Ocean) extends is very great; from which fact it is clear that the beginning of India in the east is far distant from us and from Spain, since the distance is so great from the beginning of Arabia toward India. From the end of Spain beneath the earth the sea is so small that it cannot cover three-quarters of the earth. This fact is proved by the weight of another consideration. For Esdras states in the fourth book that six parts of the earth are habitable and the seventh is covered by water . . . therefore, according to these facts the extent of the habitable portion is great and what is covered by water must be small."

The arguments used by Bacon to establish the nearness of India to Spain were repeated by Pierre d'Ailly during the 15th century and, as Kimble remarks, " . . . however uncertain the premises and unsound the conclusions may appear to a modern reader, they were seized on by Columbus and his contemporaries and stimulated exploration to the west and south . . . "[1].

Roger Bacon is cited by Nordenskiöld[2] as the originator of the "meridian projection" which the latter describes as follows: "(This) . . . projection is characterized by equidistant, rectilinear parallels and meridians formed by arcs of circles passing the poles and dividing the equator in equidistant parts. The maps of the whole surface of the earth on this projection were generally, when it became practically employed, divided into two hemispheres . . . . " Nordenskiöld indicates that this projection was employed in several world maps contained in atlases printed during the period 1512 and 1582.

Christopher Columbus, a cartographer himself, came across this projection in Cardinal d'Ailly's "Imago Mundi," and doubtless (as did many others) believed it to be d'Ailly's own development.

John of Holywood

John of Holywood (d. 1256). or Sacrobosco, as he called himself, was born in England but became a teacher of Mathematics and Astrology at the University of Paris. He wrote a four chapter work *Tractatus de Sphaera* which was essentially a manual of principles of astronomy and cosmography. More precisely, the work treated on the terrestrial globe, of circles great and small, of the rising and setting of the stars, and of the orbits and movements of the planets. While the work was entirely unoriginal, being based on Ptolemy and Arab astronomers, its simplicity of presentation and its respect for authority made the work remarkably popular. Twenty-five printed editions appeared before 1500 and another 40 after that date. Sacrobosco included in his *climata* a coverage of Alfragan's estimates of the degree and the earth's circumference, including Ptolemy's "overstatements" in these premises (as seen by Alfragan)[3]. It is possible that this is where Cardinal d'Ailly learned of Alfragan's views which he later included in his *Imago Mundi.*

Sacrobosco's inclusion of Alfragan's geodetic conclusions, as indicated in the foregoing, was not necessarily a sign that he subscribed to those conclusions. Kimble tells us[4] that Sacrobosco used the Macrobian interpretation of Eratosthenes, i.e., 25,200 Roman miles, or 252,000 stadia, for the earth's circumference and 70 miles or 700 stadia, for the length of a degree. The stadium was taken to equal 500 Roman (Palesgic) feet, of 0.2963 metre = 148.15 metre.

Sacrobosco's *De Sphaera* became one of the chief nautical textbooks of 14th and 15th century Portuguese budding mariners.

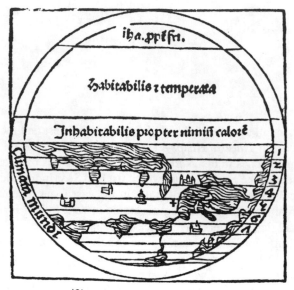

(3) JOH. DE SACROBUSTO,

OPUSCULUM

SPERICUM.

LIPSIÆ

S.A.

<u>Figure 1a:</u> <u>A Map of the World by Sacrobosco</u> <u>(Sacrobosto), also known as John of Holywood</u> <u>(1265-1321).</u> Contained in his work "Tractatus de Sphaera", it is believed by Nordenskiöld to be founded on "pre-Ptolemaic originals". <u>In</u> this map south is at the top, north at the bottom, west to the right, and east to the left. Since Sacrobosco considered any place south of the equator as uninhabitable, the map for this region is blank except for the tropic of capricorn and what served as his antarctic circle. Note also his climatic zones for the northern hemisphere of which 7 of the 12 are numbered.

<u>Credit</u>

This map is taken from A.E. Nordenskiöld's "Fac-simile-Atlas", plate <u>XXXI</u> (3), Dover Pubs, N.Y., 1973, first published in Stockholm, in 1889. It is reproduced here by permission of Dover Publications.

Durante Alighieri

Another writer of the 13th-14th century who was widely read among the educated classes was the Italian, Durante Alighieri, (1265 - 1321) author of *The Divine Comedy* and better known as Dante. None of Dante's work are strictly geographical, yet they contain in the aggregate, much geographical information of the type that the well-educated Italian in the early 14th century might possess.

In his *Quaestiao de aqua et terra,* a Latin lecture delivered in Verona in 1320, he indicates his adoption of Alfragan's estimates of 20,400 miles for the earth's circumference, 56 2/3 miles for the length of a degree, and 4,000 cubits for the mile. Our source for this information, Kimble, by indirection, would appear to believe that the cubits in question were the 15 Roman unciae, or 0.3705 metre size since the miles he refers to are both Arabian and Roman.

In Table 3, Chapter 4, eight estimates of the metric length of the Arabian mile were presented. All the authorities agreed the Arabian mile had a length of 4,000 cubits, but because each attributed a different metric length to the cubit (See column I, Table 3), there resulted eight different metric lengths for the Arabian mile (See column II, Table 3). When the length of the cubit was taken as 0.3705 metre (or 15 unciae), the Arabian mile had the same length as the Roman (Italian) mile-1482 metres (See Row A, Table 3).

MARCO POLO (1254-1324)

His Impact Upon Christopher Columbus, per Morison.

We come now to that prince of travelers, Marco Polo, and begin with a quotation from the foremost Columbian biographer of our time, Samuel Eliot Morison[5]:

"The important thing that Columbus obtained from Toscanelli apart from the prestige of having an eminent scholar approve his enterprise, was the Florentine's approval of Marco Polo. For the Venetian traveler had added some 30° of longitude to the the easternmost point of China described by Ptolemy. And beyond Mangi, Cathay, Quinsay, and Zaitun, 1500 miles out to sea, Marco Polo placed the fabulous wealthy island of Cipanju (Japan) with its gold-roofed and gold-paved palaces . . . "

The Toscanelli-Columbus relationship will be described in its chronological spot. We take up, here, Marco Polo and his effect upon geography and exploration.

Background, per Skelton

R. E. Skelton[6], the renowned map historian and Superintendent of the Map Room of the British Museum sets the tone for our treatment of Marco Polo with the following:

> The history of geographic knowledge records no more violent illumination of a practically unknown continent and civilization than the discovery of Central and Eastern Asia to European eyes at the end of the 13th century. From Hellenistic times, indeed, contact between Europe and the Far East had been gradually developed by sea

and land. By the 2nd century, A.D., Roman merchants were making voyages with the monsoons from the Red Sea to India, Malaya, and up the Chinese coast, and had travelled by the old silk road overland to markets of the Pamirs where they trafficked with Indians and Chinese, and as far as China itself, where in the 7th century Syrian missionaries introduced Nestorian Christianity. Although the world map in manuscripts of Ptolemy's 'geographia' . . . showed a landlocked Indian Ocean, the Greek geographer Cosmas Indicopleustes, writing in the 6th century, knew that Tzinitza' (China), the 'country of silk,' could be reached by a sea voyage, first east, then north; but he added that 'one who comes by the overland route from Tzinitza to Persia makes a very short cut.' There were, in fact, two sea routes, starting respectively from the Red Sea and the Persian Gulf; and two principal land routes, to the north and south of the Caspian sea.

These lines of communication were abruptly cut by the conquests of the Moslem Arabs who from the 7th century spread through the Near and Middle East to the gates of Byzantium; and the curtain which fell between Europe and Asia was not to be drawn back for six centuries . . .

In the early part of the 13th century a Neapolitan Franciscan, John de Plano Carpini, was sent by Pope Innocent IV to visit the Great Khan in Tartary (the rather indefinite but large area in Eastern Europe and Western Asia over-run by Genghis Khan and his Asian tribesmen--chiefly Mongol and Turkish--during the Middle Ages). The Franciscan's "Book of the Tartars," published upon his return to Europe, created quite a stir among those intellectual groups who could gain access to it. A few years later William de Rubruquis made a similar trip to the court of the Great Mogul[7] as an emissary of Louis IX of France (the Louis who was ultimately canonized). Upon his return to Europe, he mesmerized his audiences with his tales of Oriental wonders. Sir Henry Yule, perhaps the very best of all the Marco Polo editors, considers the report on Friar Rubruquis' mission published subsequent to his return, a "*Book of Travels* with few superiors in the whole Library of Travels."

The Book of Ser Marco Polo

Introduction

The event which really stirred things up was the publication[8] of *The Book of Ser Marco Polo* sometime after its dictation in a Genoese prison in 1298. Polo who had participated in a naval war between his native city-state, Venice, and its arch rival, Genoa, had been taken prisoner and was languishing in jail when he met a Pisan "romancer," also a prisoner, by the name of Rustigielo (Rustichello, Rusticiano, Rusticien). The latter fascinated by the story of Polo's experiences in the Orient, prior to his capture by the Genoese, persuaded him to dictate (first from memory and later augmented by his own notes forwarded from Venice) a full and complete account of what he had seen and heard. This dictation was done in Italianate French, the language, also, of the first manuscript. The popularity of the book is attested by the survival of 138 manuscripts and by its translation, before 1500, into Latin, Italian, German, and Spanish (and into Parisian French).

The Original Trip of the Elder Polos to the Orient

The book starts with an account of a trading journey made by the Venetian jewel merchants Nicolo and Maffeo Polo to the Mongols on the Volga via Constantinople, which they left in 1260. Encountering local wars in the area, they were advised to push on across the steppes to the south-east and Bokhara (in the present-day Uzbek S. S. R.), about 300 kilometres east of Samarkand and about the same distance north-west of the Afghanistan border at its nearest point. The Polos stayed in Bokhara three years, learning and becoming proficient in the use of the Tartar language. An ambassador of the court of the Great Khan, Kublai, passing through Bokhara, met the Polos and, charmed by their worldliness and mastery of the Tartar tongue, persuaded them to accompany him to Peking and meet his lord. Travelling over the northern silk road, they reached the capital sometime in the period 1264-1265.

Kublai was delighted with these Venetians, never having fallen in with European gentlemen before, and listened with great interest to all they had to say about the Latin world. Sometime around 1266 he dispatched the Polo brothers as his ambassadors to the Pope with a request that the Pope send him a large body of missionaries to instruct his people about Christianity and answer their many questions about it. As Yule puts it, " . . It is not likely that religious motives influenced Kublai in this, but he probably desired religious aid in softening and civilizing his rude kinsmen from the Steppes . . . "

The brothers arrived at Acre in April 1269 to find that Pope Clement IV had died the year before and no successor had been elected. Returning to Venice, they found that Nicolo's wife was no longer among the living, but Marco, his son, was now a fine lad of fifteen.

The Elder Polos Return to the Court of the Great Khan, This Time with Marco

No Pope having been elected during their two year wait in Venice, the Polos set out again in 1271 for Kublai's court without the requested missionaries, but accompanied by Marco, now seventeen. Travelling by sea to Acre (and Jerusalem, where they were to meet and receive letters--for the Khan--from the next Pope) and then to the Gulf of Issus and Lower Armenia, they then proceeded overland to Ormuz at the mouth of the Persian Gulf, north through Persia to Arbre Sec, and then along the southern silk road over the Pamirs to Kashgar, Lop Nor, and finally to Chandu (K'ai-p'ing-fu), Kublai's summer palace northwest of Peking, arriving there in 1275.

The Great Kahn, delighted to see them again, even though they had not brought the desired missionaries, put them into his personal service to help him administer his great empire. They were to remain in that service for seventeen years.

Kublai took a particular liking to young Marco who had a great facility for mastering the many tongues spoken within the vast empire. In additon the Khan noticed quickly the keen ovservational powers and the facility for explaining the 'customs and habits of foreign countries' that the young Venetian had been blessed with. Thus, Marco soon became entrusted with "all the important and distant missions." So far, in so many directions, did Polo travel before the Great Khan reluctantly permitted the three Polos to return to their native Venice, that "The Book of Ser Marco Polo," when published, elicited much incredulity--more, however, because of the things he related that had been told him than the things he saw.

Yule's Summary of Marco Polo's Travels

Yule has summed up Marco Polo's travels incomparably and we present it here:

He was the first Traveller to trace a route across the whole longitude of ASIA, naming and describing kingdom after kingdom which he had seen with his own eys: the Deserts of PERSIA, the flowering plateaus and wild gorges of BADAKHSHAN, the jade-bearing rivers of KHOTAN, the MONGOLIAN Steppes, cradle of the power that had so lately threatened to swallow up Christendom, the new and brilliant Court that had been established at CAMBALUC: The first Traveller to reveal CHINA in all its wealth and vastness, its mighty rivers, its huge cities, its rich manufactures, its swarming population, the inconceivably vast fleets that quickened its seas and its inland waters; to tell us of the nations on its borders with all their eccentricities of manners and worship; of TIBET with its sordid devotees; of BURMA, with its golden pagodas and their tinkling crowns; of LAOS, of SIAM, of CHOCHIN CHINA, of JAPAN, the Eastern Thule, with its rosy pearls and golden-roofed palaces; the first to speak of that Museum of Beauty and Wonder, still so imperfectly ransacked, the INDIAN ARCHIPELAGO, source of those aromatics then so highly prized and whose origin was so dark; of JAVA the Pearl of Islands; of SUMATRA with its many kings, its strange costly products, and its cannibal races; of the naked savages of NICOBAR and ANDAMAN; of CEYLON the Isle of Gems with its Sacred Mountain and its Tomb of Adam; of INDIA THE GREAT, not as a dreamland of Alexandrian fables, but as a country seen and partially explored, with its virtuous Brahmans, its obscene ascetics, its diamonds and the strange tales of their acquistion, its sea-beds of pearl, and its powerful sun; the first in medieval times to give any distinct account of the secluded Christian Empire of ABYSSINIA, and the semi-Christian Island of SOCOTRA; to speak, though indeed dimly, of ZANGIBAR with its negroes and its ivory, and of the vast and distant MADAGASCAR, bordering on the Dark Ocean of the South, with its Ruc and other monstrosities; and, in a remotely opposite region, of SIBERIA and the ARCTIC OCEAN, of dog-sledges, white bears, and reindeer-riding Tunguses.

Skelton's Assessment of the Impact of Marco Polo's Book

It matters not whether Polo was right or wrong. It is the impact of the information he provided which is important. While this was not profound immediately, with the advent of printing and particularly among cartographers, map-makers, and would-be explorers (like Columbus trying to sell his "Project of the Indies"), it could not be and was not ignored. Skelton reports that:

. . . a world map . . . (delineating) . . . Marco Polo's . . . route was apparently painted on the walls of the Ducal Palace at Venice in the early 14th century. The earliest surviving map to show his influence is the world map on eight panels, now known as the Catalan Atlas, drawn about 1375 by the Majorcan Jew Abraham Cresques, cartographer to the King of Aragon, and presented to King Charles of France in 1381 . . . On the Catalan map, the itineraries of the Polos may be traced in lines of towns; Marco's southwest journey to Burma is the source of many names; the great cities of Chambalech (Cambaluc on Yule's map), Cansay (Kinsay), Zayton (all from his Book) are marked; the trend of the coastline of China is correctly indicated; and the descriptive legends are taken from Marco Polo.

Cartographers of the 15th century, notably Fra Mauro, continued to draw their geography of Asia from Marco's Book, with some additions for India and Indo-China from the narrative of Niccolo di Conti, who between 1419 and 1444 had coasted India and visited Burma, Java, and perhaps south China. Conti, the first European to report on the Spice Islands, may have supplied information personally to Fra Mauro, whose world map completed in 1459 is the earliest to name 'Zimpagu' (Japan)--Marco Polo's "Chipangu.' The representation of Central and Eastern Asia by mapmakers of the 16th and 17th centuries, often curiously reconciled with new data, continued to be drawn from Marco Polo, whose topography of the mainland of China was not superseded until the Jesuit surveys in the 17th and 18th centuries. In 1561, five years after the publication of Ramusio's Italian edition of Marco's Book, Jacopo Gastaldi made a map of Asia in which he attempted to reproduce all the localities and names mentioned by Polo. Ortelius acknowledged his debt to Marco Polo for the maps of Asia in his atlas of 1570 . . . in 1577 John Frampton made the first English translation of the Book.......

Again we quote from Skelton:

In the search for other sea routes to Cathay, Marco Polo's Book remained the driving force. From Toscanelli's letter Columbus learnt of the wealth of Cathay and Cipangu, and of their longitudinal distance from Europe; all this came from Marco Polo. Later Columbus owned a copy of a Latin edition (1483-5) of Marco's Book, which was freely annotated in his hand . . . (preserved at the Bibliotheca Colombina, Seville) . . . His journals and letters abound in identifications of his discoveries with places recorded by Marco Polo, and in Cuba in 1493 he even sent an embassy in search of the Great Khan.

The first Spanish version of Marco's Book, that by Rodrigo de Somegalla, although not published until 1503 was - perhaps significantly - completed in August 1493 . . . .

But it was the Portuguese who, opening a seaway into the Indian Ocean, were the veritable heirs of Marco Polo. It is certain that his work was known to them, and probable that it influenced their plans for discovery to the south and east. About 1426, Prince Pedro (elder brother of Prince Henry 'the Navigator'), visiting Venice, was presented with a copy of Marco Polo's Book and a map said to be by Marco. About the same date, Jafuda (or Jacme) Cresques, the son and assistant of Abraham who drew the Catalan Atlas, was summoned from Majorca to Portugal by Prince Henry. In 1502, a Portuguese translation of Marco Polo was printed with an address to those who were . . . 'going out to India.' After the Portuguese expedition to the Spice Islands in 1511-12, their knowledge of China grew rapidly . . .

Text continued on page 148

143

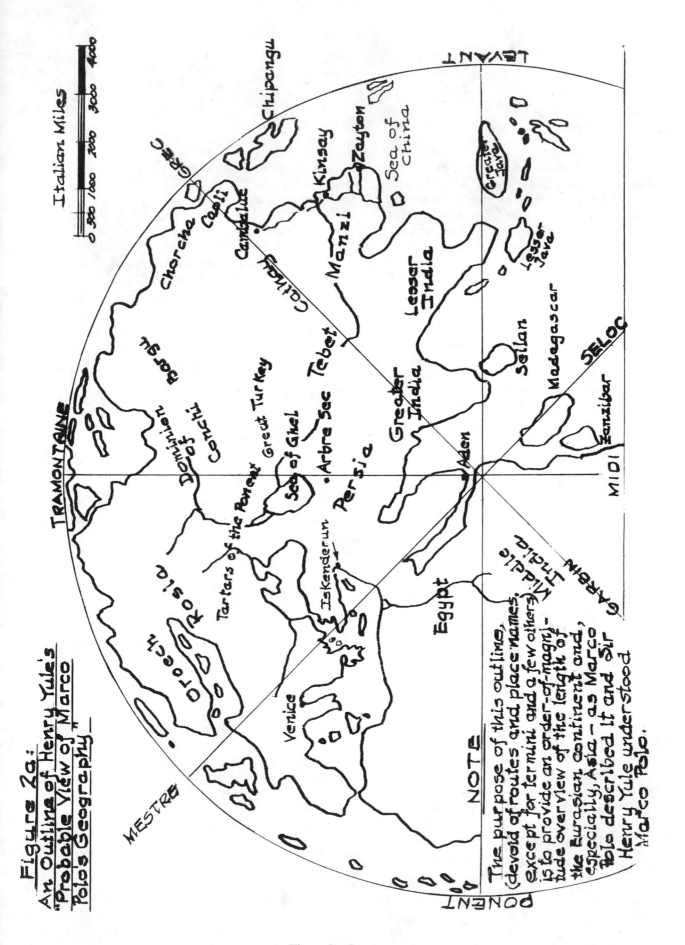

Figure 2a. *See note and credit on page 145.*

144

Additional Comment, including Source

The original Yule map of the "Habitable Earth" was based as far as possible on Marco Polo's own data. The Elder Polos travelled from Constantinople to Peking, 1255-1266; all the Polos from Venice to Peking, 1271-1275; all the Polos, Zayton to Venice, 1292-1295. They were resident in and travelled throughout Kubla Khan's empire, 1275-1292. (For their travels and visits, see Figure 2b.) Kinsay (Hangchow) is at 30°15' N. latitude and 120°10' E. longitude. It is now called Hangzhou. Zayton (Tswan-chou) is at 24°54' N. and 118°35' E. It is now called Quanzhou. The basic source for this outline map is Yule's map, contained in the 3rd edition, revised and edited by Henry Cordier, 1903, of Henry Yule's "The Book of Ser Marco Polo", J. Murray, London, 1871.

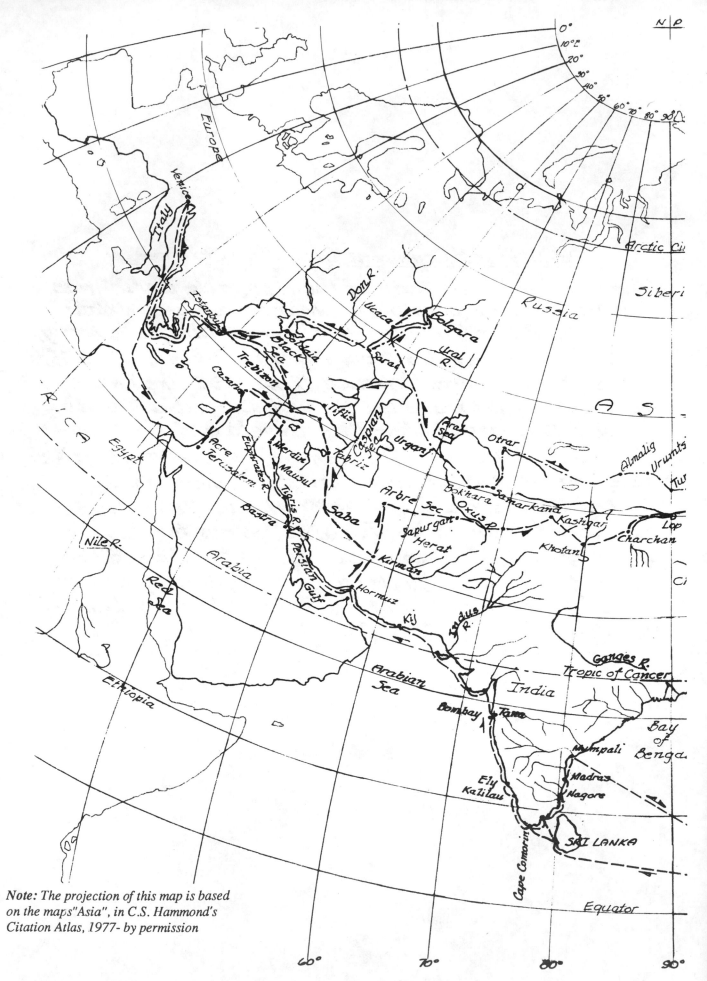

Note: The projection of this map is based
on the maps "Asia", in C.S. Hammond's
Citation Atlas, 1977- by permission

Figure 2b, part 1. Travels of the Polos in 13th Century Asia.

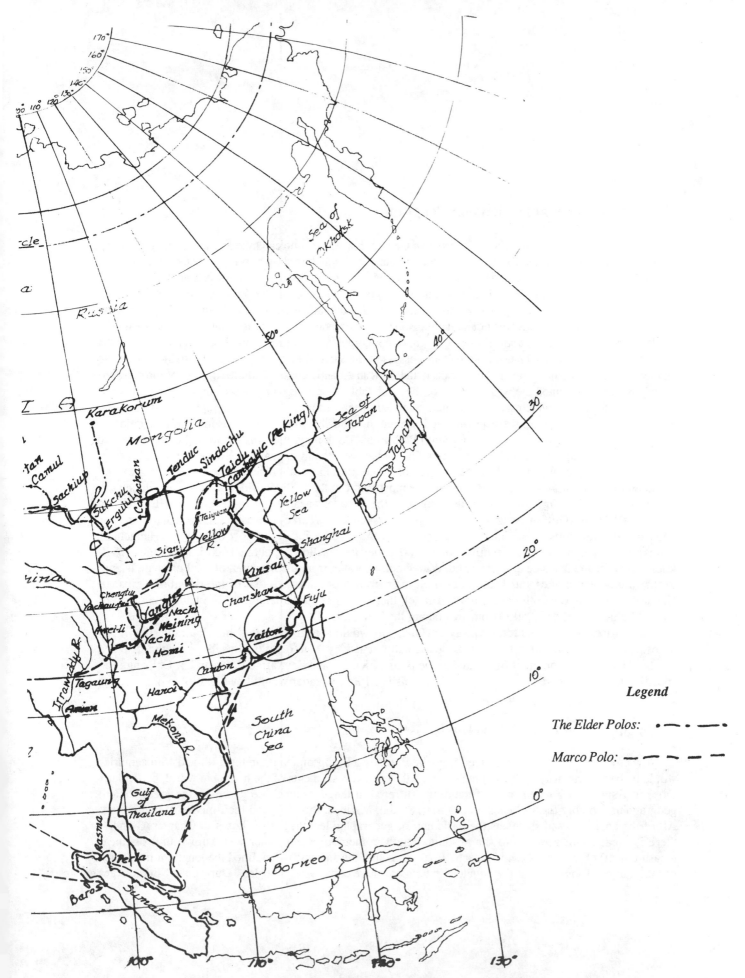

Figure 2b, part 2. Travels of the Polos in 13th Century Asia.

Marco Polo's Probable Views on Geography

In attempting to determine Marco Polo's Own Geographic Views, one can take a giant stride forward by referring to Yule's interpretation of Marco Polo, an outline of which is shown in Figure 2a.

At the outset, it should be understood that just as Marco Polos' Book is a mix of personal experiences and stories and information passed along to him by others, so Yule's *Probable View of Marco Polo's Own Geography* is a mix of geographic information actually contained in the Book and that which Yule supposes Polo possessed, but was not actually detailed, or even mentioned, in his work. The great value of the Book and of Yule's *Probable View . . .* map, geographically, is pretty much confined to Asia and eastern Europe. Whether or not Marco Polo possessed the kind of knowledge which would support the delineation of the Mediterranean, the European and African Atlantic coasts, and the spotting of the offshore islands which represented the western limit of the "Habitable World" is conjectural. In any event, it is not the sort of information which would have impressed European cartographers, cosmographers, and mariners, whose ideas as to such areas were already well-formed, if somewhat inaccurate. It is in Yule's depiction of Asia that new information provided by Marco Polo in his Book may, with justification, be considered to have had its real impact. This should be borne in mind as we attempt to evaluate Yule's map.

In the abscence of any grid system of parallels and meridians other than the equator and a central meridian taken through Aden and Arbre Sec, we must resort to the scale provided in this map which is in Italian miles[9]. Scaling Yule's map, the length of the Mediterranean from the Pillars of HERCULES (Gibraltar)[10] to the Gulf of Issus (taken as the port of Iskenderun, Turkey) measures 4500 Italian miles, or about 6670 kilometres. The "airline" distance from the latter point to the port of Kinsay (Hangchow, China) scales about 7600 Italian miles, or 11,260 kilometres. Allowing about 1600 Italian miles (2370 kilometres) for the distance along the 36th degree north parallel from the meridian of the Fortunate Islands to the Pillars of Hercules and 1500 miles from the meridian of Kinsay to that of the island of Chipangu (Japan), also along the 36th degree north parallel, there emerges a total length for the habitable world of 15,200 Italian miles, or 22,500 kilometres (along the 36th parallel).

The last segment of this total-Kinsay to Chipangu--would appear to scale somewhat more than 1500 miles. However, the latter figure was specifically stated by Marco Polo himself [11] as the distance "eastward" from the continent (intimated as the ports of Kinsay and Zayton) to the Island of Chipangu (no specification as to where on the island). Since this is all Polo's reader would have had to go on, we have departed in this instance from the scaled distance.

Geographic Impact of the Book

Now any geographic impact derived from the Book by a European reader in the 14th and 15th centuries would be based primarily on the degree to which the east-west segments of the habitable world, particularly those in Asia (as evaluated in the foregoing), differed in length from those given by the great ancient geographers, notably Eratosthenes, Posidonius, Strabo, Marinus, Ptolemy, and their medieval interpreters. All of the ancients used stades as the basic unit of geographic length, but as shown in Chapters 3 and 4, there is great controversy as to the absolute length in metres (or any other well-known unit of linear measurement) of the stades used by each. Further, each of the ancients had defined the length of his degree of latitude (or of longitude at the equator) in terms of these controversial stades. During the Middle Ages the use of the stade was replaced by the Roman or Italian mile and in Islamic countries by the Arabian mile, whose relationship to the Roman mile is questionable (see Table 3, Chapter 4.)

Still, it is possible to construct a table which compares the most likely interpretations of the ancient's on the habitable world and its major segments so as to highlight Marco Polo's contribution on Asia. Table 1 purports to do this. While the notes in this table are reasonably explanatory, the reader will find the text which follows helpful.

Longitudinal lengths, in degrees and stades are taken for Eratosthenes and Ptolemy from Figures 2d and 5, Chapter 3, respectively. While Posidonius drew no world map, he had expressed himself on the length of the habitable world (180°) and the distance westward from Spain to the Orient (180°). He also is credited by Strabo as reducing the size of the world by reducing the number of stades in a degree of latitude from Eratosthenes' 700 to 500. Strabo, himself, despite considerable criticism of Eratosthenes, finally accepted the older geographer. However, he specifically related the stade, 700 to a degree of latitude, to the Roman mile--8 stades to the mile. This fixed his stade length at 185.2 metres and that of his degree at 87.5 miles or 129.6 kilometres. While there is no unanimity as to the length of Eratosthenes' stade, the most credible estimates are 148.2-148.8 metres (10 to a Roman mile) and 157.5 metres (9.4 to the Roman mile).

Marinus accepted Posidonius' degree of 500 stades, but lengthened the habitable world from Posidonius' 180° to 225°. He was quite hazy about the character of the eastern end of the habitable world. The easternmost places which had been reported to him by mariners and traders were Sera, the Capital of Serica; Thine (Sina), the capital of the Sinae; and Cattigara, the most important port of the Sinae, all three of which he placed close to the 225th degree east meridian. There is wide speculation as to the identity of the places and provinces named, not alone because of the grossly inaccurate conclusions Marinus reached as to the distances and bearings between places which had been reported to him, but also because he landlocked the Indian Ocean. Beyond the 225th meridian was Terra Incognita and while he rejected the idea that the Atlantic coast lay somewhere within this unknown land, this was precisely the conclusion reached by later cartographers, cosmographers, and mariners--except for Ptolemy.

Ptolemy accepted Marinus, including the 500 stade degree, but reduced the length of the "known" world to 180°, as Posidonius had held it to be. His habitable world west of the Euphrates was identical to Marinus', including a 62° length for the Mediterranean, a prime meridian just west of the Fortunate Islands, the Sacred Promontory of Spain (Cape St. Vincent)--considered to be the westernmost point of continental Europe--put at 2 1/2 degrees east of the prime meridian, and the Pillars of Hercules (Gibraltar) 5 degrees east of that. Like Marinus, Ptolemy considered the Indian Ocean (Indicum Mare) to be landlocked. (See Figures 3a, 3b, and 3c).

Marinus had given the length of "known" Asia, from the Gulf of Issus to the 225th degree east of the Fortunate Islands as 155 1/2 degrees and Ptolemy reduced this to 110 1/2, which at 400 stades to the degree at their "main parallel," 36° N, equalled 62,200 and 44,200 stades, respectively. Marco Polo (per Yule's map) made the comparable length to the Mangi coast at Zayton about 8300 Italian miles and to Kinsay, somewhat farther north, 7600 miles (See Fig. 2b for identification of Zayton and Kinsay).

As has been indicated previously, there have been at least three credible interpretations of the length of Ptolemy's stade, 185.2, 197.6-198.4, and 210 metres. However, to simplify the following comparisons, only the smallest equivalency, that of 185.2 metres per stade, will be employed (which, incidentally, tends to minimize the impact of Marco Polo's Book). Thus, Marinus' 155 1/2 degrees equate to about 11,500 kilometres, or 7800 Roman or Italian miles. Ptolemy's 110 1/2 degrees equate to about 8200 kilometres, or 5500 miles.   *(Text continued on page 154)*

	Fortunate Islands (or Las Palmas de Gran Canaria) to Pillars of Hercules (or East end Strait of Gibraltar)			Pillars of Hercules to Gulf of Issus Length of the Mediterranean Sea (Gulf of Issus taken as vicinity Port of Iskenderun, Turkey)			Gulf of Issus to Easternmost Point in Continental Asia (or to Port of Kinsay); Length of Continental Asia along 36° N. Parallel		
Column No. →	I			II			III		
Geographer or Other Authority ↓	Degrees of Longitude	Stades or Italian (Roman Miles)	Kilometres	Degrees of Longitude	Stades or Italian Miles	Kilometres	Degrees of Longitude	Stades or Italian Miles	Kilometres
Actual Distances	10	600 Ital.Mi.	900	41⅔	2550 Ital.Mi.	3750	84	5100 Ital.Mi.	7550
Eratosthenes [1]	9	5000 stades	a.750 b.790	48	26,800 stades	a.3970 b.4220	78½	44,000 stades	a.6520 b.6930
Posidonius per Strabo [2]	—	—	—						
Strabo [3]	—	—	—	48	26,800 stades	5000	78½	44,000 stades	8150
Marinus [4]	7½	3000 stades	560	62	24,800 stades	4600	155½	62,200 stades	11,500
Ptolemy [5]	7½	3000 stades	560	62	24,800 stades	4600	110½	44,200 stades	8,200
Marco Polo, per Yule's Map (Fig.) [6]	—	1600 Ital. Mi.	2370	—	4500 Ital. Mi.	6670	—	7600 Ital. Mi.	11,260
Marinus and Ptolemy from Fortunate Is. to Gulf of Issus and Polo/Yule from Gulf of Issus to E.end Asia [7]	7½ (See Note No. 7)	3000 stades or 375 Ital. Mi.	560	62	24,800 stades or 3100 Ital.Mi.	4600	150	7600 Ital. Mi.	11,260
Morison/ Raisz on Marco Polo [8]	—	—	—	—	—	—	140½	7100 Ital.Mi.	10,500

1. Eratosthenes (276-194 B.C.), the first mathematical geographer, referred all longitudinal distances on his map to the parallel of Athens which he believed to be the longest in the "habitable world". He believed Athens to be at 36°55'43" N. latitude. His degree of longitude at the equator had a length of 700 stades. At the parallel of Athens, the degree had a length of 560 stades (approx.). The two most credible lengths which metrologists claim for E's stade are 148.2 metres (10 to the Roman mile) and 157.5 meters (9.4 to the Roman mile).

2. Posidonius drew no map of the world, but did express himself on the length of the habitable world and the distance westward from Spain to the Orient. The length of his stade is in dispute. Strabo said Posidonius had reduced the length of a degree from E's 700 stades to 500, but even this is questionable. However, Marinus took Strabo at his word, accepting P's size for the degree, rather than E's.

3. Strabo accepted E, generally, except that he specified his stade to be 1/8 th the Roman mile, hence 185.2 metres. While he reported E reasonably faithfully, he indicated his unhappiness with the older geographer's attempts to locate the Fortunate Islands and those off the coast of the eastern end of the Eurasian continent. He preferred to say this continent had a length of 70,000 stades between Cape St. Vincent and Cape Comorin (Cape Coniaci on E's map of the world).

4. Marinus accepted P's 500 stades to the degree and (presumably) Strabo's stade of 8 to the Roman mile. Whatever the reason, he set the length of the Mediterranean at 62 degrees, instead of its actual 41 2/3. He grossly misinterpreted reports on distances in Asia, placing the eastern end of theknown world about 225° east of the Fortunate Islands. The areas east of this meridian, he designated "Terra Incognita". While he rejected, as did Ptolemy after him, the idea of an Atlantic coast somewhere within Terra Incognita, this is precisely what those who followed him-except Ptolemy-did conclude. He considered the Indian Ocean to be landlocked, thus further distorting reports on the Far East.

Table 1, part 1: Marco Polo's impact on 15th century cosmography.

Easternmost Point in Continental Asia (or Port of Kinsay (Hangchau)) to Offshore Asian Islands (or Port of Tokyo/Yokohama) IV			Length of "Known" or "Habitable" World Fortunate Islands (or Las Palmas) to Easternmost Point in "Known" Asia (or Ports of Tokyo/Yokohama) V = I + II + III + IV			Fortunate Islands westward to Easternmost Point in "Known" Asia 360° (or equivalent) − V		
Degrees of Longitude	Stades or Italian Miles	Kilometres	Degrees of Longitude	Stades or Italian Miles	Kilometres	Degrees of Longitude	Stades or Italian Miles	Kilometres
19 1/3	1150 Ital.Mi.	1750	155	9400 Ital.Mi.	13,950	205	12,450 Ital.Mi.	18,450
3 1/2	2000 stades	a.b. 300	139	77,800 stades	a. 11,500 b. 12,250	221	123,600 stades	a. 18,300 b. 19,500
—	—	—	180	70,000 stades	?	180	70,000 stades	?
—	—	—	125	70,000 stades	13,000	235	—	—
—	—	—	225	90,000 stades	16,700	135	54,000 stades	10,000
—	—	—	180	72,000 stades	13,350	180	72,000 stades	13,350
—	1500 Ital.Mi.	2220	—	15,200 Ital.Mi.	22,520	—	—	—
29 2/3	1500 Ital.Mi.	2220	249	12,575 Ital.Mi.	18,640	111	5600 Ital.Mi.	8300
29 2/3	1500 Ital.Mi.	2220	239 2/3	12,100 Ital.Mi.	17,930	120 1/3	6100 Ital.Mi.	9000

5. Ptolemy followed Marinus in all of the assumptions and estimates described above, save one. Sensing the Tyrians over-statement of the length of Asia, he reduced (arbitrarily) the distance from the Euphrates to the east end of the "Known" world from 153 1/4 degrees (equivalent to 61,300 stades, or 11,350 kilometres) to 108 1/4 degrees (equivalent to 43,300 stades, or 8000 kilometres). Even so, he was still almost 32% too high. He thus had re-established the length of the "Known" world at Posidonius' 180 degrees. The most puzzling of his endorsements of Marinus was that of the length of the Mediterranean which was confirmed at 62°, about 49% too long. Marinus' and Ptolemy's equivalent length in kilometres was, however, only about 23% too great, and, if one were to apply the same metric lengths to the stade as conjectured for E, the estimate of 24,800 stades would have been almost perfect. Conversion from stades to kilometres has been made based upon the 185.2 metres per stade equivalency, the most popular among writers on ancient geography, but not some 20th century metrologists (including this reviewer.)

6. The distances in Italian miles shown here result from applying the scale shown on Sir Henry Yule's map, "Probable View of Marco Polo's Own Geography". For the reasons stated in the text, i.e. the inflated conversion factor from days of a journey to Italian miles, and the unlikelihood of anyone accepting Polo's views on Mediterranean geography, this approach represents too unrealistic a scenario.

7. In this approach, the Marinus/Ptolemy estimates are employed from the Fortunate Islands to the Gulf of Issus, and scaled distances from Yule's map for distances east of the Gulf. Conversions from Italian miles of 1482 metres to degree is at Ptolemy's 400 stades to the degree at the 36th degree, north, parallel and the most popular estimate of 185.2 metres/stades.

8. See Morison's "Admiral of the Ocean Sea, A Life of Christopher Columbus", pp. 64-66, Little, Brown and Co., Boston, 1942.

Table 1, part 2.

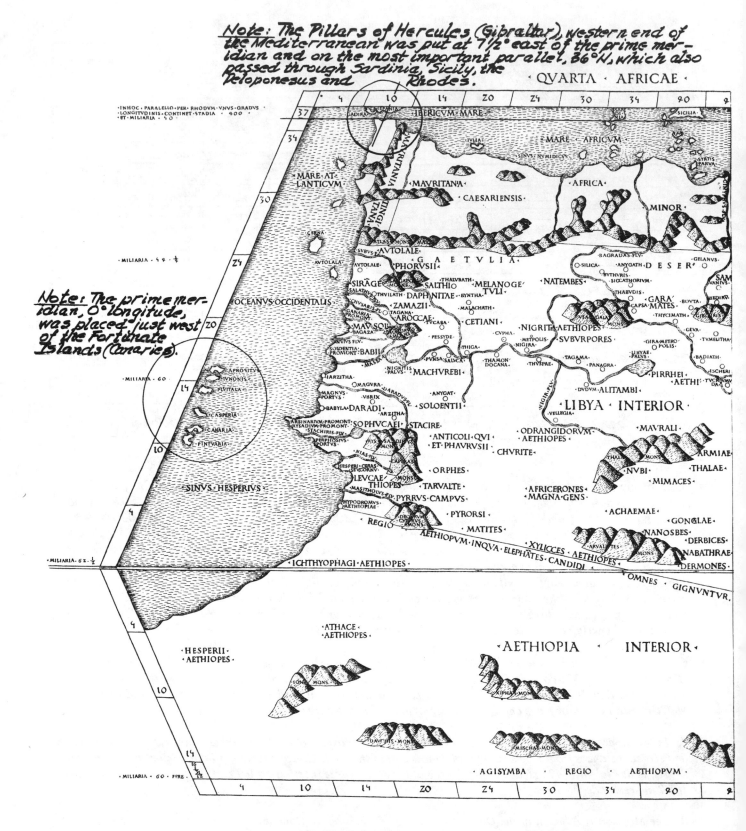

Figure 3a, part 1.

Figure 3a: XV. Ptolemæus Romæ 1490. (A.E. Nordenskiöld's "Facsimile Atlas", Dover Pub., 1973). Marinus' and Ptolemy's concepts of the prime meridian (the Fortunate Islands,) the most important parallel (36°N), the length of the Mediterranean (62°), and its eastern and western ends (the Gulf of Issus and the Pillars of Hercules, respectively). See above left and top of map.

152

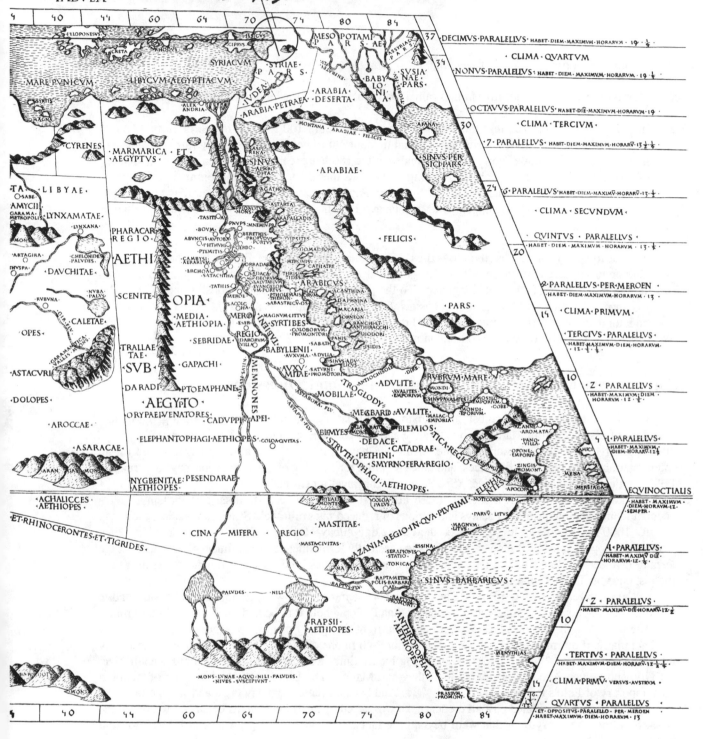

Note: The eastern end of the Mediterranean was put at the shore end of the Gulf of Issus, 69½° east of the prime meridian, making the length of the Mediterranean 62° (instead of the actual 41½°).

Figure 3a, part 2.

Three things are apparent from the foregoing:

1. Marco Polo tended to confirm Marinus rather than Ptolemy as to the length of continental Asia;

2. In doing so, he indicated that the Asian continent was from 2100 to 2800 miles, or from 42 to 56 degrees of longitude, longer than Ptolemy had made it out to be; (To be perfectly fair, Ptolemy had indicated that the "known world," only, extended to the 180th degree, east, meridian, "Terra Incognita," of unknown dimensions and character, lying east of this meridian. Thus, the geographic information provided by Marco Polo was considered by many, Paolo Toscanelli, Martin Behaim, and Christoper Columbus among them, to supplement Marinus and Ptolemy, particularly the former, rather than to refute them.).

3. The westward distance, "as the crow flies," from the meridian of the Fortunate Islands to that of Chipangu appeared (to those with cosmographical bents of mind) to be no more than about 5600 Italian miles, or about 111° of longitude. To the port of Kinsay, add 1500 miles or about 30°. The distance between the port of Lisbon (considered in the 15th century to lie 40 1/2 degrees north of the equator and about 5 1/2 degrees east of the Fortunate Islands) and the port of Kinsay appeared to be no more than 146 1/2° or about 7400 Italian miles.

An Assessment of the Book (Exaggerations and Omissions)

Now the foregoing exercise must be understood for what it is--an attempt to develop a geographical summary of distances from Marco Polo's narrative and compare this summary with the most popularly held standards of Polo's (and Columbus') time. And, we have utilized Yule's map which is based on an interpretation of the bearings, distances, and times required for each leg of the various journeys described in Polo's meticulously detailed narrative. (See Figure 2b for travel itineraries.)

Unlike previous itineraries, such as those utilized by Eratosthenes, Marinus, and Ptolemy, Polo's account identified and quantified delays and periods of reduced rates of travel. Still, the scale in Yule's map equates 1000 Italian miles with 42 days or an average day-in/day-out rate of 23.8 Italian miles[12] per day, which does appear somewhat high considering the absolutely forbidding nature of much--if not most--of the terrain traversed by Polo and his confreres. Indeed, others have assessed Polo's statements differently--see the last entry in Table 1, that for Morison/Raisz.

But anyone reading Marco Polo's Book in the two centuries following its initial publication had to come to the conclusion that Asia was a lot longer than Ptolemy held it to be and more nearly like Marinus' estimate of its length --or, that Polo was, if not an outright liar, guilty of gross exaggeration.

That many, if not most, of the Book's readers put Polo in one or both of these categories is attested by the nickname "Milioni" which he acheived during his lifetime--a reference to the enormous wealth, size, activity, and power Polo had attributed to Cathay and Mangi. The Dominican Friar, Jacopo of Acqui, a contemporary of Polo's is quoted by Yule[13]: " . . . And because there are many great and strange things in that Book, which are reckoned past all credence, he was asked by his friends on his death bed to correct the Book by removing everything that went beyond the facts. To which his reply was that he had not told one-half of what he had really seen!"

(Text continued on page 158)

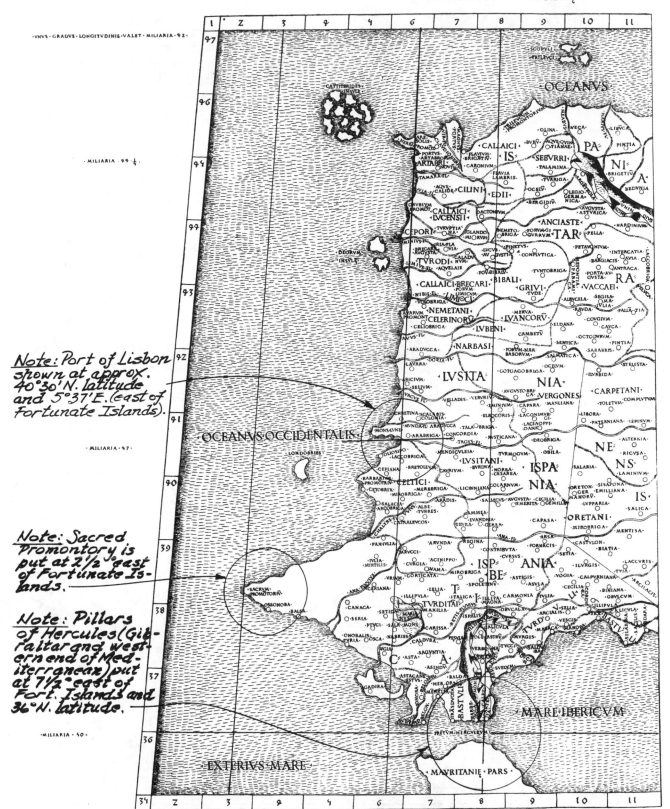

·VNVS·GRADVS·LONGITVDINIS·VALET·MILIARIA·42·

· MILIARIA·44 ½·

· MILIARIA·47·

· MILIARIA·40·

Note: Port of Lisbon shown at approx. 40°30' N. latitude and 5°37' E. (east of Fortunate Islands).

Note: Sacred Promontory is put at 2½° east of Fortunate Islands.

Note: Pillars of Hercules (Gibraltar and western end of Mediterranean) put at 7½° east of Fort. Islands and 36° N. latitude.

OCEANVS OCCIDENTALIS

EXTERIVS MARE

OCEANVS

MARE IBERICVM

Figure 3b: Ⅲ. PTOLEMÆUS ROMÆ 1490. (A.E. Nordenskiold's "Fac-simile-Atlas", Dover Pubs. N.Y., 1973.) The Sacred Promontory, considered by Marinus and Ptolemy to be the westernmost point in continental Europe, and the Pillars of Hercules (Gibraltar), the western end of the Mediterranean Sea. Also encircled is the port of Lisbon on the Tagus.

155

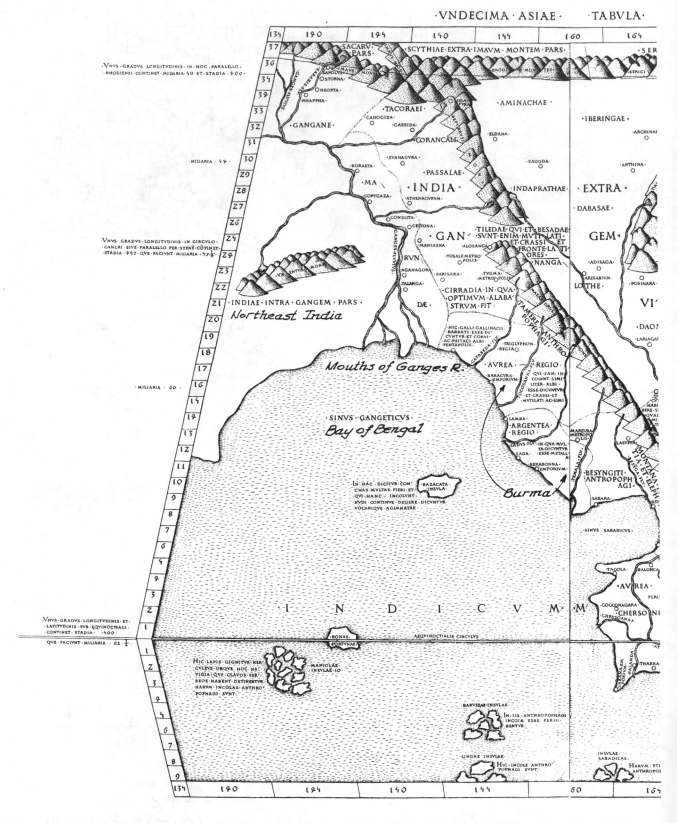

Figure 3c: XXVI. Ptolemæus Romæ 1490. (A.E. Nordenskiöld's "Facsimile Atlas", Dover pub. NY, 1973) Ptolemy's concept of the eastern end of the "Habitable World" (right border of map), presumably based on Marinus' ideas but placed at 180° east of the Fortunate Islands, reduced from Marinus' 225°. The impression gained by many was that the Chinese coast lay some unspecified distance east of the 180th meridian (or of the 225th.)

156

164 170 174 180

SERICAE · P · A · R · S · 37

36 · DECIMVS · PARALELLVS · PER · RHODVM ·

·DIFFERT · AB · AEQVINOCTIALI · HORIS · 2 · ½ · HABENS · MAXIMVM · DIEM · HORARV · 19 · ½ ·

34 · CLIMA · QVARTVM ·

SERICI · MONTES

SEMANTHINI

39 · NONVS · PARALELLVS ·

33 ·DIFFERT · AB · AEQVINOCTIALI · HORIS · 2 · ¼ · HABENS · MAXIMV · DIEM · HORARV · 19 · ¼ ·

IGAE

ARCHINARA

ASANAMARA

VRATHENE

SINA

32

32

31 · OCTAVVS · PARALELLVS · PER · ALEXANDRIAM ·

·DIFFERT · AB · AEQVINOCTIALI · HORIS · 2 · HABENS · MAXIMV · DIEM · HORARV · 19 ·

CACOBAE

ANTHINA

30

29 · CLIMA · TERTIVM ·

SALATHA

RHANDAMARCOTZA · IN · QVA · NARDVS

RA

28 · SEPTIMVS · PARALELLVS ·

·DIFFERT · AB · AEQVINOCTIALI · HORA · 1 · ¾ · HABENS · MAXIMVM · DIEM ·

BASANARE

27 HORARVM · 15 · ¾ ·

26

FLV·

M·

CHALCITIS · REGIO

ACADRAE

24 · SEXTVS · PARALELLVS · PER · SYENEN ·

·DIFFERT · AB · AEQVINOCTIALI · HORA · 1 · ½ · HABENS · MAXIMVM · DIEM ·

CIMARA

RVM·

POSINARA

PANDASSA

23 HORARVM · 15 · ½ ·

22 · CLIMA · SECVNDVM ·

SIPIBERIS

SPIORAE

ACATHRA

21 · QVINTVS · PARALELLVS ·

VI·

VM·

CVDVTE

20 ·DIFFERT · AB · AEQVINOCTIALI · HORA · 1 · ¼ · HABENS · MAXIMVM · DIE ·
HORARVM · 15 · ¼ ·

DAONE

AGIMOETHA

LARIAGARA

19

RHINGIBERI

TOMARA

18

BARRE

AGANAGARA

ASPITHABA

ASPITHRA

17 · QVARTVS · PARALELLVS · PER · MEROEN ·

·DIFFERT · AB · AEQVINOCTIALI · HORA · 1 · HABENS · MAXIMV · DIE ·

Probably Thailand

AMBASTAE

16 HORARVM · 15 ·

14 · CLIMA · PRIMVM ·

SI·

DAONA

QVI · HANC · REGIONEM
INCOLVNT · ET · IN · SPECVBVS
HABITARE · ET · PELLEM · HA-
BERE · SIMILEM · HIPPOPOTAMIS
QVAE · SA · MINIME
TRAICI · POTEST

PAPRA
SA

GITTIS

SINDI

SYNDA

BRAM-
MA

13 · TERTIVS · PARALELLVS ·

·DIFFERT · AB · AEQVINOCTIALI · HORAE · ¾ · HABENS ·
MAXIMVM · DIEM · HORARVM · 12 · ¾ ·

LASYPPA

BAREVAG-
RA

CORTATHA · ME-
TROPOLIS

SYNDA
ZABE

LABASTVS · FLV

12

11

GITI-
POPH-
AGI

MONTANA · TIG-
RES · HABE · ET · ELEPHANTES

DAONA · FLV·

Probably · Gulf of Thailand

10

MAGNVS · SINVS

RHABANA

9 · SECVNDVS · PARALELLVS ·

BSYGA

THROANA

8 ·DIFFERT · AB · AEQVINOCTIALI · HORAE · ½ · HABENS ·
MAXIMVM · DIEM · HORARVM · 12 · ½ ·

BARICVS

BERO
BAE

REGIO
LESTORV·

BALONGA
METROPOLIS

7

6

AMARADA

PAPRASA

ACADRA

MAGNVS
PROVINTIA

4 · PRIMVS · PARALELLVS ·

ACOLA

BALONCA

SAENVS · F·

·DIFFERT · AB · AEQVINOCTIALI · HORAE · ¼ · HABENS ·

AVREA

PENIMVLA

NOTIVM · PRO-
MONTORIVM ·

MAXIMVM · DIEM · HORARVM · 12 · ¼ ·

3

CONAGARA

CHERSONESVS

A·

AR·

E·

THERIODIS ·
SINVS ·

2

SINVS ·
INTERIOR ·

ICTHIOPHA-
GI

SATYRORVM ·
PROMONTORIVM ·

V·

· CIRCVLVS · EQVINOCTIALIS ·

CALIPOLIS

Maylaysia

· HABENS · DIEM · HORARVM · 12 · CONTINVE ·

1

ATTABA · FLV·

SATYRORVM ·
INSVLAE

SINARVM ·
SINVS ·

SINAE

KANAGA-
RA

2

THARRA

PALANDA

QVI · HAS · INHABITANT · CAVDAS
HABERE · DICVNTVR ·

THIINEME-
POLIS

3

SATRATA

1 · PARALELLVS · VERSVS · AVSTRV ·

·DIFFERT · AB · AEQVINOCTIALI · HORAE · ¼ · HABENS ·
MAXIMVM · DIEM · HORARVM · 12 · ¼ ·

4

5

6

COTTIABIS · FLV·

S·

7

LABADII · HOC · EST · ORDEI · INSVLA · FERACISSIMA · ENIM ·
HEC · INSVLA · DICITVR · ET · PRETEREA · MVLTVM · AVRI ·
EFFICERE

8

2 · PARALELLVS · MERIDIONALIS ·

DICAE

ARGENTEA
METROPOLIS

CATTIGARA · SINA-
RVM · STATIO

9 ·DIFFERT · AB · AEQVINOCTIALI · HORAE · ½ · HABENS ·

HARVM · ETIAM · ICOLAE
ANTHROPOPHAGI · SVNT ·

· MAXIMVM · DIEM · HORARVM · 12 · ½ ·

164 170 174 180

Note: Thine was considered to be the capitol of the Sinae. While the Sinae has been variously identified, the most convincing is that of Cambodia or Vietnam.

Note: Cattigara, in the Sinae, was the most easterly port known to Marinus and Ptolemy. It has been interpreted to be Kao-chi, Saigon, Hanoi, Canton, the mouths of the Yangtse R., and as far north as Tsingtao.

Figure 3c, part 2.

157

Another aspect of the book seized-upon by his detractors and troubling to Polo's many friends is his failure to mention certain things intrinsically Chinese such as the Great Wall, the use of tea, the compressed feet of the women, the employment of the fishing cormorant, artifical egg-hatching, printing of books, the great characteristic of Chinese writing, or a score of other remarkable arts and customs. Yule concludes Polo never learned Chinese although he evidently mastered the tongues of the people at Kubla Khan's court, Mongols, Tartars, and Persians.

With regard to Marco Polo's geographic itinerary, it is clear that some elements of distortion exist. Whether this is due to faulty assessments of distances, bearings, and durations of the many segments of his many travels, or whether his chroniclers and biographers have distorted his narrative as originally related to Rusticiano--probably both--the net result is a gross inflation of east-west distances. Not only Yule, but every other geographer of record who has attempted to evaluate the length of Asia from Polo's book has come up with inflated results, although no two appear to be the same. For whatever reasons, travellers' itineraries throughout recorded history have been over-stated, no matter how sophisticated the recorders.

For that school of thought, in the 15th century, which wanted to believe the distance westward from Iberia to the Orient was less than Ptolemy had made it out to be, Marco Polo's book was a God-send. In Columbus' mind, Marco Polo confirmed and extended Marinus who made the Eurasian continent very long. Alfragan had made the earth significantly smaller so that the extended Eurasian continent occupied even more of the earth's suface, longitudinally. Clearly, the width of the Atlantic Ocean between Europe and India was shrinking.

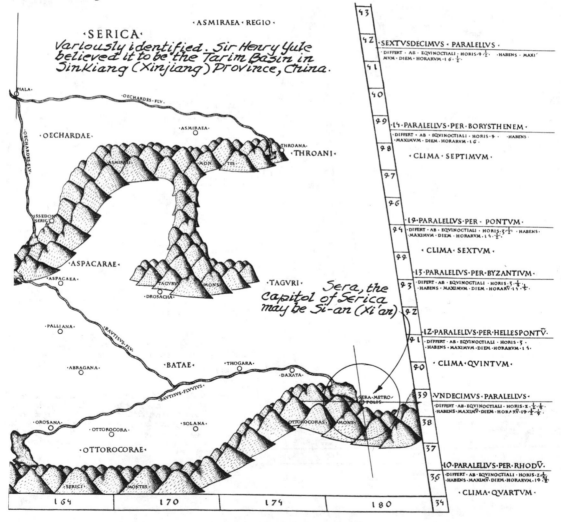

Figure 3d: A Section of XXIII Ptolemaeus Romae.
Ptolemy's concept of the eastern end of the known world north of the 35th degree N. parallel.

Chapter 6

14th CENTURY
COSMOGRAPHERS AND WRITERS

INTRODUCTION

This chapter covers the cosmographic views of three 14th century writers. It is important to our story because, inter alia, it continues to demonstrate the difficulty of assessing with any assurance the significance of numerical pronouncements as the the size of the earth.

Whether the writers were western or oriental, laymen or churchmen, sensationalist or scientific, they continued the practice-since earliest recorded history-of using units of linear measurement poorly defined and, therefore, subject to a variety of interpretations. Even in the case where intent seems clear enough, it becomes clouded by other pronouncements of the same writer in gross contradiction.

Of greater importance, we come to grips-finally-with the direct source of Christopher Columbus' cosmographic views in the coverage devoted to Cardinal Pierre d'Ailly.

JEAN d'OUTREMEUX

During the 14th century, the most famous piece of vernacular writing in the English language was a document describing a good part of the known world purportedly visited by its main character. *The Travels of Sir John Mandeville* was the English counterpart to *The Book of Ser Marco Polo*. The major difference is that there was no Sir John Mandeville, all but a few pages of his so-called "travels" having been plagiarized by its author Dr. Jean d'Outremeux from works of other authors. The work has some value for us because it illustrates the 14th century layman's idea of the world[1].

On the earth's circumference, d'Outremeux first says it is 20,425 miles, but then (without indicating he has any quarrel with this figure) he develops Isidore's theory, based on Macrobius: 700 furlongs = 1 degree = 87 miles + 4 furlongs, or 87-1/2 miles. Thus, the earth's circumference = 31,500 miles. And, he says, "The miles are after our country."

The English mile ultimately evolved into the statute mile of 5,280 English feet, 1760 yards, or 1609.34 metres. The legislation which made this "the mile" of the realm was "35 Elizabeth, c. 6 (1592)".

A furlong was defined as equal to 40 poles of 16-1/2 feet. Thus, the furlong, the counterpart of the Greek and Roman stade (stadium) was set equal to 660 English feet and the mile of 8 furlongs equalled 5,280 feet. But what was the length of the mile during the period of *The Travels of Sir John Mandeville* (first half of the 14th century)?

Klein calls *Arnold's Customs of London* , which appeared about 1500, one of the earliest tables of English linear measure. We quote from a portion of that table he has excerpted[2]: "The length of a barley corn 3 times makes an ynche and 12 ynches make a fote and 3 fote make a yerde and 5 quaters of the yerde make an elle: 5 fote make a pace. 125 pace make a furlong and 8 furlongs make an English myle."

The inch, foot, pace, furlong, mile relationship was the same as the earlier Roman sytem in use in Britain for the four hundred years of the Roman occupation and quite a period beyond. But there were Anglo-Saxon, Norman, and other influences as well. Petrie[3] shows that there were at least four feet in use in England during the 12th to 14th centuries: a version of the Roman foot of 11.66 English inches, or 0.2945 metres; the Olympic foot of 12.17 inches, or 0.309 metres; a version of the Babylonian foot measuring 12.47 inches; and a version of the Drusian foot measuring 13.22 inches, or 0.3357 metres.

Thus, the 5000 foot mile of Arnold's "tables could have measured anywhere between 1472.5 and 1678.5 metres.

Berriman[4] however, tells us that "on the Gough (Bodlein) map of England (C. 1335), the marked distances, according to Petrie ('Proceedings of the Royal Society of Edinburgh, 1883') are all in units that exceed the statute mile; his examination suggested the use of three different standards:

2220	to	2240 yards for most of England
2499	to	2567 yards for Wales
2534	to	2886 yards for Cheshire and Shropshire."

From this bewildering array, it is yet possible to make an intelligent guess as to d'Outremeux's mile. His furlong to mile relationship was 1 to 8, this calls to mind the Olympic stade to Roman mile relationship which employed the Olympic foot of 0.30864 metres, 600 of which made the stade of 185.18 metres. Taking the slightly enlarged English version of the Olympic foot, 0.309 metres and multiplying this by 5000, we obtain a possible d'Outremeux mile of 1545 metres. 87-1/2 of such miles would yield a degree of 135,187.5 metres and 360 such degrees would give an earth's circumference of 48,667.5 kilometres.

In a footnote to his commentary on Sir John Mandeville's Travels, Kimble tells us that "the 'new' (medieval) astronomers", such as Honorius of Autun and William of Conches, used figures for the earth's circumference which approximated 20,250 miles (presumably Roman).

CARDINAL PIERRE d'AILLY

Pierre d'Ailly (d. 1422), Cardinal of Cambrai in the early 15th century, published his *Tractatus de Imagine Mundi* (Imago Mundi) in 1410, 140 years after the time of Roger Bacon, yet it contained very little of the new geographic knowledge which had been gained by the world's explorers and geographers since Bacon's time. Four years later he felt impelled to bring out his "Cosmographiae Tractatus Duo" to make up for the shortcomings of his Imago Mundi, for by this time he had seen a Latin copy of Ptolemy's "Geography" which represented a veritable mine of "new" material for him.

D'Ailly was a respector of traditional sources but he did on occassion differ with them as he did with Ptolemy in his second work. The bone of contention was the size of the habitable earth, i.e., Eurasia and North Africa from west to east and from the equator to the Arctic Circle. Here, he attempts to demonstrate that this length is greater than that postulated by Ptolemy and that the easternmost part of India (China's coast) must not be as far distant from the western extremity of Africa as the great Greek geographer and astronomer had indicated. On this argument he relied heavily on Roger Bacon's views previously quoted. Further, unlike Ptolemy, he believed the equatorial regions could be temperate and possess impressive economic resources, ideas which Columbus heartily endorsed and which he saw confirmed by Portugese discoveries.

D'Ailly, besides being a prelate of the church, a cosmographer, and a university teacher, was an astrologer of some note, for he foretold, the French Revolution 350 years before its outbreak. We quote Crone[5]: "After referring to the seven great conjunctions of the planets Saturn and Jupiter and the political upheavals which had accompanied them, he wrote of the eighth, forecast for the year 1789: 'This said, if the world lasts until that year, which only God knows, there will be great, many, and astonishing changes in the world, principally in law and religion.'"

On the size of the earth, he seemed to hold with Alfragan's 20,400 mile circumference and 56-2/3 mile degree, yet he wrote" . . . if a man set out walking at a steady rate of 20 miles a day he would complete the circuit in four years, sixteen weeks, and two days . . . "[6] This, of course, leads to a size for a spherical earth far greater than he had indicated for Alfragan. Assuming the man was traversing a meridian or the equator, he would have travelled:

$$20 \text{ m. p. d. } \times (4 \times 365\text{-}1/4 + 16 \times 7 + 2) \text{ days} = 31,500 \text{ miles.}$$

This was the estimate of Isidore of Seville and far larger than the 20,400 mile circumference and 56-2/3 mile degree which he quoted for Alfragan. The largest of the Arabian mile possibilities we had examined, Table 3, Chapter 4, was 2161.6 metres, which, when multiplied by 20,400 equalled 44,096.6 km. or 29,765 Roman miles.

On the assumption, which seems far-fetched today, but may have been a more realistic possibility in the strict religious climate of early 15th century France, that d'Ailly's "man" did not walk on the Sabbath (Sunday), the number of days he walked would reduce from 1575 to 1350 (approximately) and the miles travelled from 31,500 to 27,006. This is a very respectable estimate for 27,006 Roman miles equals 40,022 kilometres, less than 0.06 of a percent from our self-adopted standard of 40,000 km.

Columbus ignored this part of d'Ailly's writings, choosing the 56-2/3 mile degree in which the miles were interpreted as Roman. Columbus annotated several passages of the "Imago Mundi where mention is made of Alfragan, and of his reduction of the degree to 56-2/3 miles. These annotations are preserved in the *Raccolta Columbiana, Postille ai di P. d'Ailly.* Vignaud to whom we are indebted for this information, says further that it was Humboldt who claimed d'Ailly had copied Roger Bacon on this, but it was John of Seville who originally translated Alfragan's *Chronologica et Astronomica Elementa* in his *Hispalensis,* in the 12th century[7].

The *Tractatus de Imagine Mundi* was a collection of 12 short treatises written by the Cardinal, four of which deal with cosmography. The first, and most important, is titled *Imago Mundi* (which name appears to have overpowered the overall title). The second treatise was titled *Epitome Mappe Mundi.* Vignaud[8] tells us "These were the two treatises which drew Columbus' attention, and literally he has covered their margins with notes, the reading of which reveals that it was there that he had found his geographical system . . ." Las Casas[9] confirms this conclusion with the remarks "this doctor has had greater influence in the decisions of Columbus than . . . (any of) . . . the authors we have previously cited. . . "

Vignaud traces the one printed edition of *Imago Mundi* to Louvain, the printer to John of Westphalia, and the time to the period between 1480 and 1487. Then, he concludes that Columbus did not read *Imago Mundi* until after his first and possibly second voyages so that the document could have played no role in

Columbus' decision to seek support for his project to reach India by sailing westwards.

Morison, however, indicates that *Imago Mundi* seems to have been his bedside book for a number of years. Unlike the postils Columbus placed in the margins or in the blank leaves at the back of other books, such as the 1477 edition of the *Historia Rerum Ubique Gestarum* by Aeneas Sylvius (Pope Pius II), those he adorned *Imago Mundi* with were undated. Thus, the best argument Morison musters to prove Columbus read this document before his first voyage of discovery of 1492 is the following convincing question: "Can anyone for a moment believe that Columbus could have made these annotations after 1492 and not inserted a single fact from his infinitely more marvelous experience in America?"[10]

Humboldt, too, is quite definite on this point: " . . . the work entitled 'Imago Mundi' exercised a greater influence on the discovery of America than did the correspondance (of Columbus) with the learned Florentine Toscanelli. All that Columbus knew of Greek and Roman writers, all those passages of Aristotle, Strabo, and Seneca, on the proximity of Eastern Asia to the Pillars of Hercules, which as his son Fernando says, were the means of inciting him to discover the Indian lands . . . were gathered by the admiral from the writings of the cardinal."[11]

The following statements, which Columbus found in *Imago Mundi*, impressed him enough to copy and comment upon them. About the sea between Europe and the Orient:[12]

> The end of the habitable earth toward the Orient and the end . . . toward the Occident are near enough, and between them is a small sea.

> Between the end of Spain and the beginning of India is no great width.

> An arm of the sea extends between India and Spain.

> India is near Spain.

> The beginnings of the Orient and Occident are close.

> From the end of the Occident to the end of India by land is much greater than half the globe.

> Water runs from pole to pole between the end of Spain and the beginning of India.

> Aristotle (says) between the end of Spain and the beginning of India is a small sea navigable in a few days . . . Esdras says six parts (of the globe) are habitable and the seventh is covered with water. Observe that the blessed Ambrose and Austin and many others considered Esdras a prophet.

> Julius (Solinus) teaches that the entire sea from India up to Spain behind (per dorsum) Africa is navigable.

> The end of Spain and the beginnings of India are not far distant but close, and it is evident that this sea is navigable in a few days with a fair wind.

And on the great extent of Asia:[13]

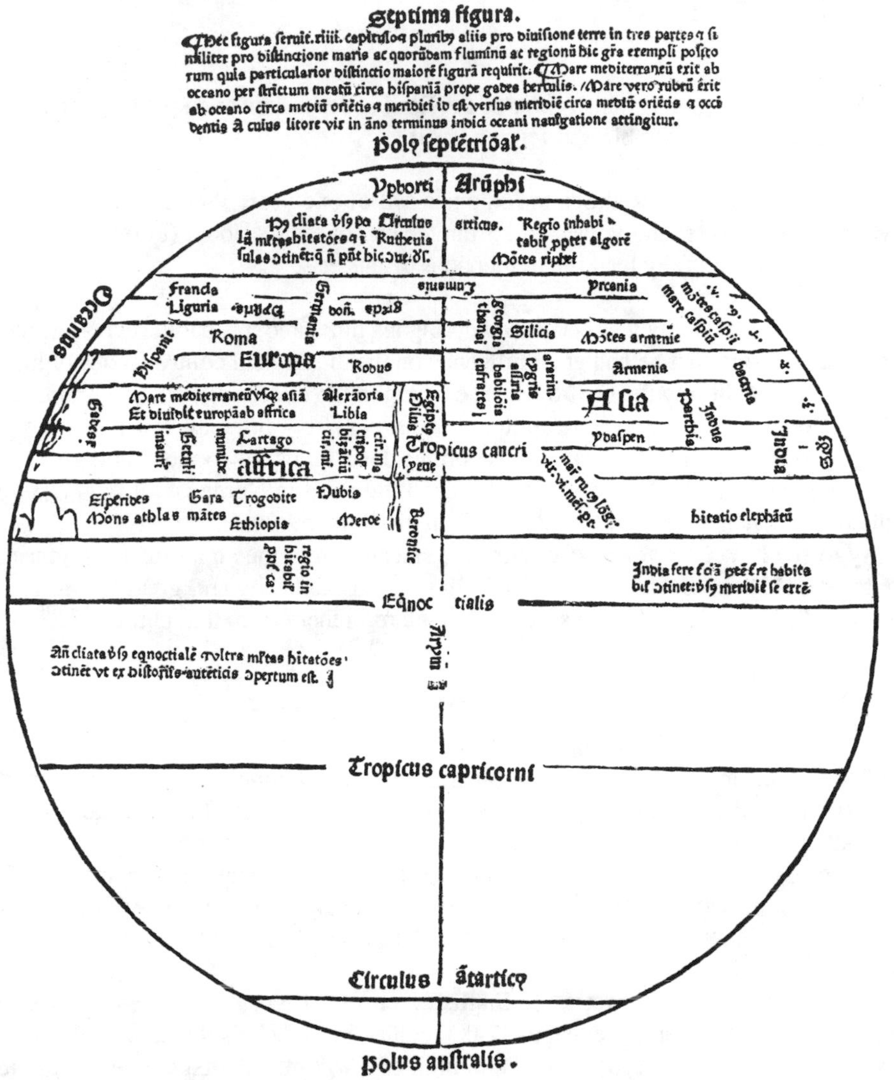

Hec figura feruit. riiii. capituloq plarib aliis pro diuifione terre in tres partes q fi
militer pro biftinctione maris ac quorûdam fluminû ac regionû hic gra erempli pofito
rum quia particularior biftinctio maiorê figurâ requirit. Mare mediterraneû erit ab
oceano per ftrictum meatû circa hifpaniâ prope gades herculis. Mare vero rubrû erit
ab oceano circa mediû oriêtis q meridiei id eft verfus meridiê circa mediû oriêtis q occi
dentis A cuius litore vir in âno terminus indici oceani naufgatione attingitur.

Poly feptêtriôal.

Yphorti Aruphi

Europa

Afia

affrica

Tropicus cancri

Eqnoc tialis

Tropicus capricorni

Circulus ftarticq

Polus auftralis.

From A.E. Nordenskiöld's "Facsimile-Atlas", first
published in Sweden, in 1889, republished in English
by Dover Pub's, New York, in 1973. "Imago Mundi" was
published in manuscript form in 1410, but not printed
until c.1483, according to Nordenskiöld. It is reproduced here by permission of Dover Publications, N.Y.

Figure 1: Map of the world from "Ymago Mundi ," by Petrus De Aliaco, c. 1483.
(Original diameter 202 m.m.)

Note that the king of Tarshish came to the Lord of Jerusalem and spent a year and 13 days on the way, as the blessed Jerome has it.

From a harbor of the Red Sea to India is a sail of one year. And Solomon took three years to make the round voyage . . . From the end of the Occident (Portugal) to the end of the Orient (India) by land is a tremendous distance.

Nordenskiold tells us that d'Ailly was celebrated during his time (and, probably several generations after his death) for his learning and that he had great influence on the theological controversies of his age. His writings provide an insight into the controversies existing in the middle ages on geographical questions. D'Ailly, while seldom providing positive opinions of his own, " . . . often allows a doubt to glimpse forth as to the geographical dogmas of the church. He brings doubtful points to view, and gives as guidance to their answer, citations from . . . " the ancients. Figure 1, taken from Nordenskiold's *Facsimile-Atlas*, is d'Ailly's rather rudimentary "Map of the World" from his "Imago Mundi." [14]

It is easy to see how Christopher Columbus, a sincerely religious man (as most mariners of his time were likely to be), (15) and deeply reverent of both religious authority and great learning, could adopt the highly respected d'Ailly's views as he was moved to interpret them from that churchman's works.

ALA-ED-DIN AL KUSGI

Dreyer/Stahl indicates[16] that a report in Shah Cholgii *Astronomica* . . . credits Ala-ed-din Al Kusgi (one of Uleg Begh's astronomers at the Samarkand Observatory during the first half of the 15th century) with the estimate of 8000 parasangs for the circumference of the earth, or 22.22 parasangs for the degree. This is the same as saying 24,000 Arabian miles for the earth's circumference and 66 2/3 miles for the degree, which was Ibn Rustah's 10th century evaluation of the 9th century Arab measurements on the plains of Sinjar. This is also similar to the "By Air" estimates attributed to 7th century Ananias of Shirak by Aubrey Diller. (See Chapter 4.)

As to the absolute value, in metres or English feet, of Al Kusgi's estimate, this is somewhat of a mystery. Assuming the parasang is equal to 12,000 cubits, inasmuch as the cubits which might have been employed vary from a low of 0.3705 metres to a high of 0.798 m. (the Isfahan elle), the resultant earth's circumference could have varied from 35,568 kilometres to 76,608 Kms. A more likely range of variation would be from 42,664 to 47,880 Kms, based upon fingers or digits varying from 0.01852 to 0.02078 m., cubits from 0.4444 to 0.49875 m., and parasangs from 5333 to 5985 metres. However, this is sheer conjecture. In reality, no conclusions can be drawn from the meagre data available.

Columbus probably never heard of Ala-Ed-Din Al Kusgi. Even if he had, he would have ignored this Arab's estimate. He had accepted Alfragan whose views had been endorsed by Cardinal d'Ailly, his oracle. His mind was made up on the size of the earth.

Chapter 7

TOSCANELLI, BEHAIM, AND COLUMBUS ON THE SIZE OF THE EARTH AND THE DISTANCE FROM PORTUGAL TO INDIA

PAOLO TOSCANELLI

Introduction

Paolo dal Pozzo Toscanelli, 1397-1482, who lived and worked in Florence all of his life except for the period he studied medicine at the University of Padua, was a physician, humanist, astronomer, mathematician, and geographer. In his day, many of the artifices of medicine were derived from astrology, thus, it was not surprising to find physicians with a working knowledge of astronomy, the science which supposedly underlaid astrology. He was considered by his contemporaries such as Nicholas of Cusa and Regiomontanus to be one of the most learned living mathematicians[1]. In the field of astronomy, Toscanelli is best known for his observations of comets which appeared in the years 1433, 1449, 1456, 1457, and 1472, and for the gnomon which he attached to the dome of Santa Maria del Fiore in Florence. This gnomon, some 300 feet above ground level, permitted Toscanelli to solve several astronomical problems more accurately than had previously been possible. It also permitted the regulation of ecclesiastical dates.

Toscanelli, however, achieved a measure of fame (after his death) when it beacme known that he had corresponded with Columbus and offered him both advice and encouragement as regards his "project" to sail westward to the Orient from the Iberian peninsula. In Chapter 1, it was pointed out that some writers, Henry Vignaud among them, have questioned the authenticity of the Toscanelli-Columbus correspondence. However most writers on Columbian subjects who have analyzed Vignaud's charges have not been persuaded that the correspondence was fraudulent. That the geographical import of Toscanelli's views as contained in his letter to Fernam Martins-a copy of which had been sent to Columbus-was susceptible to a variety of interpretations should surprise no one who has faithfully-if laboriously-plodded his way through this book.

It is proposed now to analyze the Toscanelli letter so as to assess its geographic significance. In the process, we shall lean heavily on the impressive scholarship of none other than the erstwhile bete noir-Henry Vignaud. There are few books about Columbus which treat, maturely, on the geographical/metrological nuances of the Toscanelli-Martins correspondence. Vignaud, in his *Toscanelli and Columbus, the letter and Chart of Toscanelli*,[2] has done it well - despite his ultimate conclusions as to its authenticity.

Toscanelli's Geographical Concept[3]

The map which accompanied Toscanelli's letter to Fernam Martins, although referred to as a sea chart, was probably not. Sea charts of the day lacked grids of latitude and longitude lines, instead using strategically placed rosettes of loxodrome-lines of value in getting from one port to another. Tucked away in a corner of the chart was a scale relating miles, leagues, and degrees of latitude, or of longitude at the equator. The rosettes and the harbor entrances were spotted with great care as regarded distance and relative bearings between them. Toscanelli's map, which has not survived, is known olnly by what he said about it in the "postscriptum" of the original letter from him to Canon Martins. According to the letter, the map represented the portion of the globe between the China coast and that of Portugal.

Quoting now from the Spanish version of the letter as found in Las Casas' "Historia":[4]

> And from the city of Lisbon straight toward the west, there are on the said map 26 spaces and in each one of them there are 250 miles, to the most noble and great city of Quinsay; this city is 100 miles in circumference, which are 25 leagues, and in it there are ten marble bridges. The name of this city in our language means City of Heaven wonderful things are told of this city in regard to the magnificance of the workmanship and of the revenues [this space is almost the third part of the sphere].[5] It is in the province of Mango near the city of Cathay, in which the King resides most of the time - and near the island Antilia, which you call the Seven Cities, and of which we have knowledge. There are ten spaces to the most noble island of Cipango which are 2500 miles, that is to say 625 leagues . . .

The parenthetical remark " . . . almost the third part of the sphere" is taken by most reviewers to mean "about" rather than "not quite." This lends to the general interpretation that Toscanelli's sphere had 72 (and not 78) "spaces" in the entire globe. These meridional divisions were then 5° each, 72 making 360°. Thus, Toscanelli had indicated that the sea portion of the globe, from Lisbon due westward to Quinsay, had a length of 6500 miles, or 26 x 5° = 130° of longitude[6]. The Eurasian continent, from Lisbon eastward to Kinsay, stretched over some 230° of longitude. Marinus had made it 225° from the Fortunate Islands to the eastern end of the known world. He and Ptolemy both had placed Lisbon about 5° 37' east of the Fortunate Islands. Clearly, Toscanelli had extended the Eurasian continent some 10 1/2° over Marinus' estimate and 55 1/2° over Ptolemy's.

Along the parallel of Lisbon,[7] Toscanelli's earth's circumference would be 72 x 250 miles = 18,000 miles. Vignaud tells us that in Toscanelli's day Lisbon was considered to be placed at the 40th or 41st parallel[8]. Indeed, Map No. 3 of the Ptolemaeus Romae of 1490 shows the mouth of the Tagus River to be at about 40° 30' N. latitude (See Figure 3b, Chapter 5). The cosine of 40° 30' being 0.76, Toscanelli's degree at the equator would have a length of 50 ÷ 0.76 = 65.79 miles and his earth's circumference would be 23,684 miles. While we cannot be sure that Toscanelli placed Lisbon at the latitude indicated in the foregoing, it does appear that he believed Quinsay and Lisbon shared a common parallel (approximately)[9].

With regard to the islands of Antilia and Cipango mentioned in the postscriptum of Toscanelli's letter to Martins, some have interpreted these latitudes to be the same as that of Lisbon and Quinsay. Others, leaning on a comment Toscanelli had made in his letter about the islands from which a westward voyage should be initiated, i.e. the Canaries, believe that Toscanelli had placed Antilia and Cipango somwhere between the 25th and 30th parallels. Toscanelli did not mention how far west of the meridian of Lisbon (or of the Canaries) Antilia lay, but had clearly stated the distances from Antilia to Cipango to be 10 spaces, 2500 miles, or 50 degrees. Since Marco Polo had given the distance from Quinsay (and Zaiton) to Chipangu to be 1500 miles and Toscanelli's venture into cosmography had been inspired by Polo's book, it is logical to assume that Toscanelli accepted Polo on the length of this leg of the sea journey. This, then fixes the distance from Lisbon to the mythical Antilia as 6500 - 1500 - 2500 = 2500 miles, or 50º on the parallel of Lisbon. (Vignaud indicates that a consensus of 15th century Portolani would place Antilia about 40º to 45º west of Lisbon).

Toscanelli's Degree (Various Opinions)

We have seen that if Toscanelli's "spaces" of 250 miles, each, were taken at the parallel of Lisbon and if he took Lisbon's latitude to be about 40º 30' North, then his equatorial degree would measure 65.69 miles. Vignaud reports that the views of several of his contemporaries varied considerably on the length of Toscanelli's degree.

The 62 1/2 mile degree.

This was the standard attributed to Ptolemy by the "exclusionists," those who held that there had never been but one stade in the Greek, and later, Roman world--that which went 8 to the Roman mile. Since Ptolemy had said 500 stades made a degree of latitude, or of longitude at the equator, the length of the degree in Roman (and later Italian) miles = 500 ÷ 8 = 62.5. This had been Macrobius' interpretation of Ptolemy (with which however he did not agree, preferring Eratosthenes' estimate). Most of the learned European geographers of Toscanelli's time accepted this standard. The noted 19th century French geographer, d'Avezac, and his contemporary, the German, S. Ruge, both believed this was Toscanelli's standard. Acceptance of this figure for Toscanelli's equatorial degree would mean that he either put Lisbon at a latitude of 36º 52' 12" or that this latitude represented the mean latitude of the westward voyage.

The 66 2/3 mile degree.

Hermann Wagner believed this was Toscanelli's standard. It will be recalled that there were those who held that Ptolemy had used 500 Philetaerian stades of 197.6 metres to make his degree of 98,800 metres, which is equivalent to 66 2/3 Italian or Roman miles of 1482 metres. Alfragan had pegged Ptolemy's degree at 66 2/3 miles. Further, it appears to have been one of the recognized measures of Toscanellis' day and later. The "Suma" or geographical compendium published by Martin Fernandez de Enciso in 1518 utilizes this measure for an equatorial degree which is also shown to equal 16 2/3 leguas (leagues). The earth's circumference measured precisely 6000 leguas, according to Enciso.

Toscanelli's use of a 66 2/3 mile equatorial degree, would have been compatible with his according to Lisbon a latitude of 41º 24' 36" and putting Antilia, Cipango, and Quinsay on the same parallel.

The 56 2/3 mile degree.

This was Alfragan's degree. Columbus adopted it even though he, as most practical seamen of the time, employed the equatorial degree of 60 miles or 15 leagues.

Whether Columbus thought Toscanelli's degree had a length of 56 2/3 Roman or Italian miles is conjectural, but the odds are for the affirmative view. Certainly, a cosmographer who had outdone all of his predecessors in reducing the number of degrees of longitude between Lisbon, or the Fortunate Islands, and "The Indies" would not be so reactionary as to undo his achievement by the use of outsized degrees, especially in the face of Alfragan's very positive statements.

Further, the 56 2/3 mile degree accorded better with the assumption that Toscanelli really intended that the parallel of the Canaries (28° N.) be utilized for the trans-oceanic voyage, as Columbus intended to do, and did.

Since Vignaud believed the Toscanelli-Martins letter was a fraud, written by Bartolomeo Columbus and expressing his brother Christopher's cosmographical views, he puts the degree of the post-scriptum at 56 2/3 Roman (Italian) miles.

The 60 mile degree.

Despite being the popular choice among 15th century seamen, no geographer has indicated a belief that Toscanelli may have employed it. Such a choice, however, would have accorded with a mean parallel of 33° 33', not unreasonable for a 28° N. Canaries and a 39° N. Lisbon and Quinsay.

The 67 2/3 mile degree.

This is the standard the Italian geographer Uzielli[10] has adopted and which, as Vignaud puts it , has attracted the most attention. It stems from the discovery in 1854 at the Biblioteca Nazionale, Florence, of a manuscript in Toscanelli's hand, the only holograph of Toscanelli's in existence. At first, the manuscript was hailed for the observations, made some four centuries earlier, of the comets of 1433, 1449, 1456 (Halley's Comet), 1457 and 1472. However, there were other parts of the manuscript which were soon perceived to have great interest for geographers and metrologists.

The manuscript contained a table of latitudes and longitudes for many well-known places and the framework, in two sheets--clearly intended to be joined, of a map of the northern hemisphere of the earth. The right-hand vertical border of one sheet and the left vertical border of the other, marking latitude, are divided into 18 spaces, each having five subdivisions, each marking one degree. The lower and upper borders, marking longitude are each divided into 36 spaces, each having five sub-divisions, each marking a degree. Thus, the two sheets when joined would have 72 spaces of longitude, each of 5° , making 360° degrees and 18 spaces of latitude, each of 5° degrees, making 90° and stretching from the equator to the North Pole. After careful authentication, this seemed to satisfy most lingering doubts as to Toscanelli's intent in his references to "spaces" in his letter to Canon Martins.

There was, however, another item of interest in the manuscript. On the upper right-hand margin of one page is a note, again in Toscanelli's handwriting, which formed the basis for a memoire by Uzielli, on Toscanelli's concept of the size of the earth. The note, written in Latin, follows, "Gradus continet. 68. miliarum minus. 3ª unius. Milarium tria milia brachia. Brachium duos palmos. Palmus. .12. uncias .7. filos."

Thatcher translates this as,"A degree contains 67 2/3 (sixty-eight less one third) miles. A mile contains 3000 brachia. A brachium contains two palms. A palm contains 12 inches and 7 fili.[11]

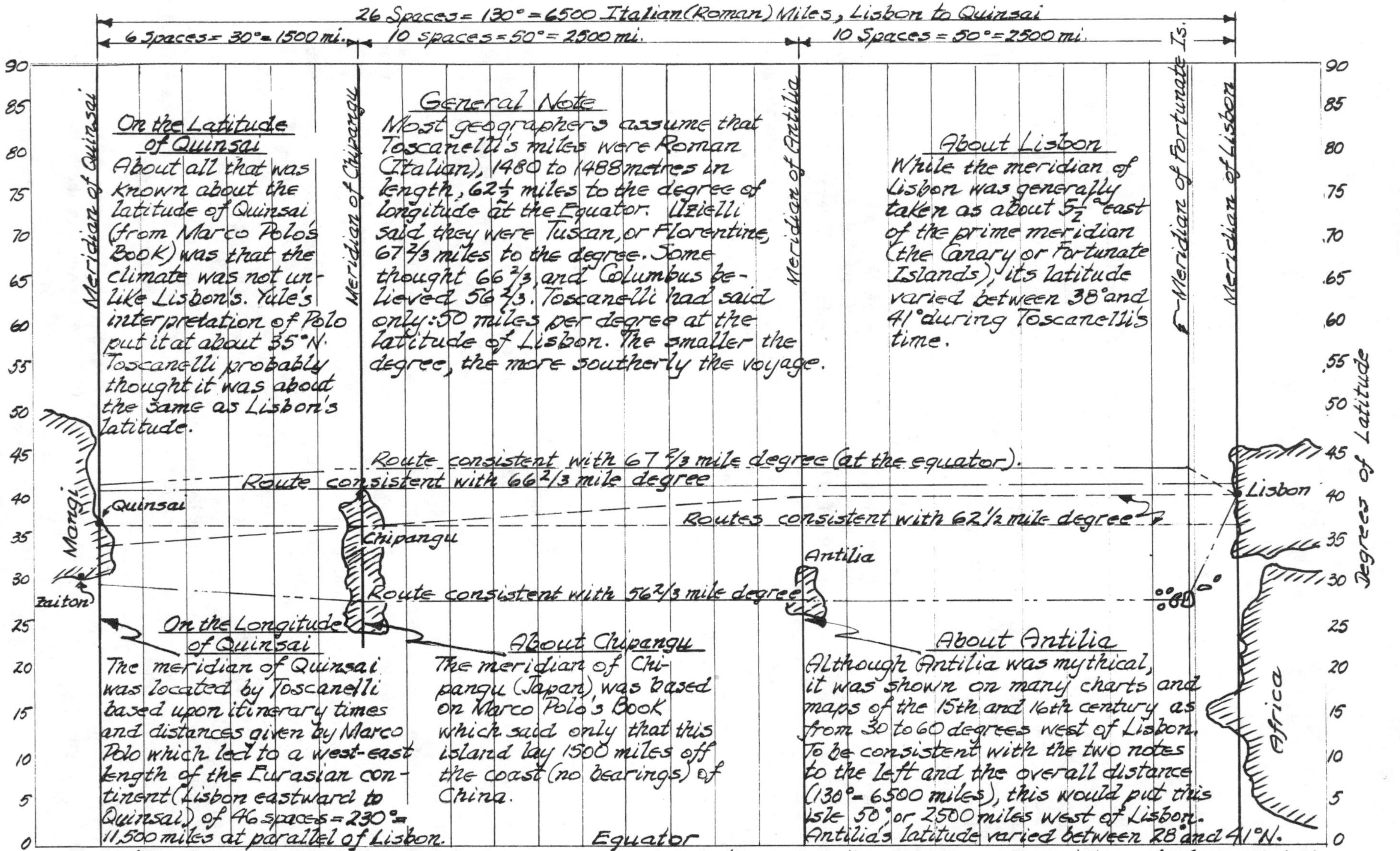

Degrees of Latitude

90 85 80 75 70 65 60 55 50 45 40 35 30 25 20 15 10 5 0

26 Spaces (30° = 6500 Italian (Roman) Miles, Lisbon to Quinsai

10 spaces = 50° = 2500 mi.   6 Spaces = 30° = 1500 mi.

Meridian of Lisbon

Meridian of Fortunate Is.

Meridian of Antilia

Meridian of Chipangu

Meridian of Quinsai

Africa

Lisbon

Antilia

Chipangu

Quinsai

Mangi

Zaiton

Equator

About Lisbon

While the meridian of Lisbon was generally taken as about 5° east of the prime meridian (the Canary or Fortunate Islands) its latitude varied between 38 and 41° during Toscanelli's time.

General Note

Most geographers assume that Toscanelli's miles were Roman (Italian) 1480 to 1488 metres in length, 62⅔ miles to the degree of longitude at the equator. Uzielli said they were Tuscan, or Florentine, 67⅔ miles to the degree. Some thought 66⅔ and Columbus believed 56⅔. Toscanelli had said only 50 miles per degree at the latitude of Lisbon. The smaller the degree, the more southerly the voyage.

On the Latitude of Quinsai

About all that was known about the latitude of Quinsai (from Marco Polo's Book) was that the climate was not unlike Lisbon's. Yule's interpretation of Polo put it at about 35°N. Toscanelli probably thought it was about the same as Lisbon's latitude.

On the Longitude of Quinsai

The meridian of Quinsai was located by Toscanelli based upon itinerary times and distances given by Marco Polo which led to a west-east length of the Eurasian continent (Lisbon eastward to Quinsai) of 46 spaces = 230° = 11,500 miles at parallel of Lisbon.

About Antilia

Although Antilia was mythical, it was shown on many charts and maps of the 15th and 16th century as from 30 to 60 degrees west of Lisbon. To be consistent with the two notes to the left and the overall distance (130° = 6500 miles), this would put this isle 50° or 2500 miles west of Lisbon. Antilia's latitude varied between 28 and 41°N.

About Chipangu

The meridian of Chipangu (Japan) was based on Marco Polo's Book which said only that this island lay 1500 miles off the coast (no bearings) of China.

Route consistent with 67⅔ mile degree (at the equator).

Route consistent with 66⅔ mile degree

Route consistent with 62½ mile degree

Route consistent with 56⅔ mile degree

Figure 1: A diagrammatic representation of the information contained in the post-scriptum of a letter dated 25 June 1474 sent by the Italian astronomer/physician Paolo Toscanelli to Canon Ferram Martins, a confidante of the Portuguese monarch, Alonso V. The letter pertained to the feasibility of a voyage westward from Lisbon to the Indies. The geographic information provided by Toscanelli was very limited and contained in the post-scriptum. This, paraphrased, said: From Lisbon to Quinsai, in the province of Mangi, there are ... 26 spaces, each space comprising 250 miles, or 6500 miles in all, making about one-third the circumference of the globe. But from the isle of Antilia to the famous isle of Cipangu, there are only 10 spaces (2500 miles). See notes above.

In evaluating the note, Thatcher tells us, "A brachium is estimated to contain 0.550637 metres. A mile or 3000 brachia would contain 1651.911 metres . . . (and) . . . a degree . . . at the equator measured 67 2/3 Tuscan miles . . . "

Thus, the degree at the equator also measured 111,779.32 metres and the circumference of a spherical earth 40,240.55 kilometres. These figures are remarkably accurate, being within 0.6 of one percent of the 111,111.11 metre degree and 40,000 kilometer circumference we have been using for a spherical earth and within 0.44 of one percent of the actual equatorial circumference of the earth.

Vignaud, in reporting the foregoing[12], indicates that Uzielli evaluated Toscanelli's degrees as 67 2/3 Florentine miles = 75 3/5 Roman miles = 605 stadia of 185 metres = 111,925 metres. Thus, Uzielli's brachium is 0.551352 metres in length, essentially the same as Thatcher's.

Uzielli, in his memoire, indicated that it was his belief that Toscanelli was referring to Tuscan or Florentine miles when he assigned 50 miles to a degree of longitude at the parallel of Lisbon. Inasmuch as this would have required an assumption that Lisbon's latitude was about 42° 22' [arc cos (50 ÷ 67 2/3) = 42.36° = 42° 22'] and that the entire westward voyage would be at that approximate latitude, few geographers of the late 19th century, or since, have endorsed this view. And, as several have remarked, to have provided the monarch of Portugal, the leading seafaring nation of Europe with a chart in which degrees were equated to a purely local measure unknown to Portugal, instead of the sea mile of the era [13], the Roman mile, would have been uncharacteristic of Toscanelli.

There is another, more tangible problem with the Uzielli/Thatcher Tuscan, or Florentine mile--it does not meet Toscanelli's own specifications.

The Tuscan (Florentine) Mile

We have already given some treatment to the brachium[14] and shown that as it varied in length throughout Italy, so did the Italian mile. In an effort to check the estimate for the value of the brachium used by both Thatcher and Uzielli (which could have been determined by working backward from the desired result of a 75.6 Roman mile degree equivalent to 67 2/3 Tuscan miles), we will revert to the basic Toscanelli note.

Toscanelli defines the brachium as being equal to two palmos and the palmus as equalling 12 uncias and 7 fili. Thatcher tells us that while he is not sure about the filium, " . . . it may have been the twenty-fourth part of an inch." ((Logic would point to a value for the filium of no greater than 1/8 uncia (inch) so that Thatcher's suggestion seems not unreasonable)).

It is clear, from Toscanelli's specification, that the palmus he was using was longer than a Roman foot, for - by definition - the Roman uncia was 1/12th the Roman foot. As explained in Chapter 2, however, there were a variety of feet in use in Italy-pre-Roman, Roman, and post-Roman. Table 4, Chapter 2, lists--besides the official Roman foot of about 0.296 m.--the Olympic foot of 0.309 m., the Babylonian foot of 0.316 m., and the Drusian foot of about 0.342 m. Moreover, Kennelly has described a Bolognese piede (foot) of 0.3801 m. which may have been used during Toscanelli's day. The shortest foot which appears to have been used in Italy, proper, was the Roman foot of 0.296 m. Therefore, Toscanelli's palmus had to be 7 fili larger than 0.296m., as a minimum, but could have been as large as 0.3801 m. plus 7 fili.

If one accepts the Thatcher suggestion that a filium may have equalled 1/24th uncia (inch) = 1/288th foot, then the palmus equalled 1 foot + 7/288ths foot = 1.0243 feet. From this, the following unit lengths would result:

1 brachium = 2 palmi = 2 x 1.0243 x (0.296 m. to 0.3801m.) = 0.6064m. to 0.7787 m.
1 Tuscan (Florentine) mile = 3000 brachia = 1819.2 metres to 2336 metres
1 degree = 67 2/3 Tuscan miles = 123,099 to 158,071 metres.
This, of course, differs from the figures Uzielli claims for Toscanelli.

Brachia, in Italy, have been known to vary considerably from as small as 0.3476 m. to as large as 0.7443 m[15]. Smyth reported 8 braccios in use in Italy (although none was for Florence) averaging 0.63675 metres in length[16]. Kennelly describes a Florentine braccio equal to 2 palmi or 0.583 metre[17]. Morison reports a Genoa braccio in Columbus' day " . . . was equivalent to 22.9 inches," or 0.58166 metres[18], virtually the same as the Florentine braccio described by Kennelly. The Florentine braccio is only about 6% greater than the Uzielli/Thatcher brachium. However, neither the Kennelly Florentine nor the Uzielli/Thatcher brachia meet Toscanelli's specification relating the brachium via the palmos to the uncia and filium.

One is struck by the similarity of Toscanelli's unit relationships with those cited by Vitruvius, Ananias of Shirak ("according to the Persians"), and others who set the mile equal to 1000 paces of 6 feet, or 6000 feet, each foot being equal to 16 digits or fingers, or 12 uncia. In Toscanelli's case, the mile equalled 3000 brachia or 6000 palmi, each palmus equal to 12 uncias + 7 fili. Clearly, the palmus was the counterpart of the foot, being 7 fili larger. Thatcher has suggested 1/24th uncia for the width of the filium, while this reviewer suggests 1/12th uncia, based on the French premetric relationship of the ligne, pouce, pied, and toise (864 lignes = 72 pouces = 6 pieds = 1 toise[19].) The toise was the counterpart of the 6 foot pace, or 6 palmi in Toscanelli's relationship. The pied equalled 12 pouces and the pouce 12 lignes.

The length of the palmus using Thatcher's ratio of 1 uncia equals 24 fili would be 12 7/24 unciae; in this reviewer's, 12 7/12 unciae. Employing the uncia of 1/12th the Roman foot of 0.2963 metre, or 0.02479 m, and Thatcher's filium, the palmus has a length of 0.3035 metre. Using the same uncia and this reviewer's filium leads to a palmus with a length of 0.3107 metre, only slightly longer than the Olympic foot of 0.30864 metre, 6000 of which made the Miglio Geografico of 1851.84 metres.

There is a distinct possibility that Toscanelli, by the time he entered the note in the margin of his manuscript on comets, was using a mile rather close in length to the Miglio Geografico. But, if he was, he was off in his assessment that 67 2/3 such miles made the degree.

On the Miglio Geografico

While doing some field research for his book *Vestiges of Pre-Metric Weights and Measures,* Kennelly found some very old stone distance markers along certain roads of the Piedmont in northern Italy in which the numerical distance to the next town was preceded by the letters "M.G." Inquiry disclosed they represented "Miglio Geografici," or geographic miles. Utilizing the odometer of his car, graduated in kilometres, he checked the stated distances on the stone markers and found the miglio grafico measured 1.853 kilometres. A reference to Bemporad's Pocket Encyclopedia revealed that the official length of the "Miglio Geografico Italiano" was 1851.85 metres. This source also indicated there had been a "Miglio Romano," or medieval Roman mile of 1489 metres, as distinguished from the old Roman mile, or "milliarum" of 1481.5 metres.

As was pointed out in Chapter 2, 5000 Roman feet, or 80,000 Roman digits, made the Roman mile and 6250 Roman feet, equal to 6000 Greek Olympic feet and 100,000 Roman digits, made the Miglio Geografico, of 1851.85 metres--equal to one minute of latitude, or of longitude at the Equator. There can be little question as to the antiquity of the 1851.85 metre mile. Whether it was realized in ancient times,

be little question as to the antiquity of the 1851.85 metre mile. Whether it was realized in ancient times, or at all before the 17th century (when the French "honed-in" on the correct length of a degree[20]) that this mile had geodetic significance, is seriously questioned. The road markers Kennelly saw in Northern Italy probably date from no earlier than the late 17th, or early 18th, century.

Summary of Toscanelli's Cosmographical Ideas

It is difficult to reconcile the only two pieces of evidence we have as to his views on the size of the earth. The information contained in the post-scriptum of the letter to Canon Fernam Martins, written in mid-1474, has been interpreted by most scholars who have researched the matter, and who believe the letter to be authentic, to be about as follows: the minimum estimate of Toscanelli's degree was 62.5 Roman miles or 92,625 metres, leading to an earth's equatorial circumference of 22,500 Roman mile or 33,345 kilometres; the maximum estimate is 66 2/3 Roman miles or 98,800 metres for the degree, and 24,000 miles or 35,568 kilometres for the earth's circumference. Vignaud, who believes the letter to be a fraud written by Bartolomeo Columbus, sees in it Christopher Columbus' cosmographical views--a degree measures 56 2/3 Roman miles and the earth's equatorial circumference 20,400 miles or 30,233 kilometres.

Uzielli, at first, maintained the postscriptum and the note in the margin of Toscanelli's manuscript on comets were mutually supportive. Later, he abandoned this position, but continued to insist that Toscanelli's (evidently later) note in the margin of the manuscript merited support. Thus he evaluates Toscanelli's degree as 67 2/3 Florentine miles = 75 3/5 Roman miles = 111,925 metres. This leads to an earth's circumference of 24,360 Florentine miles, 27,216 Roman miles or 40,293 kilometres. Most reviewers do not take Uzielli's memoire too seriously and either ridicule its thesis or ignore it. Vignaud believes that 67 2/3 miles per degree were Roman, not Florentine. This view cannot be supported for the same reason that Uzielli's evaluation of Toscanelli's Florentine mile cannot be supported--failure to meet elements of Toscanelli's own specification.

This reviewer does not concur with Uzielli in the belief that Toscanelli had estimated the size of the earth almost precisely.

Our own evaluation of Toscanelli's note in the margin of his manuscript on comets follows:

> 1 filium = 1/12th uncia = 1/144th Roman pes of 296.3 mm = 2.06 mm.
> 1 uncia = 12 fili = 1/12th Roman pes = 24.69 mm.
> 1 palmus = 12 unciae + 7 fili = 310.7 mm.
> 1 brachium = 2 palmi = 621.4 mm. = 0.6214 metre
> 1 mile = 3000 brachia = 1864.2 metres
> 1 degree = 67 2/3 miles = 126,145 metres
> earth's circumference = 45,412 kilometres

Text continues on page 176

Table 1: Various Estimates as to the Length of Toscanelli's Equatorial Degree and Earth's Circumference.

Authority	Estimate Based on		Length of Degree of Longitude at Equator in		Earth's Equatorial Circumference in	
	Ltr., Tosc. to Martins	Note in MS on Comets	Roman Miles	Metres	Roman Miles	Kilometres
D'Avezac and Ruge	Mean latitude Lisbon-Quinsay voyage taken as 36°52'12"N.		62.5	92,625	22,500	33,345
Wagner	Mean lat. L-Q voyage taken as 41°24'36"N.		66⅔	98,800	24,000	35,568
Columbus and Gignaud	Mean lat. L-Q voyage taken as 28°N.		56⅔	83,980	20,400	30,233
Vignaud		Belief that Toscanelli was using Roman miles	67⅔	100,147	24,360	36,053
Uzielli and Hatcher		Assumption that Tosc. used brachium of 0.551352m.	67⅔ Florentine miles = 75 3/5 Roman miles	111,925	24,360 Florentine miles = 27,216 Roman miles	40,293
This reviewer		Assumptions Roman pes=296.3mm filium=1/144 pes=2.06mm uncia=1/12 pes=24.69mm brachium=621.4mm Flor. mile = 1864.2 m.	67⅔ Florentine miles = 85.146 Roman miles	126,145	24,360 Florentine miles = 30,653 Roman miles	45,412

Table 1

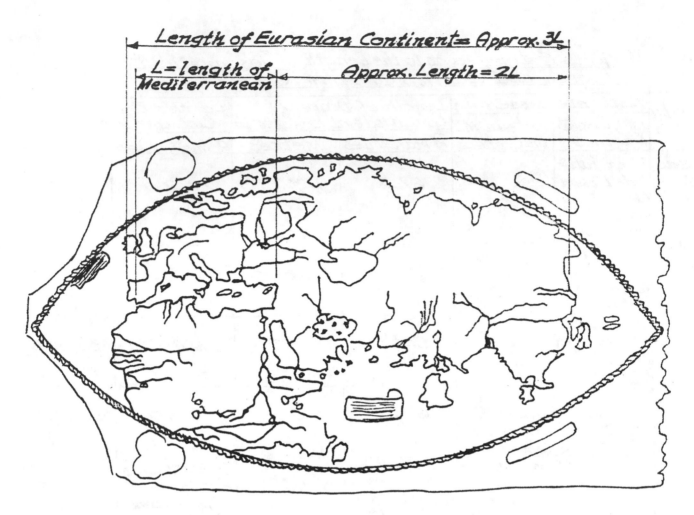

Length of Eurasian Continent= Approx. 3L

L= length of Mediterranean

Approx. Length = 2L

Figure 2a: Toscanelli's Map?

An outline of a world map in the possession of the Firenze Biblioteca Nazionale (Florence, Italy) which has been photographed and reproduced by many publications. Some claim that it is the work of the Florentine scholar Paolo Toscanelli. Others call it simply the "Genoa Map". Either way, it does not represent Toscanelli's geographic views as described in this chapter, which accords the Eurasian continent a length of about 230°. This map supports the Ptolemaic concept which made the length of the Mediterranean about 60° and the Eurasian continent (the "known world") about 3 times that length, or 180° of longitude.

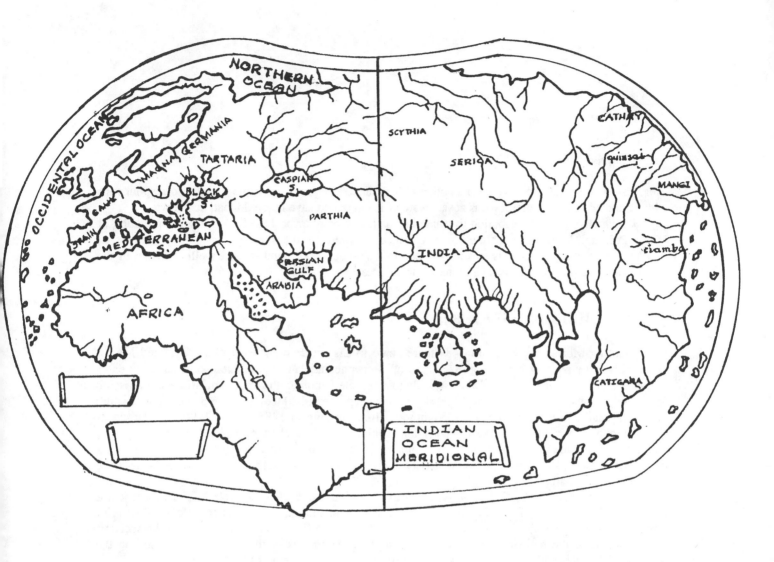

Figure 2b: An outline of a Ptolemaic Map of 1489-90 attributed to the German Cartographer, Henricus Martellius. Portugese exploration in Africa and Marco Polo's travels in east Asia and coastal China are reflected.

Extended comment and credits for Figure 2b,

Reflected in Martellus' map - otherwise Ptolemaic in appearance and proportioning - are the discoveries of the Portuguese in Africa, including Bartolomeu Dias' doubling of the Cape of Good Hope, and the reports of Marco Polo concerning eastern Asia and in particular, its "Atlantic" coastline. Not shown here, but shown on another map attributed to Martellus in the possession of the Yale University Library, is Marco Polo's Cipangu (Japan). In the absence of meridian lines, it is not possible to be precise in gauging Martellus' longitutinal extension of the Eurasian continent, but compared to the Ptolemy map shown in Figure 5, Chapter 3, which shows 178° for this dimension, Martellus appears to be indicating a length of about 255 to 265°

Outline shown is based upon a photo by Walter W. Roberts, Aldus Books, of the original map in the possession of the Royal Geographioc Society, London. The photo appeared in color on pages 10 and 11 of "The Seafarers, The Explorers" by Richard Humble and the editors of Time Life Books, Alexandria, VA, 1978. See also pages 16 and 17 of Bjorn Landstrom's "Columbus", Macmillan Co., New York, 1966, for a facsimile and photograph of a version of Martellus showing Cipangu, and in which the known world has a length of 280° of longitude.

Text continued from page 172

This, however, appears to have been a notion of Toscanelli's acquired subsequent to the correspondence with Canon Martins. Columbus was never aware of it and would have ignored it had he known.

Toscanelli's great contribution to mankind was his colossal error, based upon his interpretation of Marco Polo and Marinus, in estimating the distance westward from Lisbon to Kinsai at 6500 miles, or 130 degrees of longitude. This was a God-send to Columbus who found in Toscanelli, a savant of the highest rank, vindication of his own cosmographical ideas.

MARTIN BEHAIM (1459-1506)

Quoting Morison[21] " . . . In 1484 there came to Lisbon a young Nuremberger named Martin Behaim, who by passing himself off as a pupil of the mathematician Regiomontanus managed to enter the most learned and courtly circles. D. Joao appointed him to the royal maritime commission (apparently he was absent when Columbus' scheme came up); he visited the Azores and married a daughter of the captain of Fayal, and received knighthood from the king in 1485 . . . in 1490 he returned to Nuremberg and there in 1492 constructed his famous globe . . . The scale, the eastward extension of Asia, and the narrow ocean in this globe are so similar to the false geographical notions on which Columbus based his voyage, as to suggest Columbus and Behaim were collaborators. But there is no positive evidence of their trails ever crossing."

Vignaud[22] comments, "It has become almost a religious belief that the globe of Behaim is just a duplicate of the so-called Toscanelli map . . . it is evident this map . . . and the globe of Behaim represented the ideas which the cosmographers of the close of the fifteenth century had formed respecting the smallness of the maritime space separating the two extremities of the ancient world, and on the position of the isles and countries that were to be found at the remoter part of that space . . . "

"Behaim's globe and the so-called Toscanelli map are cartographical documents closely related, inasmuch as they are the expression of the same conception and cosmographical ideas . . . Columbus and his brother Bartholomew must certainly have known Behaim's works, for Behaim dwelt in Portugal at the same time as they, took an interest in the same occupations, was like Columbus in personal relations with the king, and, if we are to believe Herrera, was his personal friend. Under such circumstances it is almost certain Columbus must have been acquainted with the cosmographical ideas of Behaim . . . "

" . . . The one important item common to both documents is the reduction to 130 degrees of the maritime space stretching from the West away to Asia. But the idea of thus reducing this space was special neither to the author of the letter nor the constructor of the globe; it came . . . from Marinus of Tyre; and Ptolemy, through whom it was known, had been printed six or seven times before the construction of the globe. Cardinal d'Ailly had also mentioned it."

Vignaud must be considered an authority on the Behaim globe inasmuch as he "was commissioned, in 1892, to get a reproduction of the Paris copy of this globe for the National Museum of the United States . . . (and) . . . had to make a study of all existing copies and reproductions . . ."

He describes it as follows: " . . . This globe has a diameter of 530 millimeters, is constructed of papier-mache, covered with a coat of plaster, which again is covered with vellum on which are depicted the coasts and legends in ink, colour, and gold . . . .It is mounted on an iron stand, with a movable meridian and a brass horizon which have been subsequently added. It is not graduated. A single meridian, the equator, the Arctic Circle, the two tropics, and the ecliptic are the only circles represented."

176

The last specification, pertaining to the ecliptic, explains the cryptic convex circle on the Morison/Greene gore map of Behaim's globe, Figure 4. Vignaud also indicates that the original Behaim globe did not show Quinsay and Zaiton which do appear on the gore map, with Quinsay somewhat misplaced (see Figure 2a, Chapter 5.)

Johan Gabriel Doppelmayer, in 1730, published a planispheric reproduction of the Behaim globe in two hemispheres. It was part of a work covering the accomplishments of great Nurnbergers of the past. This map, taken from Nordenskiold's *Facsimilie-Atlas*, is shown in Figure 3 [23].

Although the original globe had neither a grid of latitude and longitude lines nor a scale relating distances on the map to miles or leagues, both Doppelmayer and Morison/Greene have obliged their readers by providing latitude and longitude scales, conveniently located. Thus, distances along parallels and meridians in degrees may be determined by scaling. To convert to miles or leagues, it is necessary to make an assumption as to the length of Behaim's equatorial degree.

Morison assumed Behaim's length of a degree to be the same as Columbus'. Vignaud's account certainly does not rule out this possibility and both von Humboldt and Nordenskiold, who are evident admirers of Behaim and provide much data on his life and other accomplishments, are silent on this matter. A clue is, however, provided by Nordenskiold's statement [24]: "On a closer examination of the drawings and legends on Behaim's globe we shall find it to be based first on Ptolemy's atlas; and on the narratives of the travels of Marco Polo and other medieval travellers in Asia; third on the Portuguese voyages of discovery; and fourth on the map of the northern countries of Europe in the Ptolemaeus Ulmae 1482." The strong Ptolemaic influence points toward adherence to the 62-1/2 mile, 500 stade degree.

On the foregoing basis, Behaim's degree of longitude at a latitude of 36° has a length of 50.56 Italian miles, quite close to Toscanelli's 50 miles at the un-named latitude at which his voyage from Lisbon to Kinsay would have been conducted. At 28° N., the latitude in which he places Grand Canaria and the northern tip of Cipangu, a degree of longitude would have a length of 55.184 Italian miles, or 81.78 kilometres.

In attempting to scale Dopplemayer's planispheric representation of Behaim's globe, so as to establish distances between important geographic features, it was found expedient to construct a limited grid based on that which Dopplemayer must inevitably have used in his reproduction. Thus, meridians are circular, pole-to-pole, and spaced evenly at 10 degrees. Only two parallels have been added to Doppelmayer's equator, tropics of Cancer and Capricorn, Artic and Antarctic circles. These are the parallels of 28° N. and 42° N., and are drawn so as to intersect each meridian 28/90 and 42/90 respectively of the distance along the meridian from equator to pole.

Text continued on page 181.

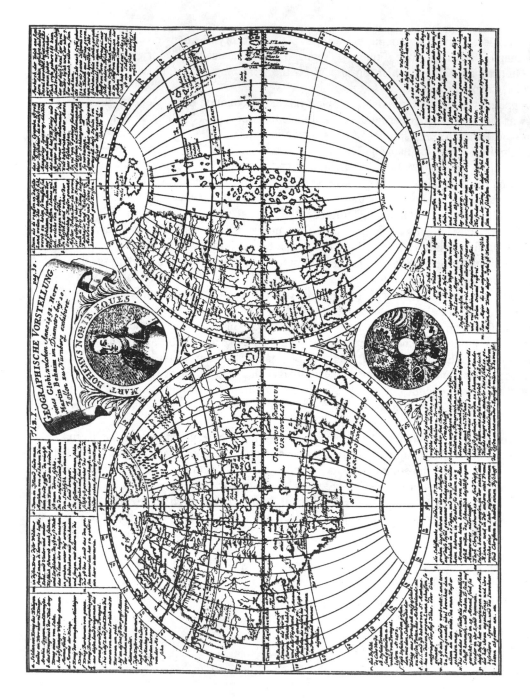

Figure 3 Martin Behaim's globe of 1492 from J.G. Doppelmayer. (Size of his drawing 414 x 302 m.m.)

Superimposed on this Doppelmayer (1730) reproduction of Behaim's globe is a limited grid consisting of meridians on a 10° spacing and the parallels of Lisbon (42°) and the northern shores of Grand Canaria and Cipangu (28°). It can be seen that Behaim considered a) the meridian of Grand Canaria to be about 10° west of that of Lisbon; b) the meridian of Cipangu to be about 88-90° west of Grand Canaria; and c) the meridian of Quinsai (Kinsay) – which is assumed to be on the coast of "India" (China) somewhere – between the 28th and 42nd degree, north, parallels – to be about 30° west of that of Cipangu. He made the meridian of Quinsai a total of 128° west of that of Lisbon (other reproductions have it at 130° west of Lisbon) and about 118° west of Grand Canaria.

This reproduction does not show the (mythical) island of Antilia. That shown in Jomard's Atlas ("Monuments de la Geographie", Paris, 1854) shows it in the same position as St. Brandon's Island on this map. Vignaud says the original globe placed Antilia 50° west of Lisbon and at about the Tropic of Capricorn (see figure 4). It is located variously in other reproductions of Behaim and on portolani of that era, as is the precise location of Cipangu (Japan). However, the Lisbon-Quinsay and Canaries-Quinsay distances are about as given in the foregoing.

Credits and Sources

The basic reproduction is taken from A.E. Nordenskiöld's "Facsimile-Atlas", pp. 72 a&b; Dover Publications, N.Y., 1973; first published in Stockholm in 1889. See also same publication, pp. 65 b, 74 a, 100 b, and 101 a; Morison's "Admiral of the Ocean Sea", pp. 76-8 and illustration 66-8, Little, Brown & Co., Boston, 1942; and Vignaud's "Toscanelli and Columbus...", pp. 174, 175, 180, note 172, pp. 191, 204, 206, and note 205, Books for Libraries Press, Freeport, N.Y., 1971, first published in 1902.

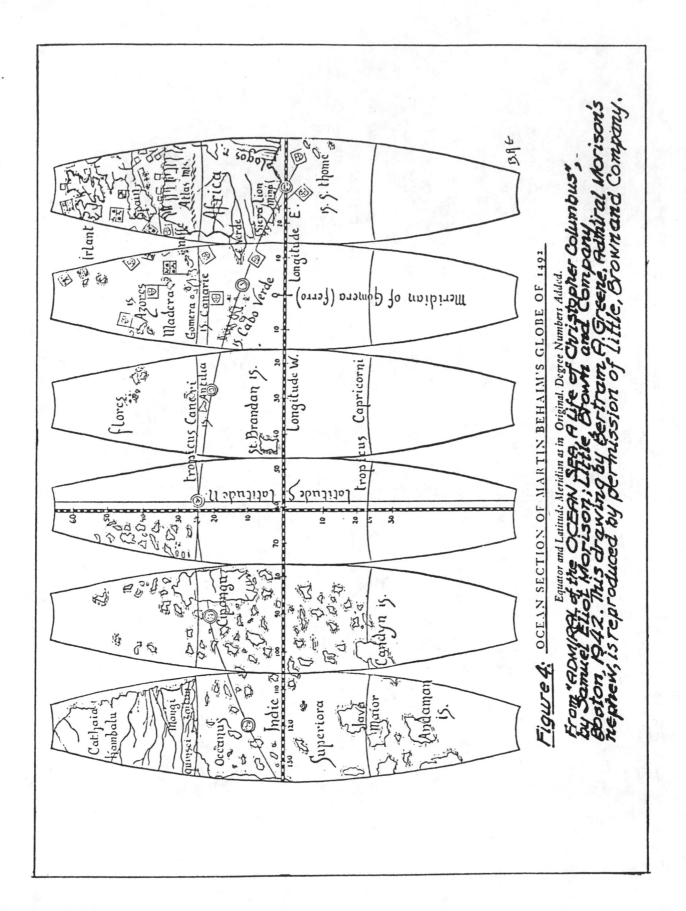

Figure 4: OCEAN SECTION OF MARTIN BEHAIM'S GLOBE OF 1492

Equator and Latitude Meridian as in Original. Degree Numbers Added.

From "ADMIRAL of the OCEAN SEA, A Life of Christopher Columbus", by Samuel Eliot Morison; Little, Brown and Company, Boston, 1942. This drawing by Bertram A. Greene, Admiral Morison's nephew, is reproduced by permission of Little, Brown and Company.

Text continued from page 177.

Behaim evidently believed the meridian of Grand Canaria to be about 10° west of that of Lisbon which appears to be located close to the 42° N. parallel. The northern tip of Cipangu (Japan) is placed 88°-90° west of Grand Canaria. While he does not designate Kinsay (Hangchau), he puts the coast of China, between the 28° and 42° N. parallels, at about 30° west of the northern tip of Cipangu, about 118° west of Grand Canaria, and about 128° West of Lisbon [25]. Antilia, the mythical isle which had managed to find a somewhat variable spot on many portolani of Behaim's day, is not shown on Dopplemayer's reproduction although it was, on the authority of Vignaud[26,] shown on the original Behaim globe at about 50° west of Lisbon.

The Behaim globe is considered to be one of the most valuable testimonials to the geographical thinking of those mapmakers of the end of the 15th century who did not dismiss Marco Polo, who tried to keep abreast of the latest discoveries in the Atlantic, the current gossip of the mariners, and who were quite aware of Ptolemy, and through him, Marinus.

CHRISTOPHER COLUMBUS

We come now to Columbus. He, of course, could not know about the note in the margin of the Toscanelli manuscript on comets because it would not be discovered until 1854. However, even if he had known of its contents, he could be depended upon to ignore it. Clearly, it was at odds with the data in the postscripture of the Toscanelli letter of 1474 to Canon Martins which he heartily endorsed. This information seemed to match almost perfectly Alfragan's 56 2/3 mile degree, for 50 miles per degree of longitude along the 28th parallel (that of the Canaries) was almost precisely equivalent to 56 2/3 miles per degree at the equator. This was even more favorable than the scales he was used to--4 miles per league, 15 leagues (or 60 miles) to the degree of latitude, or of longitude at the equator. And the miles--Vignaud, Nordenskiold, and Morison agree--were Roman (Italian) miles of 1480-1482 metres, defined at the time in terms of palms, braccia, feet, or stadia, depending on the occupation and location of the definer.

It must be admitted we do not know precisely how Columbus defined the composition of the sea mile he used. Earlier, we cited Yules' identification of the Genoa palm of 9.725 English inches or 0.247 metre. This could be the "building block" of a Roman (Italian) mile of 1482 metres if we used Toscanelli's stated relationship of 2 palms per brachium and 3000 brachia per mile. However, Morison cites[18] at least one instance in which Columbus estimated the height of a wave in "braccia," thus indicating it was a measurement in use in Genoa in the 1470's. In a footnote, Morison then says "A Genoese braccio was equivalent to 22.9 inches." 22.9 English inches is the equivalent of 0.58166 metre. Since the palm/braccio relationship seems to have been reasonably steady and widespread in Italy, two palms equalling one braccio, it is evident there must have been a time lapse between Yule's small palm of 0.247 metre and Morison's large one of 0.29083. 3000 Morison/Genoese braccia would make a rather large mile of 1745 metres. We doubt that Columbus or other seamen of his day used a mile this large.

The inference in both the Yule and Morison instances cited is that the lengths for the units defined were those in vogue in Genoa at the times of Marco Polo, in one case (c. 1300) and Christopher Columbus in the other (c. 1477). Such is not expressly stated, would have been difficult to determine with certainty, and would have required a greater sensitivity about the subject of metrology than either source has demonstrated.

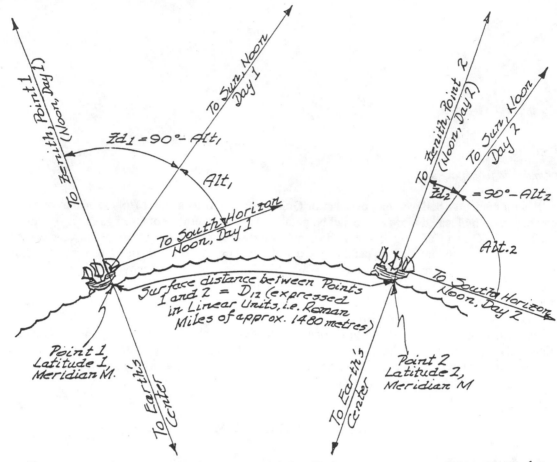

Note: Zd_1, Zd_2, Alt_1, and Alt_2 are zenith distances and altitudes of Sun at Noon on days 1 and 2. Ship is assumed to be proceeding due South, so that Points 1 and 2 are on same meridian, a distance D_{12} apart.

$$\text{Latitude 1} - \text{Latitude 2} = (Zd_1 + \text{declination}_1) - (Zd_2 + \text{decl.}_2) \quad \} \text{ all in degrees}$$
$$= (Alt_2 - \text{decl.}_2) - (Alt_1 - \text{decl.}_1) \quad \}$$

D_{12} is an estimate by ship's pilot of distance travelled between noon on Days 1 and 2, expressed in Roman Miles

$$\underline{\text{Length of 1 degree of Latitude}} = \frac{D_{12}}{(Alt_2 - \text{decl.}_2) - (Alt_1 - \text{decl.}_1)}$$

$= 56\frac{2}{3}$ Roman Miles, according to Columbus

$= 75$ Roman Miles (approx. 111,111 metres), actually

Figure 5: A Rationalization of Columbus' "Measurement" of the Length of a Degree of Latitude at Sea En-Route to Guinea (1482-1484).

Comment

The sun's declination (zenith distance or angle at the equator) is constantly changing as is the declination at noon - from day to day. The latter varies from 24 minutes per day on the days near the equinocti to only about one minute per day on the days near the solstices. The end points of the measurements may have been as much as 1000 kilometers, 675 Roman miles, or 9° of latitude apart. For ships of Columbus' day, this could have been anywhere from a 4 to 8 day sail, depending upon the wind. Such a variable time span could involve a total difference in declination of as little as 4 minutes or as much as 3 1/2°, depending upon the dates of the measurement. (See Table 1, Chapter 1.) Whether Columbus even considered declination is moot considering the almost impossible conditions for taking the sun's altitude at noon at the points 1 and 2.

182

Most of the learned geographers of the day (but not all, as we shall see) stood with Ptolemy and accorded an equatorial degree of longitude 62 1/2 of the same Roman (or Italian) miles. The seagoing community, of which Columbus became a part, had little in common with the geographers who looked upon them (and their chart-makers) as a bunch of ruffians who knew how to sail a ship (mostly in sight of land) and somehow - by the Grace of God - get it from port to port. Yet the portolani of Columbus' day were far more faithful in representing Mediterranean and Atlantic shorelines and port locations than the various editions of the Ptolemy atlases which began to appear in the second half of the 15th century with the introduction of printing in Europe.

During Columbus' period of employment in Portugal, he sailed several times to Guinea during the years of 1482 and 1483. He claims (through the writings of his son Ferdinand) to have measured the length of a degree at different times by taking readings of the sun's altitude (while his ship was proceeding on an essentially meridional course). Thus he "found" Alfragan's standard of 56 2/3 miles to a degree of meridional arc to be correct.

Details of Columbus' measurement are lacking but can be easily rationalized. Columbus made his observations of the sun's altitude at noon on successive days, utilizing a mariner's astrolabe or a quadrant. The difference in the sun's altitude on these successive noons was essentially due to the change in latitude as the ship sailed southward (or northward). Nevertheless, some part of this difference was due to the change in the sun's declination from one noon to the next. This could be negligible (as during the period of the summer and winter solstices) or quite significant (as much as 24 minutes daily change during the periods of the equinocti).

Columbus obtained the surface distance between the points at sea (believed to be on the same meridian) at which he made his noon observations of the sun's altitude on successive days from the ship's chart, entries having been made by the ships navigator. Such entries, based on estimates of the ship's speed everytime there is a change in wind direction or velocity, or whenever the ship changes course, are heterogeneous and generally exaggerated as to distances covered.

A degree of latitude measures 75 Roman miles (111,111 meters ÷ 1481.5 metres/Roman mile). Neglecting for the moment the effect of declination change, for Columbus to have obtained 56 2/3 Roman miles/ degree, the difference in the zenith distances he observed on successive noons had to be about 75 ÷ 56 2/3= 1.324 times as great as it should have been. Considering that distances shown in the log were probably inflated, Columbus's errors in reading his quadrant or astrolabe were even greater. Marine historians attest to the extreme difficulty in obtaining reliable readings of altitude from the deck of a small, 15th century ship undergoing, simultaneously, varying degrees of heave, surge, sideslip, roll, pitch, and yaw. Columbus must have taken dozens of observations (the altitude of the sun varies comparatively little between 11:30 a.m., and 12:30 p.m.). In this melange, he evidently found at least one pair (for successive days) which led to his degree length of 56 2/3 Roman miles. These he kept and discarded all the others. The final result, 56 2/3 miles/degree, dictated which observations would be considered valid and which otherwise. It also governed whether any allowance (and how much) would be made to account for the change in the sun's declination from day to day. Figure 5 illustrates our concept of the measurement.

His precise mission on these voyages is not known, although it is known that on one of these trips he saw, for the first time, gold being mined for the Portuguese by black slaves. This was the prototype for his later, less successful effort to do the same with the Indians of Hispaniola.

Columbus' precise geodetic views are perhaps best described by Ferdinand Columbus, his illegitimate son and biographer. In 1523-24, meetings were held at Badajoz between representatives of Spain and Portugal on the continuing problem of defining the line of demarcation between Spanish and Portuguese areas of influence. At this meeting, Ferdinand Columbus, the leading member of the Spanish junta, presented his views pertaining to the matters at hand. Harisse describes Ferdinands's testimony as follows[27]:

> He commences by declaring that, above all, it was necessary to determine the size of the globe, the length of the league, and how many Spanish leagues of 4 miles constitute a degree. But that being a most difficult thing to ascertain, a choice should be made between the dicta of cosmographers. In accordance with that suggestion, he proposes to reject the opinion current on the subject in the time of Aristotle, as well as the opinions of Strabo, Macrobius, Eratosthenes, Maximus,and Ptolemy. The estimate which must be adopted, he says, is that of 'Tebit, Almeon, and Alfragan, which was followed by Pedro de Aliaco, Juan de Pecan, and Christopher Columbus, as shown in many of his writings; all of whom give to each degree 56 miles and 2/3, equal to 14 leagues and 2/3 of a mile, and assign to the globe a circumference of 5100 leagues. (A degree = 14 1/6 leagues or 56 2/3 miles; the earth's circumference = 5100 leagues or 20,400 miles.)

In the foregoing, the Pedro de Aliaco referred to by Ferdinand is, of course, Cardinal Pierre d'Ailly, the source of almost all of Christopher Columbus' cosmographical views.

As to the distribution of land and sea in the east-west direction, there are various views as to Columbus' position in the matter. Table 2 illustrates two such views, those of Morison and Skelton. Also shown are the concepts held by Toscanelli and Behaim, as developed in this chapter. For purposes of comparison, the interpretations of the views of earlier geographers and travelers, notably Marinus and Marco Polo, shown in Table 1, Chapter 5, are repeated. It was, of course, Marinus' 225° Eurasia and Marco Polo's extensions which underlay the sharply reduced estimates of the westward distance from Europe to Asia propounded by Toscanelli, Behaim, and Columbus.

There is no document in the literature which records Columbus' precise views on the distances involved in a westward voyage from Lisbon or the Fortunate Islands to the Orient. Historians and geographers, such as Morison and Skelton, have pieced together many of Columbus' statements, made in isolation, in order to arrive at estimates such as are detailed in the four major columns of Table 2. That Morison and Skelton appear to differ widely is not surprising. Each of these estimates, and several others, can be justified depending on the weight given to such factors as Columbus' estimate of Marco Polo's additions to the length of the Eurasian continent (unknown) and whether he actually converted Marinus' 225° length of the "known" world to 265° to conform with Alfragan's statement that Ptolemy (and therefore Marinus) made the degree too large in the ratio 66 2/3 to 56 2/3. If he did, as Morison suggests, and extended the Asian mainland 30° eastward to conform with some estimates of Marco Polo's descriptions of Asia, and if he allowed 30° or 1500 miles for the distance from the Asian mainland to Chipanju, a length from the Fortunate Islands eastward to Chipanju of 325° would result and not even Morison attributes this distorted an estimate to Columbus.

An analysis of the data in columns IV, VI, and VII shown for Skelton would indicated that Columbus considered the meridian of the Fortunate Islands to lie 20° east of that of Lisbon, which is surprising for Ptolemy (and Marinus) had made it only about 5 2/3 degrees. (The meridian of the western-most of the Canary Islands is actually about 9° west of that of Lisbon.) Still, Columbus may have come across certain mariners' reports which justified this inflated estimate.

While Table 2 confirms that Columbus made a westward voyage from Lisbon to the Orient a more attractive one than either Toscanelli or Behaim made it out to be, all three managed to shrink the distance to about one-half of that held by Ptolemy's followers and one-third of the actual.

The general concept of Toscanelli, Behaim, and Columbus was to gain adherents in the first few decades after Columbus' initial voyage of exploration and discovery. This will be demonstrated and the events leading up to a more realistic view of the earth will be described in the next chapter.

Notes for Table 2, on pages 186 and 187.

1. Eratosthenes (276-194 B.C.), the first mathematical geographer, referred all longitudinal distances on his map to the parallel of Athens which he believed to be the longest in the "habitable world". He believed Athens to be at 36°55'43"N latitude. His degree of longitude at the equator had a length of 700 stades. At the parallel of Athens, the degree had a length of approx. 560 stades. The two most credible lengths which metrologists claim for E's stade are 148.2 metres (10 to the Roman mile) and 157.5 metres (9.4 to the Roman mile).

2. Posidonius drew no map of the world, but did express himself on the length of the habitable world and the distance westward from Spain to the Orient. The length of his stade is in dispute. Strabo said Posidonius had reduced the length of a degree from E's 700 stades to 500, but even this is questionable. However Marinus took Strabo at his word, accepting P's size of the degree, rather than E's.

3. Strabo accepted E, generally, except that he specified his stade to be 1/8 the Roman mile, hence 185.2 metres. While he reported E reasonably faithfully, he indicated his unhappiness with the older geographer's attempts to locate the Fortunate Islands and those off the coast of the eastern end of the Eurasian continent. He preferred to say this continent had a length of 70,000 stades between Cape St. Vincent and Cape Comorin (Cape Coniaci on E's map of the world.)

4. Marinus accepted P's 500 stades to the degree and (presumably) Strabo's stade of 8 to the Roman mile. Whatever the reason, he set the length of the Mediterranean at 62 °, instead of its actual 41 2/3 . He grossly misinterpreted reports on distances in Asia, placing the eastern end of the known world about 225 ° east of the Fortunate Islands. The areas east of this meridian, he designated "Terra Incognita." While he rejected, as did Ptolemy after him, the idea of an Atlantic coast somewhere within Terra Incognita, this is precisely what those who followed him - except Ptolemy - did conclude. He considered the Indian Ocean to be landlocked, thus further distorting reports on the Far East.

5. Ptolemy followed Marinus in all of the assumptions and estimates described above, save one. Sensing the Tyrian's overstatement of the length of Asia, he arbitrarily reduced the distance from the Euphrates to the east end of the "Known" world from 153 1/4 degrees (equivalent to 61,300 stades, or 11,350 kilometres) to 108 1/4 degrees (equivalent to 43,300 stades, or 8,000 kilometres). Even so, he was still almost 32% too high. He thus had re-established the length of the "Known" world at Posidonius' 180°. The most puzzling of his endorsements of Marinus was that of the length of the Mediterranean which was confirmed at 62°, about 49% too long. Marinus' and Ptolemy's equivalent length in kilometres was, however, only about 23% too great, and, if one were to apply the same metric lengths to the stade as conjectured for E, the estimate of 24,800 stades would have been almost perfect. Conversion from stades to kilometres has been made based upon the 185.2 metres per stade equivalency, the most popular among writers on ancient geography, but not some 20th century metrologists (including this reviewer).

185

Table 2: Distances along the 36°N. Parallel, or as otherwise indicated in Table or Notes, between the Meridians

Geographer or Other Authority	I. Fortunate Islands (or Las Palmas de Gran Canaria) to Pillars of Hercules (or East end Strait of Gibraltar)			II. Pillars of Hercules to Gulf of Issus; Length of the Mediterranean Sea (Gulf of Issus taken as vicinity Port of Iskenderun, Turkey)			III. Gulf of Issus to Easternmost Point in Continental Asia (or to Port of Kinsay); Length of Continental Asia along 36°N. Parallel		
Column No. →	Degrees of Longitude	Stades or Italian (Roman Miles)	Kilometres	Degrees of Longitude	Stades or Italian Miles	Kilometres	Degrees of Longitude	Stades or Italian Miles	Kilometres
Actual Distances	10	600 Ital. Mi.	900	41 2/3	2550 Ital. Mi.	3750	84	5100 Ital. Mi.	7550
Eratosthenes [1]	9	5000 stades	a.750 b.790	48	26,800 stades	a.3970 b.4220	78 1/2	44,000 stades	a.6520 b.6930
Posidonius per Strabo [2]	—	—	—	—	—	—	—	—	—
Strabo [3]	—	—	—	48	26,800 stades	5000	71	40,200 stades	7,445
Marinus [4]	7 1/2	3000 stades	560	62	24,800 stades	4600	155 1/2	62,200 stades	11,500
Ptolemy [5]	7 1/2	3000 stades	560	62	24,800 stades	4600	110 1/2	44,200 stades	8,200
Marco Polo, per Yule's Map (Fig.) [6]	—	1600 Ital. Mi.	2370	—	4500 Ital. Mi.	6670	—	7600 Ital. Mi.	11,260
Marinus and Ptolemy from Fortunate Is. to Gulf of Issus and Polo/Yule from Gulf of Issus to E.end Asia [7]	7 1/2 (See Note No.7)	3000 stades or 375 Ital. Mi.	560	62	24,800 stades or 3100 Ital.Mi.	4600	150	7600 Ital. Mi.	11,260
Morison/Raisz on Marco Polo [8]	—	—	—	—	—	—	140 1/2	7100 Ital. Mi.	10,500
Toscanelli, based on Marco Polo [9]	—	—	—	—	—	—	—	—	—
Martin Behaim [10]	16	At 36°N: 810 It.Mi. 28°N:885 " "	At 36°N: 1200 28°N: 1310	62	At 36°N: 3135 It.Mi. 28°N:3420 " "	At 36°N: 4645 28°N: 5070	164	At 36°N: 8290 It.Mi. 28°N:9050 " "	At 36°N: 12,290 28°N: 13,415
Christopher Columbus, per Morison [11]	—	—	—	—	—	—	—	—	—
Christopher Columbus, per Skelton [12]	—	—	—	—	—	—	—	—	—

Table 2, part 1

6. The distances in Italian miles shown here result from applying the scale shown on Sir Henry Yule's map, "Probable View of Marco Polo's Own Geography." For the reasons stated in the text, i.e., the inflated conversion factor from days of a journey to Italian miles, and the unlikelihood of anyone accepting Polo's views on Mediterranean geography, this approach represents too unrealistic a scenario.

7. In this approach, the Marinus/Ptolemy estimates are employed from the Fortunate Islands to the Gulf of Issus, and scaled distances from Yule's map for distances east of the Gulf. Conversion from Italian miles of 1482 metres to degrees is at Ptolemy's 400 stades to the degree at the 36th degree, north, parallel and the most popular estimate of 185.2 metres/stade.

8. See Morison's "Admiral of the Ocean Sea, A Life of Christopher Columbus", pp. 64-66, Little, Brown and Co., Boston, 1942.

Easternmost Point in Continental Asia (or Port of Kinsay (Hangchau)) to Offshore Asian Islands (or Port of Tokyo/Yokohama) IV			Length of "Known" or "Habitable" World Fortunate Islands (or Las Palmas) to Easternmost Point in "Known" Asia (or Ports of Tokyo/Yokohama) V = I + II + III + IV			Fortunate Islands westward to Easternmost Point in "Known" Asia VI = 360°; or equivalent –V			Lisbon, westward to Kinsay (length of Ocean Section between European and Asian mainlands) VII = VI + IV + Dist. Lisbon to Fort. Is.		
Degrees of Longitude	Stades or Italian Miles	Kilometres	Degrees of Longitude	Stades or Italian Miles	Kilometres	Degrees of Longitude	Stades or Italian Miles	Kilometres	Degrees of Longitude	Stades or Italian Miles	Kilometres
19 1/3	1150 Ital.Mi.	1750	155	9400 Ital.Mi.	13,950	205	12,450 Ital.Mi.	18,450	230 3/4	14,000 It.Mi.	20,750
3 1/2	2000 stades	a.b. 300	139	77,800 stades	a.11,500 b.12,250	221	123,600 stades	a.18,300 b.19,500	228 1/4	127,580 st.	a.18,900 b.20,095
—	—	—	180	70,000 stades	?	180	70,000 stades	?	236 1/3	133,868 st. 16,733 It.Mi.	24,800
—	—	—	225	90,000 stades	16,700	135	54,000 stades	10,000	—	—	—
—	—	—	180	72,000 stades	13,350	180	72,000 stades	13,350	—	—	—
—	1500 Ital.Mi.	2220	—	15,200 Ital.Mi.	22,520	—	—	—	—	—	—
29 2/3	1500 Ital.Mi.	2220	249	12,575 Ital.Mi.	18,640	111	5600 Ital.Mi.	8300	146 1/3	7400 Ital.Mi.	11,000
29 2/3	1500 Ital.Mi.	2220	239 2/3	12,100 Ital.Mi.	17,930	120 1/3	6100 Ital.Mi.	9000	155 2/3	7870 Ital.Mi.	11,660
30	1500 Ital.Mi.	2220	265 2/3	13,300 Ital.Mi.	19,700	94 1/3	4700 Ital.Mi.	7000	130	6500	9530
30	At 36°N: 1520 It.Mi. 28°N: 1655 " "	At 36°N: 2250 28°N: 2455	272	At 36°N: 13,750 It.Mi. 28°N: 15,010 " "	At 36°N: 20,380 28°N: 22,250	88	At 36°N: 4450 It.Mi. 28°N: 4860 " "	At 36°N: 6595 28°N: 7200	128	At 36°N: 6470 It. 28°N: 7065 Mi.	At 36°N: 9590 28°N: 10,470
28 3/4	At 36°N: 1320 It.Mi. 28°N: 1440 " "	At 36°N: 1955 28°N: 2130	300	At 36°N: 13,750 It.Mi. 28°N: 15,010 " "	At 36°N: 20,380 28°N: 22,250	60	At 36°N: 2750 It.Mi. 28°N: 3000 " "	At 36°N: 4080 28°N: 4450	97 3/4	At 36°N: 4480 It. 28°N: 4890 Mi.	At 36°N: 6640 28°N: 7250
30	At 36°N: 1375 It.Mi. 28°N: 1500 " "	At 36°N: 2040 28°N: 2225	280	At 36°N: 12,840 It.Mi. 28°N: 14,010 " "	At 36°N: 19,025 28°N: 20,765	80	At 36°N: 3670 It.Mi. 28°N: 4000 " "	At 36°N: 5435 28°N: 5930	130	At 36°N: 5960 It. 28°N: 6505 Mi.	At 36°N: 8832 28°N: 9642

Table 2, part 2.

9. Paolo Toscanelli set the distance from Lisbon to Kinsay at 6500 miles, which most, but not all, reviewers believe to be Italian miles of 1480 to 1488 metres. He estimated that a voyage between these ports would be made at an average latitude in which the length of a degree of longitude was 50 miles. Since the length of his equatorial degree of longitude is a matter of some dispute, the precise latitude of Lisbon in Toscanelli's day varied with the geographer, and the latitudes assumed for Antilia, Chipangu, and Kinsai are not stated it is not possible to indicate what the distances along the 36°N. parallel between the meridians of the indicated sites would be. The distances shown refer to the unknown average or mean latitude at which the Lisbon-Antilia-Chipangu-Kinsay voyage would be made. See section on Toscanelli in text and Figure 2, Chapter 6. See also "Toscanelli and Columbus, the Letter and Chart of Toscanelli," by Henry Vignaud, pp. 22-29, 188-202, and Appendix A, Books for Libraries Press, Freeport N.Y., reprinted 1971, first published 1902; and " Christopher Columbus, His Life, His Work, His Remains," by John Boyd Thatcher, pp. 301-107, Ames Press/Krause Reprint, N.Y.

10. *Martin Behaim constructed a globe in 1492, a reproduction of which in the plane, in two hemispheres, is shown in Figure 3. Gores of the ocean section of a reproduction of Behaim's globe are shown in Figure 4. Behaim, being a cartographer rather than a mariner, probably set his equatorial degree equal to 62.5 Italian miles. At 36° N, his degree of longitude had a length of 50.56 It. miles and at 28°N, 55.18 miles. In an age, when most mariners had a more realistic view of the length of the Mediterranean, Behaim clung to the erroneous figure of 62° handed down by Marinus and Ptolemy. Where Marinus/Ptolemy had made the longitudinal distance between the Fortunate Islands and the Pillars of Hercules 7 1/2°, Behaim provided much more detailed information about the islands, but overstated the distance (from Grand Canaria) to the Pillars by 6°, making it 16°. Toscanelli had interpreted Marco Polo 1500 miles from the China coast to Cipangu as equivalent to 30° of longitude. To be consistent, we have chosen the northern tip of Cipangu as the location for its representative meridian since it is just about 30° east of the China coast between latitudes 28° and 42° (the latitude of Lisbon). See Figure 3.*

11. *Most reviewers of the cosmographical views of Toscanelli, Behaim, and Columbus are struck by their general similarity despite the controversy about the length of Toscanelli's equatorial degree, the assumption that Behaim used an equatorial degree of 62.5 Italian miles, and the general agreement that Columbus had interpreted (and adopted) Alfragan's 56 2/3 mile degree as referring to Italian, or Roman, miles.*

Morison introduces an element of Columbian cosmographical thinking that goes beyond this general thrust when he attributes to Columbus an Eurasian continental length of 300 degrees of longitude and an Ocean span, from the Fortunate Islands to Cipangu of only 60 degrees. Morison's descriptions of Columbus' arguments are described in text and in his "Admiral of the Ocean Sea", pp. 64-67. Morison's references are general enough in character to support an array of interpretations.

12. *R. E. Skelton, the highly respected Superintendent of the Map Room of the British Museum, attributes to Columbus a length for the Ocean Section of the globe, Lisbon to Kinsay, in degrees, quite similar to Toscanelli's and Behaim's estimates. Columbus' employment of the inordinately tiny degree of 56 2/3 Italian miles led to an exaggerated Lisbon to Fort. Islands longitudinal distance, in degrees, and hence a smaller Fort. Is. to Cipangu longitudinal distance, in degrees, vis-a-vis Toscanelli and Behaim. Skelton, agreeing with G. E. Nunn, believes that Columbus and Behaim drew on a common map-source and this was Toscanelli. See Skelton's The Cartography of the First Voyage, pp. 217-219, an Appendix to "The Journal of Christopher Columbus", translated by Cecil Jane, Clarkston N. Potter, New York, 1960.*

Chapter 8

COSMOGRAPHY AND CARTOGRAPHY IN THE 16TH CENTURY

INTRODUCTION

Background

The publication of Marco Polo's travels reawakened in the western mind an interest in the Orient. The land routes between East and West which Polo had described were soon to be utilized by missionaries carrying western Christianity into Asia and caravans bearing the spices and textiles of Persia, India and China to European markets. When the Moslem Turks planted themselves across these routes by the end of the 14th century, the Genoese and the Venetians were able to reach a rapprochement with them whereby they were permitted to establish agencies on the Black Sea and in the Levant for the transfer of goods from caravan to ships bound for the ports of Europe. For such exclusive rights, the Italians paid a stiff price, which, of course, was passed on to the rest of Europe.

To break the monopoly of the Italian maritime states, the nations of Europe looked to the sea, for Polo had also described the advantages of the sea route between Hormuz and China via India and Java. Portugal took the lead among European nations in opening a sea route to India. The route which it finally elected to follow was that leading southward along the west coast of Africa. As the sea captains of Prince Henry the Navigator methodically worked their way down the coast, the Infante sought and obtained Papal Bulls which would protect and insure Portugal's right to whatever it discovered and exclude other Christian nations from such territories.

The second Bull, signed by Nicholas V in January 8, 1454, was in response to the intention expressed by Prince Henry " . . . to discover a route at the south and east as far as the countries of the Indians". In this Bull, the Pope granted to Alfonso V (the king of Portugal) " . . . all the regions discovered and to be discovered south of (the African) Capes Bojador and Noun (Nun), towards Guinea, and all those which are 'on the south coast and on the east side'". This Bull was affirmed by that granted by Sixtus IV in June 21, 1481 and that of Innocentius VIII, dated September 12, 1484.

The effect of the Papal Bulls was to preclude Spain, had it wanted to, from pursuing exploration in the wake of Portugal. Thus, when finally in 1492, the last Moslem stronghold in Spain had been captured and the Catholic monarchs of Spain were free to entertain proposals of exploration, Columbus' ideas which involved a route for reaching India other than that being pursued by the Portuguese had a degree of fascination for them.

Christopher Columbus was able, ultimately, to sell his "Project of the Indies" to the monarchs of Spain because he firmly believed a westward voyage from Spain to the Orient was achievable by the sailing ships of the day and was shorter than the Portuguese route around Africa. Utilization of offshore islands (real and mythical) as "stepping stones" or way stations, and favorable trade winds, were important factors in making this projected voyage feasible. The most important consideration, however, was that the distance between continents, and especially between offshore islands of both continents, was, as he (and others) figured it, much shorter than the ancients had estimated it to be. (Implicit in all this was that there lay no continental barrier between Europe and Asia in the Atlantic. The ancients had never mentioned such a barrier nor had anyone else up to Columbus' time).

As was shown in Table 2 of the last chapter, Eratosthenes' estimate of distance between continents (remarkably close to the actual distance) had been enormously reduced by Marinus and restored only partially by Ptolemy. Then, it was reduced again enormously by Toscanelli, Behaim, and Columbus based upon their interpretations of Marco Polo's descriptions of the geography of Asia and Alfragan's remarks about Ptolemy's degree being oversized.

An important factor in making this strange scenario possible was the repeated misinterpretation of cosmographers, geographers and cartographers of the itinerary distances and units of length found in the literature and reported by seamen, travelers, surveyors, and others.

The League Enters the Picture

It will be recalled that in the last chapter there was quoted a part of the letter written by Paolo Toscanelli to Canon Fernam Martins pertaining to the distance between Lisbon and Quinsay (Kinsai). In this quotation, the league was twice related to the mile (presumably the Roman or Italian mile of 1480-1482 meters): 4 miles = 1 league.

Later in our coverage of Toscanelli, the views of several authorities on the likely length of Toscanelli's degree of longitude at the Equator were presented. These included, in the order of their magnitude, 56 2/3, 62 1/2, 66 2/3, 67 2/3, and 75 3/5 Roman miles per degree. At 4 miles per league, the relationship between the league and the degree attriubted to Toscanelli was (from smallest to largest) 14 1/6, 15 5/8, 16 2/3, 16.92 and 18.9 leagues per degree. Columbus was described as holding Alfragan's view that 56 2/3 (presumed Roman) miles, or 14 1/6 leagues, made the degree.

Following Columbus' return to Spain at the end of his first voyage, he recommended to the monarchs Isabella and Ferdinand that a Papal Bull be sought to protect Spain's rights to the lands he had discovered and might discover on subsequent voyages. There were strong signs that King John II of Portugal planned to contest Spain's right to Columbus' discoveries based upon prior rights granted to Portugal by the Papal Bulls of 1454, 1481, and 1484.

Isabella and Ferdinand acted promptly and there resulted three Papal Bulls, all issued in May 1493 by Alexander VI. These Bulls granted to Spain just about the same rights previously granted to Portugal by the prior Bulls. The only restriction placed on Spain's rights was that its dominion in the Atlantic was made to start " . . . west of a meridian 100 leagues west and south of the Azores and of Cape Verde . . ." These terms have puzzled geographers for they are vague and contradictory, there being a difference of at least 22 degrees of latitude and 7 degrees of longitude between the cape and the Azores. India was not excluded from the lands which might be gained by Spain.

Subsequently, Ferdinand and Isabella importuned Pope Alexander to remove the restrictions imposed upon Spain by the May Bulls. The Pope acquiesced and a new edict, issued in September 1493, did so. This put the Spanish sovereigns in a better position to bargain with King John, both sides having realized the desirability of reaching an amicable agreement between the parties, one which could later receive Papal approval.

190

The Treaty of Tordesillas, 1494

Following months of haggling, a Line of Demarcation separating Portuguese and Spanish areas of pre-eminence in the Atlantic was agreed to by the monarchs of both countries. The line was set at 370 leagues west of the Cape Verde Islands. Everything east of the line not already owned (or claimed) by a Christian nation was to be in Portugal's sphere, everything west in Spain's. The line was to run from Pole to Pole, but whether it was to continue on the "other side" of the earth, i.e. along the meridian 180° east and west of the Line of Demarcation, if discussed, was not mentioned in the Treaty. The Treaty was signed June 7, 1494 at Tordesillas, a small town near Vallodolid in Castile.

Soon thereafter, in an effort to obtain the most authoritative expression available as to what meridian the Line of Demarcation would fall on, Ferdinand and Isabella consulted Jaime Ferrer, a Catalan cosmographer highly regarded in his day.

COSMOGRAPHY, CARTOGRAPHY, AND EXPLORATION IN THE PERIOD A.D.1494-1571

Ferrer's Geodetic Views - 1494

Ferrer's geodetic views as applied to the determination of the longitude of the Line of Demarcation are taken from Henry Harisse's *The Diplomatic History of America*[1]. We quote from Harisse's interpretation of Ferrer's response to Ferdinand and Isabella:

1st.   "The 370 leagues must be counted from the most central of the islands in the group of the Cape Verde islands.

2nd.   "Each degree in that parallel (15°) comprises 20 leagues and 5/8.

3rd.   "It is necessary to count each degree as equal to 700 stades[97], according to Strabo, Alfragano, Teodoci, Macrobi[98], Ambrosi, Euristenes[99].

4th.   "The 370 leagues [counted from the middle island in the Cape Verde archipelago] comprise westward 18 degrees.

5th.   "Each degree in the Tropics is equal to 20 leagues and four parts of 360.

6th.   "In the equinoctial circle, each degree is equivalent to 21 leagues and 5/8.

7th.   "According to Strabo, Alfragano, Ambrosius, Theodosius, Macrobius, and Eratosthenes, the circumference of the earth is 252,000 stades, which 252,000 stades, at the rate of 8 stades per mile, equal 31,500 miles, which, in counting 4 miles for each league, equal 7,875 leagues."

Footnotes 97 through 99 are reproduced below:

Note 97: "The stade of Macrobius, Strabo, etc., which Ferrer takes

191

as a basis for his calculations is the Olympic stade, now mathematically ascertained to have been equal to 192.27 meters. *Die Ausgrabungen zu Olympia, V. Ubersicht der arbeiten und funde vom Winter and Frujahr, 1879-1880 and 1880-1881. XLIII tafeln, herausgaben von E. Curtius, F. Adler, G. Treu, and W. Dorpfeld,' Berlin, 1881, folio, p. 37, and Plate XXXI.-XXXII."*

Note 98: "Viz.: Aurelius-Theodosius Macrobius."

Note 99: "Eratosthenes. Let us remark that the great Greek mathematician did not count each degree as equal to 700 stades, as the circumference of his sphere was not divided in 360 degrees, but in 60 parts, each of 4200 stades; which , however, amounted to the same thing."

As Harisse points out, the seven specifications of Ferrer are mutually contradictory: the 2nd leads to the length of a degree at the Equator of 21.353 leagues which contradicts specification 6 (21.625 leagues) and specification 7 (7875 ÷ 360 = 21.875 leagues); the 5th specification leads to an equatorial degree of 21.813 leagues; and specifications 2 and 4 do not match, precisely. Harisse logically selects from among these that part of the 7th specification which sets the earth's circumference at 252,000 stades - Eratosthenes' figure. Then based upon Note 97, he evaluates Ferrer's circumference of the earth as 252,000 x 192.27 m. = 48,452,040 meters, which he properly judges to be some 21% too great.

As pointed out in Chapter 4, the Olympic stade is considered by most metrologists today to be about 185.2 metres, which meets the criterion of 1/8th the Roman (Italian) mile of 1482 metres. Harisse, using a stade of 192.27 metres, by indirection, ascribes to Ferrer's mile a length of 1538.16 metres. Ferrer's league, based on the Roman mile, has a length of 4 x 1482 m. = 5928 metres. According to Harisse, it is equal to 4 x 1538.16 m., or 6152.64 metres (see Table 1).

 Ferrer is wrong, however, in ascribing to Eratosthenes and Macrobius the stade of 8 to the mile. Eratosthenes, himself, never related his stade to the Roman mile. As pointed out in Chapter 4, Macrobius set Eratosthenes' stade at 1/10th the Roman mile, or about 148.2 metres. It was Strabo who, while accepting Eratosthenes' degree of 700 stades,nevertheless believed those stades to have a length of 1/8th Roman mile.

The foreoging comments aside, Ferrer calculated 370 leagues at the latitude of Fogo, the central island of the Cape Verdes, 15° N, to be equivalent to about 18° of longitude. Taking the island of Fogo at 24° 25' west of Greenwich, this put Ferrer's Line of Demarcation at about 42° 25' W. Harisse, however, correcting for Ferrer's oversized degree, puts the line at 45° 37' W[2].   This meridian cuts through Brazil over 700 miles west of Recife, on the "Bulge," and 150 miles west of Rio de Janeiro, on the south coast.

Enciso's Geodetic Views - 1518

The matter of the precise location of the Line of Demarcation continued to engross the thoughts of the Spanish government and the views of experienced pilots were solicited from time to time. These were people who had been to the New World either on one of the numerous voyages of exploration which took place in the wake of Columbus' first two voyages, or who had participated in re-supply voyages to newly established colonies there. While up to date "padrons," navigational charts fortified with accurate sailing directions, were closely held, there is nothing to indicate that any information bearing upon the size of the earth or the length of a degree was ever with-held from Spain's own cosmographers.

Unit →	Stade	Mile	League	Degree	Earth's Circumference
Column No. →	①	②=8×①	③=4×②	④=700×① =87.5×② =21.875×③	⑤=360×④
Identification ↓					
Harisse	192.27m.	1538.16m.	6152.64m.	134,589 m.	48,452 Km.
Reviewer's Conjecture	185.2 m.	1481.6 m.	5926.4m.	129,640 m.	46,670 Km.

Table 1. *Interpretations of the cosmographical views of Jaimie Ferrer, solicited by the Spanish monarchs, Ferdinand and Isabella, in 1495.*

Harisse indicates that Sebastian Cabot and Juan Vespuccius, among others, were consulted in 1515. In 1518, Martin Fernandez de Enciso, a prominent Spanish cosmographer of the time, published at Seville his "Suma," or geographical compendium. The only geodetic data Harisse quotes from the compendium follows:

> 1st. "The Equator contains in longitude three hundred and sixty degrees of sixteen leagues and a half each[109].

> 2nd. "As each degree is estimated to be in length sixteen leagues and a half and one-sixth, the circumference of the entire globe is three hundred and sixty degrees, amounting to six thousand leagues.

Footnote 109 (above) is quoted hereunder:

> "It must be noted, however, that in the windrose added to the "Suma," the difference between two points of the compass seems to have been calculated on the basis of 17 1/2 leagues (to the degree) . . . "

Thus, it would appear that Enciso has suggested--directly or indirectly--that the degree contains 16 1/2, 16 2/3, and/or 17 1/2 leagues. Harisse has chosen 16 2/3 leagues as Enciso's intended equivalency. Then, Harisse says:

"The probability is that the league, which is always a unit usual and fixed, was the same for Enciso and for Ferrer; that is at the rate of 32 stades for one league. We shall therefore adopt the same value for the league of both cosmographers and ascribe the difference in the valuations which they give to the equatorial degree only to their different valuations of the dimensions of the earth."

To assume that ". . . the league was always a unit usual and fixed . . .," i.e., unvarying, is naive. Within Spain, as late as 1852, there were at least 6 different leagues employed within the 49 provinces. These varied from a minimum of 5495 metres in the province of Navarra to a maximum of 6687.24 metres in the province of Cuidad-Real. 24 of the 49 provinces employed a legua de 6666 2/3 varas castellanos. Since the vara castellano was rated at 0.835905 metres, the legua equalled 5572.699 metres. Harisse ascribes a league equal to 6152.64 metres to both Ferrer and Enciso. This, of course, is a possiblity, but only one of several possibilities.

Evaluating Enciso using Harisse's proposal, one obtains:

1	stade	=	192.27 metres
1	mile	=	8 stades = 1538.16 metres
1	league	=	4 miles = 6152.64 metres
1	degree	=	16 2/3 leagues = 66 2/3 Roman miles
		=	533 1/3 stades = 102,544 metres

The earth's circumference = 36,915.84 km.

Based on Enciso's 16 2/3 leagues to a degree, 370 leagues at 15º N latitude put the Line of Demarcation 22º 59' west of Fogo or 47º 24' W of Greenwich. Harisse then, as for Ferrer, calculates where the Line of Demaraction would come on a properly sized globe, i.e., one with an equatorial degree of 111,111.11 metres and a degree at 15º N = 107,325.1 metres. This, Harisse puts at 45º 38' W of Greenwich, one minute west of Ferrer's line on a properly sized globe.

Returning now to Harisse's assignment to Enciso's league of a length of 32 (stades) x 192.27 metres/stade = 6152.64 m.:

As in the Ferrer case, it is possible to suggest other reasonable relationships. Thus, one alternative is to choose the Olympic stade of 185 metres as the starting point. Another alternative is to assume Enciso was using the legua of Cuidad-Real, composed of 8000 varas castellanas of 0.836 m. each. At 6687.24 metres, this was the largest of the Spanish leguas. With Enciso's ratio of 16 2/3 leagues to the degree, this yields a remarkably close figure for the degree-111,453.5 metres. Still another alternative assumes the legua of Huesca was used by Enciso. This unit, which in 1852 measured 6176 metres, quite close to Harisse's assumption of 6152.64 metres, produces a degree with a length of 102,934 metres and an earth's circumference of 37,056.07 kilometres.

Summing up, because Enciso did not specify the length of his degree in any other way than its relationship to the league (or vice versa), we are left with the several interpretations of his cosmographical views shown in Table 2, as well as other possible interpretations.

194

Unit →	Stade	Mile	League	Degree	Earth's Circumferene
Column No.→	①	②=8×①	③=4×② unless shown otherwise	④=16⅔×③	⑤=360×④
Identifica-tion ↓					
Harisse	192.27m.	1538.16m.	6152.64 m.	102,544 m.	36,915.8 Km.
Reviewer's Conjecture #1	185.2m.	1482m.	5928 m.	98,800 m.	35,568 Km.
Reviewer's Conjecture #2	209 m.	1671.8m.	6687.24 m (legua de Ciudad-Real)	111,453.55	40,123 Km.
Reviewer's Conjecture #3	192.88m.	1543m.	6172 m. (legua de Huesca)	102,934 m.	37,056.07Km.
Reviewer's Conjecture #4	185.2m.	1482m	5081.14 m. (3.43 × 1482m) (See Fig.19)	88,920 m.	32,011.2 Km.

Table 2. Interpretations of the cosmographical views of Martin Fernandez de Encisco, as contained is his "Suma" or geographic compendium of 1518, Seville.

European Exploration, 1494-1518

Harisse's estimate for Enciso having yielded a length for the degree and the earth's circumference somewhat closer to reality than his estimate for Ferrer, he then adds[3] " . . . the difference between Enciso's and Ferrer's valuation in this respect, shows, to a certain extent, the geodetic progress accomplished in Spain between the years 1495 and 1518."

There having been no formal measurements of a degree during this period, this could only be ascribed to voyages of exploration and re-supply and land exploration which occurred usually after the establishment of a colony or the conquest of a region. The most significant such events were:

- Christopher Columbus' 2nd, 3rd, and 4th voyages which explored all of the West Indies, Trinidad, Venezuela, Honduras, Nicaragua, Costa Rica, and Panama (1493-1504).

- John Cabot's voyages to the maritime provinces of Canada and the Eastern coast of the U.S. (1497-8).

- Amerigo Vespucci's two voyages to South America, which explored almost the entire Atlantic Coast (1499-1500 and 1501-1502).

- Vasco de Gama's and Pedro Alvares Cabral's voyages to India (1497 and 1500).

- The conquest and establishment of colonies in Hispaniola, Puerto Rico, Cuba, and Jamaica (1493-1511).

- The establishment of colonies in Panama (1510).

- Vasco Nunez de Balboa's exploration of the Isthmus of Panama and discovery of the South Sea (Pacific Ocean) (1511-1513).

- Mathias Albuquerque's conquest of Malacca, the East Indian spice center, for Portugal (1511).

- Ponce de Leon's discovery and exploration of Florida (1515).

- Juan Diaz de Solis' discovery of the mouth of the Rio de la Plata (1515).

- Francisco Fernandez de Cardoba's exploration of 400 miles of coastline of Yucatan Peninsula (1517).

- Juan de Grijalva's exploration and mapping of 1200 miles of Gulf of Mexico shoreline, from Yucatan to Panuco (Tampico) (1518).

Mapping in the Period 1494-1518

The voyages of discovery in the North and South Atlantic, the Caribbean, and the Gulf of Mexico were reflected almost immediately in the official "padrons" of Spain, as the voyages to India and the Moluccas were reflected in Portugal's padrons. Since these documents were closely held, it took some time for the chartmakers of each of these countries to learn the secrets of the other and still longer for non- Iberian cartographers to learn of them -- except as breaches of security occurred. Eventually, the cosmographers

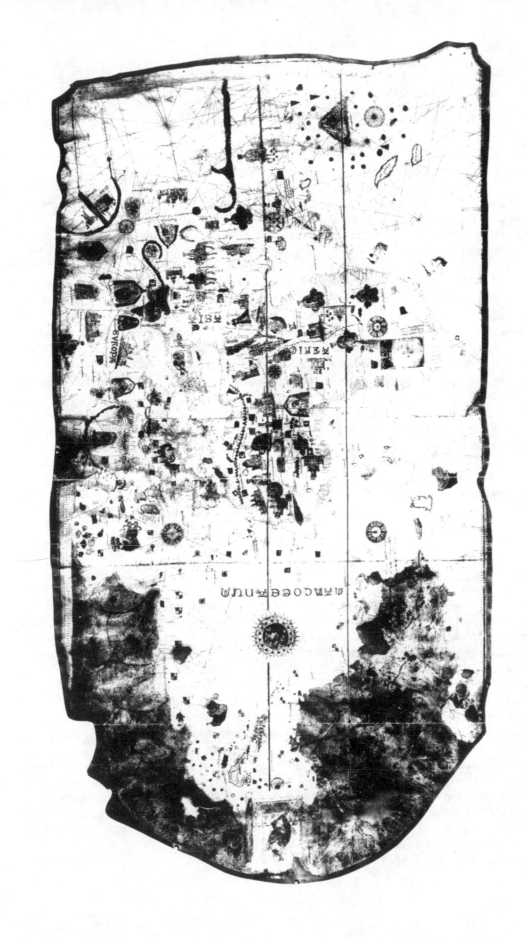

Figure 1. The oldest map (sea chart) showing islands and coasts of the New World, reproduced here by permission of the Museo Naval, Madrid. It was drawn around 1500 by Juan de la Cosa. See comment on page 198.

Comment on Figure 1, page 197

The map in Figure 1 was drawn about 1500 by Juan de la Cosa, navigator, cartographer, and cosmographer, who had accompanied Columbus on his first two voyages. The map reflects information gained during the first three of Columbus' voyages to the New World, as well as those of Cabot, Vespucci, Pinzon, de Lepe, and de la Cosa's own voyage with Hojeda to the Spanish Main. The detail provided for Africa is based, almost exclusively, upon Portuguese exploration from Prince Henry's time through Dias' doubling of the Cape of Good Hope to de Gama's voyage to India.

Sea charts of this period did not generally consist of grids of evenly spaced parallels and meridians, but usually showed the Equator, at least the Tropic of Cancer, the zero meridian, usually through the Fortunate Islands, (Canaries), and a series of carefully spotted windroses with Rhumb lines, or Loxodromes, emanating therefrom to other windroses or strategic geographic entities (entrances to harbors or straits, or land projections). De la Cosa's chart gives prominence to the Tropic of Cancer, the Equator, and what appears to be the Line of Demarcation resulting from the Treaty of Tordesillas between Spain and Portugal. The zero meridian is identifiable as running through the two windroses off the European and African coasts.

De la Cosa's placement of Hispaniola and the length he has given the Mediterranean (60° west of Lisbon and 42° of longitude respectively) are far more realistic than many of the Prolemaic based maps o the first half of the 16th century. The reader can use as a reasonably reliable scale the meridional distance between the Tropic of Cancer and the Equator, which represented 23.5° of latitude. A degree of latitude an longitude on such maps were generally shown with the same length, although a degree of longitude actually approximated the cosine of the latitude X the actual length of a degree of latitude.

Comment on Figure 2, page 199

Shown in Figure 2 is an outline of a part of an original print of the "Cantino Map" preserved at the Biblioteca Estense, Modena, Italy, the British Museum, London, and doubtless other places. Cantino, a representative of the Duke of Ferrara at the court of João II of Portugal, paid 12 ducats to have an illegal copy of the official Portuguese "padron" made for his master. The sea chart dates from 1502 and shows four prominent parallels - the Equator, Tropics of Cancer and Capricorn, and the Arctic Circle. Also shown are several compass roses and a host of rhumb lines calculated to be of assistance to navigators. As in other sea charts of the day, no formal grid of parallels and meridians is used.

It is possible by scaling a photograph of an original "Cantino Map" to obtain an order of magnitude estimate of the Lisbon to Hispaniola (port of Santo Domingo) longitudinal distance - approximately 53.3 degrees, and the length of the Mediterranean - approximately 37 degrees.

An order of magnitude estimate of the length of a degree of longitude on this map in terms of a Treaty league (of unknown length) can be obtained by scaling the map in two places: (1) between the meridian of Fogo, central island of the Verdes (shown on map) and the Line of Demarcation, which is equivalent to 383 leagues (meridians 370 leagues apart at 15°N. latitude will be 383 leagues apart at Equator), and (2) between both Tropics taken as 47°. Thus, 1°, at Equator = 71/47 x 383/35 = 16.53 leagues. At the paralle, of Fogo, 15°N, 1° = cos15°(=0.966) x 16.53 = 15.97 leagues.

The Line of Demarcation is 370/15.97 = 23.17° west of Fogo.

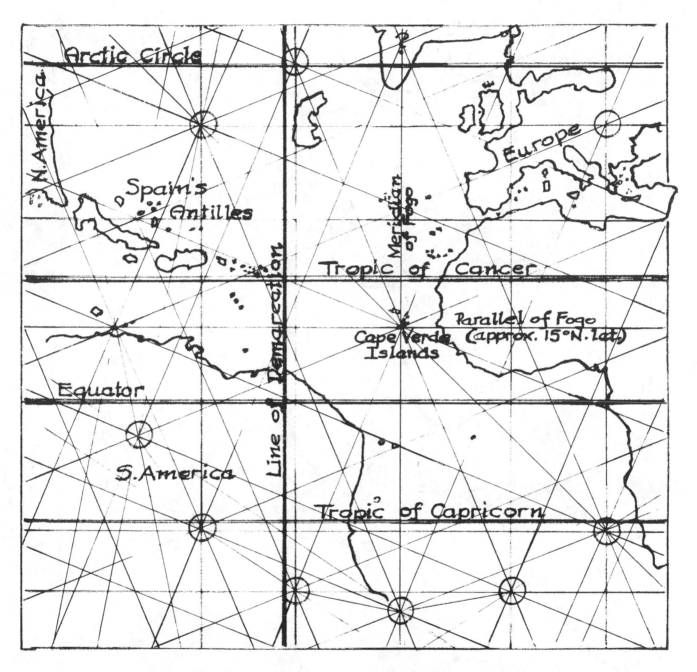

Figure 2: The First Printed Map to Show the Line of Demarcation Stemming from the Treaty of Tordesillas — "The Alberto Cantino Map.

199

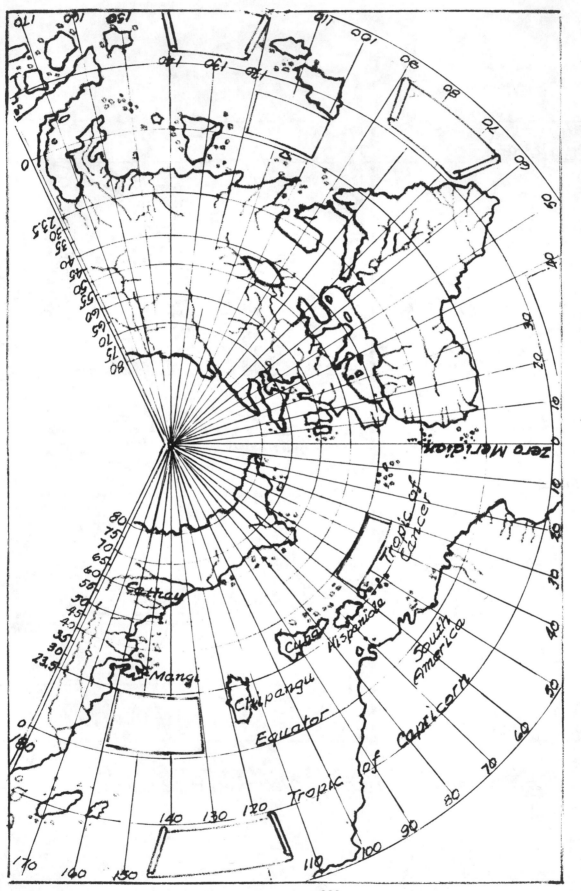

Figure 3. An Outline of G.M.Contarini's Map of the World, 1506, drawn on a conical projection. This was the earliest printed map to show America. His length of the Mediterranean is 58°; Lisbon to Hispaniola (Santo Domingo) is 72°; Lisbon west to Asian mainland is about 163°; and Canaries west to Japan is 116°.

Comments on Figure 3, page 200

This was the earliest printed map to show America. Contarini's length of the Mediterranean is 58°; Lisbon to Hispaniola (Santo Domingo) is 72°; Lisbon west to the Asian mainland is about 163°; and Canaries west to Japan is 116°. All longitudinal distances are approximate and represent the number of degrees of longitude between the meridians of the places named, measured along the parallel 36° north.

One of the original Contarini prints is owned by the British Museum, London. A photo copy has been included in R.V. Tooley's "Maps and Mapmakers", B.T. Batsford, London, 1978, as well as several other publications. The outline shown is based on such photocopies.

The projection employed in constructing the Contarini map is not a standard Prolemaic conical projection (See Figure 4, Chapter 3, or the Ruysch map, Figure 5, Chapter 8) which would produce undistorted areas in the vicinity of at least one, and possibly two, parallels, i.e. at circles of tangency or intersection between cone and globe. While degrees of latitude are everywhere equal, they are too small in relation to degrees of longitude causing the entire map to be condensed in the north-south direction and broadened in the east-west direction.

Comments on Figure 4, page 202

Waldsseemüller believed the Florentine Amerigo Vespucci (who had explored 1300 miles of Brazil's north coast and part of Guiana in 1499 - 1500) had discovered the South American continent - hence the name America which he gave to that land. Figures of Prolemy and Vespucci appear on either side of the miniature hemispheres at the top of the large 7 1/2' x 4 1/2' original map, honored as the major sources of Waldseemüller's information. Numerous notations credit other sources. Like other cartographers of the late 15th and early 16th centuries, Waldseemüller has abandoned Prtolemy's length of 180° for the "known world", favoring the sharply increased length for Asia suggested by Marco Polo and accepted by Toscanelli, Behaim, and Columbus. He shows a full 270° of longitude for the eastward span from the Canaries (zero meridian) to the eastern coast of Japan, hence only 90° for the westward span. This despite the necessity of accommodating the New World discoveries. Like most maps of the 15th and 16th centuries, his Mediterranean is 50% too long.

Waldseemüller's cordiform projection had a grid of meridians and parallels each spaced at 10°. Each degree of latitude was intended to have equal length regardless of latitude. Meridians were equally spaced along each parallel. A degree of longitude at any latitude had a length equal to the cosine of the latitude times the length of a degree of longitude at the equator.

This outline is based on a photocopy of the original Waldseemüller map used to illustrate the article "The Naming of America" by Egon Klemp in the "EXPLORERS JOURNAL", june 1977. This article was adapted from the book "America in Maps" by the same author, published by Holmes and Meier, New York, 1976.

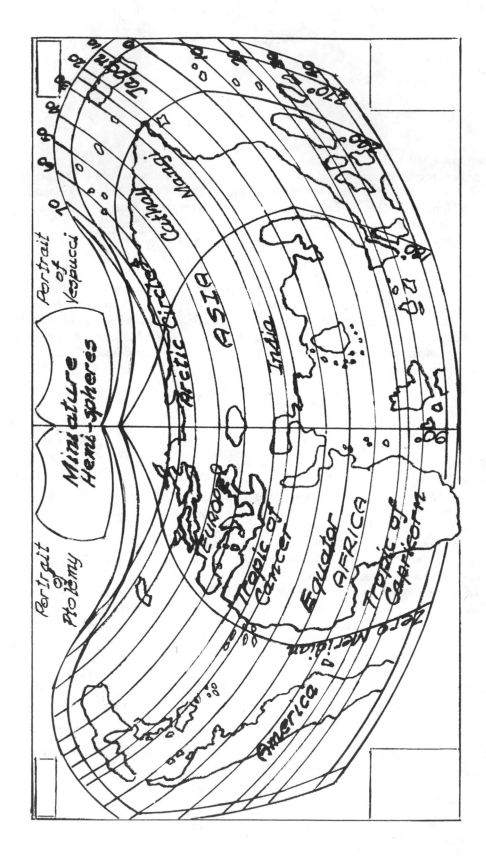

202

Figure 4. An Outline of Martin Waldseemüller's World Map of 1507.

of the world might be given to wondering if the aggregation of discoveries might not portend a world larger, or smaller, than generally accepted. But, it took something pretty iron-clad to persuade a geographer that the degree, as he understood it, would have to be changed.

Some examples of world maps produced before 1518 are shown in Figures 1 through 7, credits and comments for which are found on pages 198, 201, and 208.

The Juan de la Cosa planisphere and the Cantino map (Figures 1 and 2, respectively) are the only sea charts in the group, thus have no grid of longitude and latitude lines. Such charts, distinguished by loxodrome or rhumb lines emanating from compass roses strategically placed on the map, were intended for seamen and designed to get them from port to port by careful use of the rhumbs (i.e., by following the indicated bearings). (Such charts anticipated Gerardus Mercator whose famous grid or projection was intended for the same purpose). The sea charts did, however, carry 5 parallels, the Equator, the two Tropics, and the two "circles," Arctic and Antarctic. Thus, it is possible to do some "order of magnitude" scaling of longitudes by using the map distance between Equator and a tropic (intended to be 23.5°) as a reference.

Such scaling reveals that both the de la Cosa and Cantino maps are far more realistic than the remaining five maps in their representation of the length of the Mediterranean (actually 41.66°) and the distance between the meridians of Lisbon and Hispaniola (Santo Domingo) (actually 60.46°). All of the remaining five make the Mediterranean much too long and the distance between the meridians of Lisbon and Hispaniola too short (except Contarini, Figure 3, and Ruysch, Figure 5, who make it too long). All five of the maps show the distance between the meridians of Lisbon and the Asian mainland not unlike that shown for Columbus in Table 2, Chapter 7 (based on data obtained from Skelton).

Only one map, that by Sylvani (Figure 6) provides a legend indicating the length of a degree of latitude, or of longitude at the Equator: 1° = 62 1/2 Roman miles. Thus, the earth's circumference = 22,500 Roman miles (a figure frequently, but questionably called Ptolemaic), or 33,345 km.

One of the maps, Figure 4, is an outline of Waldeseemuller's famous World Map of 1507. This map named the continent discovered by Christopher Columbus (S. America) after Amerigo (Latin Americus) Vespucci who, Waldseemuller thought, had preceded Columbus there. (See Appendix A story.)

The Petrus Apianus Map of 1520

One of the few maps published during the 16th century which contained information as to the length of a degree of longitude at the Equator, and at 10° intervals, was that of Petrus Apianus, whose cordiform map of the world is shown in Figure 8.

Apianus was a prolific author of cosmographical works and an innovative cartographer. The first edition of his cosmography with the imposing title "Cosmographicus liber Petri Apiani Mathematici studiose collectus" was published at Landshut (Germany) in 1524. Subsequently, a number of editions in different languages were published throughout Europe and used as cosmographical manuals in the universities. From 1527, he was professor of mathematics at the celebrated academy of Ingolstadt.

One of the maps included in his "Cosmography" was the cordiform projection of 1520. On this map, the left perimeter, from the Equator to 50° S. latitude, at 10° intervals, has the equivalent length of a degree of longitude given in miliaria (Roman or Italian miles). As indicated in the caption to Figure 8, the mathematics appears contradictory, although the reproduction is sufficiently unclear and nomenclature strange to give this reviewer some pause. In any event, evaluating the apparent markings at 10°, 20°, 30°, 40°, and 50° indicates that the length of a degree of longitude at the Equator could be as small as 60.05 miliaria and as much as 63.78, the average of which is 61.76 miliaria, an unlikely number. 62 1/2 miliaria may have been intended, as it was in the case of the 1511 Sylvani map. If so, our remarks for Sylvani would hold for Apianus. In any event, a later, less equivocal, presentation on the subject by this cosmographer will be entertained in its chronological spot.

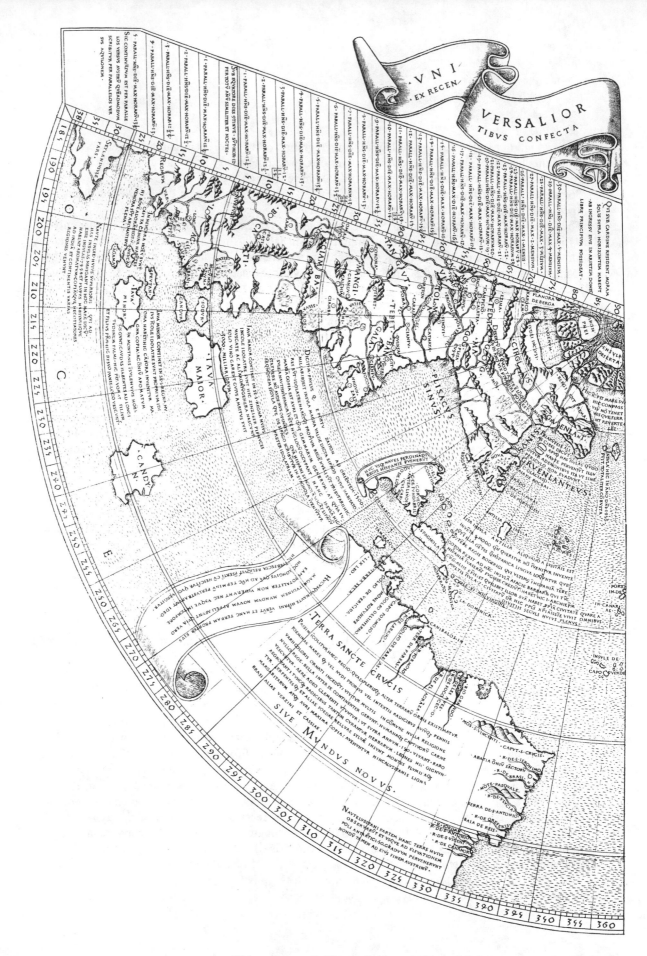

Figure 5, part 1. The conical projection of Johannes Ruysch - 1508.

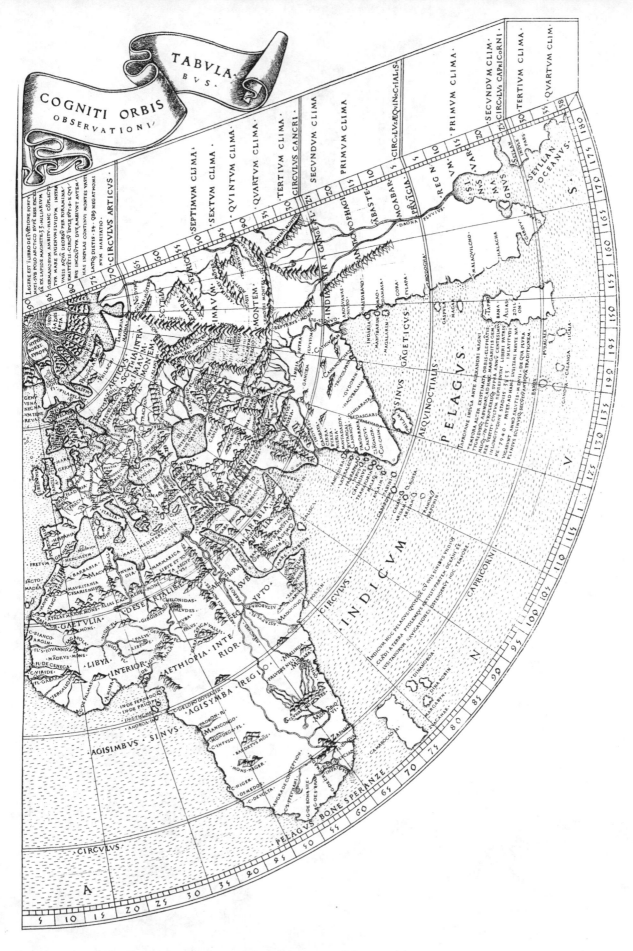

Figure 5, part 2. Comments and credit on page 208.

Figure 6 part 1. Sylvani's Map of the World.

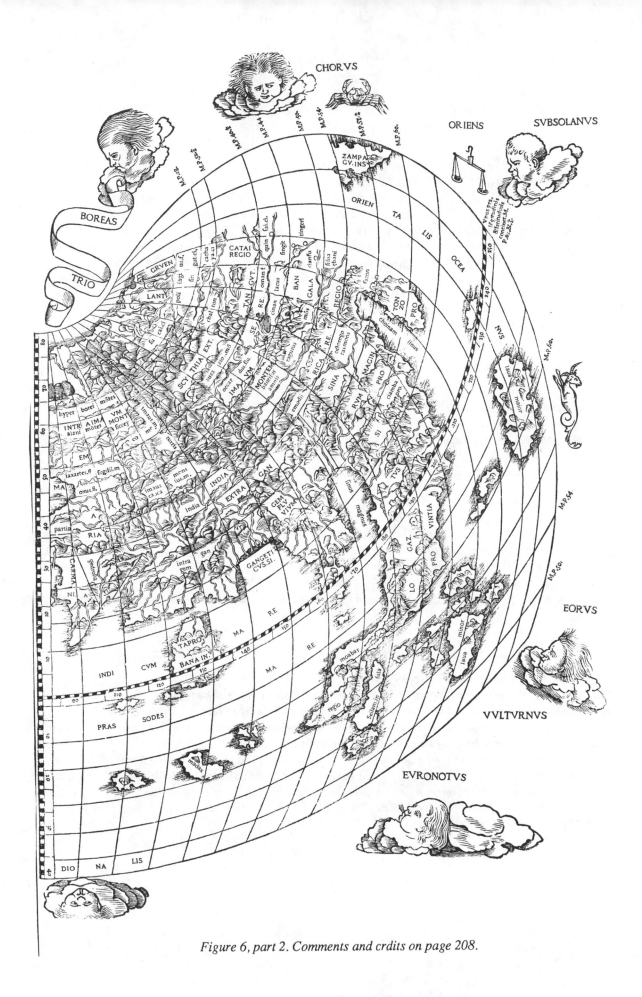

Figure 6, part 2. Comments and crdits on page 208.

Credit and comment on Figure 5, page 204, 205

This conical projection of Johannes Ruysch shows Mediterranean 58° long; Lisbon to Hispaniola, 74°; Lisbon to Asian mainland 135°; Japan (Zipanju) is not shown. Ruysch follows Contarini rather than de la Cosa and Cantino's Portuguese padron in placing Hispaniola so far offshore.

The conical projection of this map is constructed by making the radius of the equatorial circle equal to the length of a quadrant (90°) on the globe. Degrees of latitude on map are everywhere equal and are also equal to degrees of longitude at Equator (on globe and map). However, the length of degrees of longitude on map are too small at latitudes betrween 0 and 90°N and much too large in the southern hemisphere. This would be reversed if center of map were placed at the South Pole.

This map is taken from A.E. Nordenskiölds "Facsimile Atlas" Dover Pubs., NY, 1973 and is reproduced here by permission of the publisher.

Credit and Comment on Figure 6, page 206, 207

This map was the first to employ Ptolemy's homeother projection beyond 180° of longitude. He made the Mediterranean 58° long; Lisbon to Hispaniola only 48°; Lisbon to the Asian mainland (westward) approximately 144°; and Fortunate Islands to Japan 112°.

This map is taken from Nordenskiöld's "Facsimile Atlas". Note that the map extends only from 290° (70° West of Fortunate Islands) to 250°. The China coast, part of "Zampagu" (Japan), and the ocean area between 250° and 290° longitude were too indefinite to be drawn.

This projection, also known as the "cordiform", differs from Ptolemy's homeother projection in construction only in that the common centre of the parallel circles is placed 100°, rather than 181° 8' from the Equator and the parallel circles are extended to cover 360° (in the Sylvani map, only 320°) rathe than ptolemy's 180° of longitude. Ptolemy's maps being limited to his "Oukimene", or known world, restricted his grid, latitude wise, between 63° N and 16 5/12° S.

This map is reproduced here by permission of Nordenskiölds U.S. publisher, Dover Pub., NY, 1973.

Credit and comment on Figure 7, 209

This map was the first to employ two hemispheres. Strobnicza made the Mediterranean 61° long; put Hispaniola only 46° from Lisbon and the Asian mainland 129° west of Lisbon. Japan, correctly surmised to lie in an ocean other than the Atlantic, is, however, placed only 95° west of the Canaries (O° meridian). Strobnicza was also the first to draw North and South America as two continents connected by a long and narrow isthmus - and this a year before Balboa discovered South Sea (Pacific) on the Isthmus of Panama. Despite the roughness of appearance (printed from a crude woodcut), Nordenskiöld (pp. 68b, 69 a & b) considers this map "of great interest and importance to the early history of cartography..".

This map is taken from Nordenskiöld's "Facsimile Atlas", Dover Pubs. NY 1763, and is reproduced here by permission of the publishers.

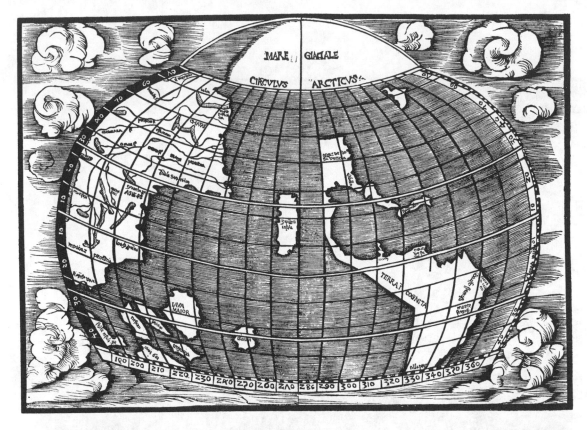

Figure 7. Johannes Strobnicza's homeother projection of the Earth's surface, the first to employ two hemispheres. See credit and comment on page 208.

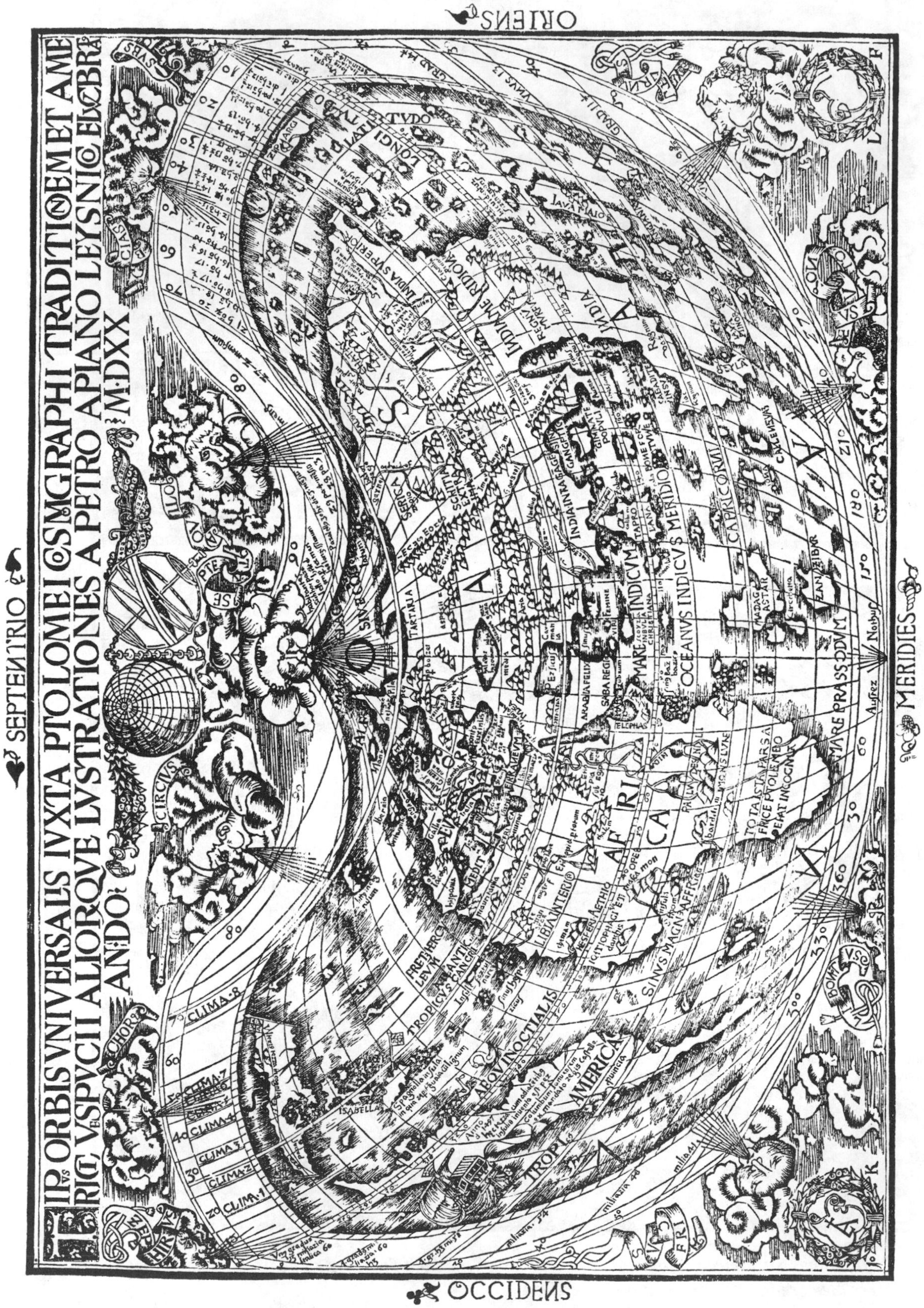

Figure 8. The cordiform projection of Pietrus Apianus of 1520.

Spanish and Portuguese Experts at the Badajoz Conference of 1523-24

It was brought out in Chapter 7 that Ferdinand Columbus, when presenting his views at the Badajoz Conference of 1523-24, indicated he stood with his famous father in estimating the length of a degree of equatorial arc of earth's surface at 14 1/6 leagues or 56 2/3 miles; the earth's equatorial (or great circle) circumference, then, had a length of 5100 leagues or 20,400 miles.

Chapter XVIII of Henry Harisse's *Diplomatic History of America*, the source for Ferdinand Columbus' testimony at the conference, also relates the views of other Spanish and Portuguese "experts" in attendance, not members of the junta, but who had been retained by that body to give advice on cosmographic and related matters. Among them were Thomas Duran, Sebastian Cabot, Juan Vespuccius, Diego Ribeiro, and the pilots of the Casa de Contratacion in 1524. (The Casa de Contratacion was the Spanish Crown agency set up during Columbus' time to handle all logistics matters pertaining to exploration, colonization, and administration in overseas territories).

We quote from Harisse:

On the proposition of Thomas Duran, Sebastian Cabot and Juan Vespuccius:

1st. The degree was to be considered equal to 17 1/2 leagues, embracing 62 1/2 miles [147].

2nd. The league contained 4 miles, one mile contained 8 stades, and there were 500 stades to a degree[148] "according to Ptolemy."[149]

3rd. The starting meridian of the 370 leagues westward was to be the centre of the island of Sant Antonio (the easternmost of the Cape Verde Islands).

4th. These 370 leagues in that parallel were calculated to be equal to 22 degrees and nearly 9 miles.

Note 147 contains this interesting comment by Harisse:"This league of 17 1/2 to a degree was stated by them (the propositioners) to be the league commonly used by the Spanish and Portuguese seamen . . ."

Note 148 contains the comment: "According to their first two data, there would have been 17 1/2 x 32 = 560 stades to a degree and 201,600 stades for the circumference."

Note 149 follows:

"Misled by Saigey (op. cit., p. 61), we first based our calculations upon his assertion that Ptolemy used the Phileterian stade. Consequently, we increased the Olympic stade, without, however, adopting Saigey's estimate of 184.8 metres for the Olympic, as it is now known to be equal to 192.27 metres. But a new study by Boeckh, Letronne, and T. H. Martin convinced us that although both kinds of stades co-existed in Egypt in Ptolemy's time, only the Olympic was resorted to for scientific mensurations and that Archimedes, Eratosthenes, Hipparchus, Strabo, Vitruvius, and Ptolemy himself must have used it to the exclusion of the Phileterian. Our computation of Ptolemy's stade is based therefore upon the present estimate of the Olympic, viz.: 192.27 metres, as we did when discussing Ferrer's estimate."

Harisse continues: "These measurements we find to give to their globe a circumference of 38,759,728 metres, which is 31 thousandths less than the real circumference of the earth (viz., 40,000,000 metres)."

Comment on the Foregoing:

The first and second statements of the experts are, of course, contradictory[4]. The first says only: 1 degree = 17 1/2 leagues = 62.5 miles. From this, 1 league = 3.5714 miles and the earth's circumference = 6300 leagues = 22,500 miles.

The second statement says only: the league = 4 miles, a mile = 8 stades and 500 stades make the degree. From this, the league = 32 stades, the degree = 62.5 miles, or 15.625 leagues, and the earth's circumference = 180,000 stades = 22,500 miles = 5625 leagues.

Harisse selects from the first statement only the ratio: 1 degree = 17 1/2 leagues, ignoring the reference to the degree also equalling 62.5 miles. He selected from the second statement the equalities: 1 league = 4 miles and 1 mile = 8 stades, ignoring the specification: 1 degree = 500 stades. Then, he applies his "Olympic" stade of 192.27 metres. Thus, he obtains:

1 mile	=	8 stades = 1538.16 metres
· 1 league	=	4 miles = 32 stades = 6152.64 metres
1 degree	=	17 1/2 leagues = 70 miles = 560 stades
	=	107,671.2 metres; and the earth's circumference = 6300 leagues
	=	25,200 miles = 201,600 stades = 38,759,728 metres

(the last is slightly in error, the figure should have been 38,761,632 metres).

There are more conventional interpretations which can be made of each of the expert's statements. With regard to the first; if the mile is taken as the standard sea-mile of the era, the Roman or Italian mile of 1482 metres, the following development is obtained:

1 league	=	3.5714 miles = 5293 metres
1 degree	=	17.5 leagues = 62.5 miles = 92,625 metres, and
		the earth's circumference = 6300 leagues = 22,500 miles = 33,345 kilometres

With regard to the second statement:

If the stade is taken as the Olympic stade of 185.2 metres,

1 mile	=	8 stades = 1481.5, say 1482 metres
1 league	=	4 miles = 32 stades = 5928 metres
1 degree	=	500 stades = 62.5 miles = 15.625 leagues
	=	92,625 metres

The earth's circumference = 180,000 stades = 22,500 miles = 5625 leagues = 33,345 kilometres

If the stade were the Philetaerian stade of 197.6 stades (see Table 2, Chapter 3):

1 mile	=	8 stades = 1580.8, say 1581 metres
1 league	=	4 miles = 32 stades = 6324 metres
1 degree	=	500 stades = 62.5 miles = 15.625 leagues = 98,813 metres, and the earth's circumference = 180,000 stades = 22,500 miles
	=	5625 leagues = 35,572.5 kilometres

If the stade were the Ptolemaic or Royal Egyptian stade of 210 metres (see Table 2, Chapter 3) which Dreyer believed Ptolemy had used, then:

1 mile	=	8 stades = 1680 metres
1 league	=	4 miles = 32 stades = 6720 metres
1 degree	=	500 stades = 62.5 miles = 15.625 leagues = 105,000 metres the earth's circumference = 180,000 stades = 22,500 miles
	=	5625 leagues = 37,800 kilometres.

There is nothing in the foregoing extrapolations which violates the conditions of the specific statement for which each is provided. Moreover, each assumption is based solidly on ancient authorities or more modern research on ancient and medieval geography and metrology.

In summation, the Spanish experts at the Badajoz Conference of 1523-24 gave testimony which because of the contradictions between the two basic statements and the failure to define the lengths of the smallest units in these statements, leads to a number of interpretations. Harisse has chosen to ignore certain of the specifications in each statement in order to arrive at a single interpretation of this testimony. An alternative approach is to consider each statement as representing the views of at least one of the Spanish experts (even though it is not known which) and provide the most plausible interpretations for the statement unencumbered by the contradictions of the other statement. The cosmographical views of the Spanish experts, as interpreted in the foregoing, are set forth in Table 3.

But what about the location of the Line of Demarcation, for which purpose the Badajoz Conference was convened and which was the centerpiece of Harisse's book?

We quote Harisse directly, "On the basis adopted by those cosmographers, the Line of Demarcation would cut the north coast of South America, on their sphere, in 47° 17' west of Greenwich.

But on our sphere their calculations for the location of the Line would correspond with 46° 36' west of Greenwich."

The Badajoz Junta, as is well known, failed to come to an agreement owing to the Portuguese experts, who could not overcome this dilemma: If the Line was pushed more to the west, Portugal would gain a greater part of Brazil; but she might lose all rights over the Moluccas, as the Line, of course, had to be carried to the other hemisphere as well.

Unit →	Stade	Mile	League	Degree	Earth's Circumference
Column No. →	①	②=8×①	③=4×② unless shown otherwise	④=62·5×② or =17·5×③	⑤=360×④
Identifica-tion ↓	✕	✕	✕	✕	✕
Harisse's Interpolation of the Two Statements	192.27 metres	1538.16 metres	6152.64 m.	107,671.2 m. =560×① =70×② =17.5×③	38,759.7 Km. =6300×③ =25,200×② =201,600×①
Reviewer's Conjecture #1, based on First State-ment	185.2 m.	1482 m.	5293 m. =3.5714×②	92,625 m. =500×① =62·5×② =17.5×③	33,345 Km. =6300×③ =22,500×② =180,000×①
R.C. #2 based on Second Statement	185.2 m.	1482 m.	5928 m.	92,625 m. =500×① =62.5×② =15.625×③	33,345 Km. =5625×③ =22,500×② =180,000×①
R.C. #3 based on Second Statement	197.6 m. Philetaer-ian stade	1581 m.	6324 m.	98,813 m. =500×① =62.5×② =15.625×③	35,572.5 Km. =5625×③ =22,500×② =180,000×①
R.C. #4 based on Second Statement	210 m. Ptolemaic or Royal Egyptian stade	1680 m.	6720 m.	105,000 m. =500×① =62.5×② =15.625×③	37,800 Km. =5625×③ =22,500×② =180,000×①

Table 3. Interpretations of the Cosmographical Views of the Spanish Experts at the Bajadoz Conference of 1523-1524.

Exploration, 1518-1522

Strangely, the testimony of the experts referred not at all to some of the momentous events which had occurred just prior to the conference which were certainly of cosmographic interest. Since Juan de Grijalva's exploration of the Yucatan and Mexican shoreline, the following events had transpired:

- Alvarez de Pineda explored the coastline of Gulf of Mexico from some point in Mexico all the way to Florida (1519).

- Francisco Gordillo explored Florida's Atlantic coast continuing northward as far as Cape Fear, North Carolina (1520).

- Hernan Cortes conquered Mexico (1519-1522) and initiated conquest of Guatemala, Honduras, and El Salvador.

- Ferdinand Magellan discovered the Straits of Magellan, crossed the Pacific, discovered Guam and the Philippines and was killed at Cebu in a clash with natives (April 1521). Sebastian El Cano assumed leadership, loaded spices at the Moluccas, passed into the Indian Ocean, rounded the southern tip of Africa and returned to Spain. The expedition in just under 3 years (1519-1522) had circumnavigated the earth.

From the charts and the log of the Magellan-Cano expedition, several things began to be evident. Among them:

a. The South Sea, or Pacific Ocean, was no mere strait, but a body of enormous width.

b. The South American continent was not only very long, but quite broad, confirming previous exploration of the Spanish Main.

c. The length of the Eurasian continent was much less than shown on current maps and padrons.

d. The earth appeared to be greater in girth than the cosmographers had said it was.

e. The route from Spain to the Spice Islands via the Straits of Magellan was longer than from Portugal, using the route around the southern tip of Africa.

f. Not only did the Moluccas lie in the Portuguese area of primacy, if the Line of Demarcation were extended across the poles, but it appeared that newly discovered Guam and the Philippines did, as well.

The last item dictated that geodetic conclusions and geographic information gained on the Magellan-Cano expedition be kept secret from the rest of the world as long as possible. Spanish padrons, however, began to reflect the new shape of the world within a few years, as we shall see.

The Diego (Diogo) Ribeiro Padron of 1529

One of the experts present at the Badajoz Conference, but who was not quoted by Harisse, was Diego Ribeiro, a Portuguese in the employ of the Casa de Contratacion. He had been working in Seville since 1519 and ultimately became "Cosmographer and Chartmaker" for the Casa. He is credited with three world maps, one produced in 1527 and two in 1529. One of his 1529 maps is reproduced in Figure 9, and is regarded as a copy of the Spanish padron real. We quote Skelton[5] " . . .

> Magellan's Track in 1519-1520 is marked by drawings of his two ships, the Vitoria and Trinidad; and the place names in the inset record his passage along the coast of Patagonia and through his strait. . . The length of Magellan's crossing of the "Mar del Zur" had already impressed cartographers, and Ribeiro represents the width of the Pacific, from Peru to the Moluccas (Gilolo), as 125 degrees of longitude. This is 25 degrees more than the width of Agnese in 1536 or Velasco in 1575 . . . but is still 25 degrees short of the true width (150 degrees), and the underestimate was perhaps prompted by the political exigency which required the Moluccas to be laid down on the Spanish (or eastern) side of the Demarcation Line in the east.

The Ribeiro map shows the Mediterranean at about its true length, but the Lisbon to Hispaniola longitudinal distance is about 5° short and that from Lisbon westward to the Asian mainland, at 208°, is about 23° short. Clearly, enormous progress is shown in depicting known lands and the seas surrounding, or bordering, them in more nearly their proper proportions. But nothing is said about the circumference of the earth or the length of a degree, about which Ribeiro and others must have speculated. And, as we shall see, because of secrecy imposed by the Spanish government at the time, maps drawn by cartographers and cosmographers not privy to this priveleged information continued in the Toscanelli-Behaim-Columbus concept for some time to come.

In 1529, Charles V of Spain reached an accommodation with the Portuguese. By the Treaty of Zaragoza of that year, Spain acknowledge the Moluccas were on the Portuguese side of the Line of Demarcation and relinquished all claims to them deriving from the Magellan expedition. Portugal paid Charles 350,000 ducats. It was agreed that the Philippines, which are just as far west as the Moluccas, would be retained by Spain. They remained Spanish possessions for the next 3 1/2 centuries.

A Bit of Geodesy in the 16th Century-Fernel Measures a Degree of Latitude[6]

Sometime during the period 1526-1528, Jean Fernel, the celebrated French physician known as "the modern Galen", undertook to measure the length of a degree of latitude. He had been studying (much as we have here) what earlier authorities had to say on the subject. Eratosthenes' 700 stades were generally considered to be equivalent to 87 1/2 Italian (Roman) miles. Regiomontanus had reduced this to 640 stades, or 80 Italian miles. Ptolemy's 500 stades were, in Fernel's time, considered equivalent to 62 1/2 Italian miles, but Fernel, himself, accorded the degree only 60 such miles. Hence, the impetus to "straighten things out".

Fernel measured the distance between two points on the Paris-Amiens road, "where the road ran true north and south", counting the revolutions of a carriage wheel as it traversed this distance. (A similar method for measuring distance was described by Vitruvius, Roman military engineer, architect, and author who flourished c. 16 B.C.) Over a period of four days, Fernel estimated that the 6 foot, 6 digit carriage measuring wheels had made 17,024 revolutions. Since the circumference of these wheels was π x 6 6/16 = 20.028 ft or 4 paces, the total distance between his end observation points was 68,096 paces, which Fernel adjusted downward to 68,095 1/4 paces.

Fernel's celestial observations at the end points of his surface measurement were made of the sun's altitude at noon, exactly four days apart. The difference in the sun's altitude at the two sites on these days was 2° 32' of which 1° 29' 25" was attributed to the sun's change in declination in this time and the remainder, 1° 2' 35", due to the difference in latitude of his end sites. Fernel must have used 2' 35" to offset certain perceived errors in his procedure for his stated length for a degree of latitude was 68,095 1/4 paces (pas). He didn't stop there but defined his result in terms of other well-known units.

Delambre, our source for Fernel's measurement, reports that Fernel equated his 68,095 1/4 paces to 68 Italian miles and 95 1/4 paces (68.09525 Italian miles), or 544 Roman stades and 95 1/4 paces, which we put at 100,883 metres-based on the 185.2 metre stade and the 1481.5 metre mile. This result is within about 9% of the mean length of a degree of latitude (at a latitude of 49°-50° N.). Fernel put the circumference of a great circle (of a spherical earth) at 24,514 Italian miles and 285 5/7 paces (24,514.2857 Italian miles) which we equate to 36,317.92 kilometres.

Of more than passing interest is the trouble to which Fernel went to insure that all the units he used were well defined. Thus, he specified the following relationships: the digit has a length of 4 grains; 4 digits make the palm; 4 palms the foot; 6 palms the cubit; 10 palms make the simple pace and 5 feet the geometric pace; the Roman (Italian) stade is equal to 125 geometric paces and the Italian mile 8 stades or 1000 paces. It is evident (from our discussion of the barley grain's length and width, in Chapter 2) that Fernel's specification of 4 grains to the digit referred to the length of that botanical standard, approximately 4.6 mm.

Despite the seeming impeccability of the foregoing, Jean Picard, the celebrated 17th century French priest, astronomer, and geodesist, evaluated Fernel's 68,095 1/4 paces as equivalent to 56,746 toises (110,600 metres)[7]. This would put Fernel within 1% of the true length of a degree of latitude. We believe Picard may have erred in according Fernel's foot the value of the French pied in his time, 0.32484 metre, instead of the value of the Roman foot as defined by Fernel which is nearer 0.2963 metre.

The Views of Alonso de Chaves in 1537 and Those of Petrus Apianus and Gonzalo Fernandez Oviedo in 1545

In 1537, Alonso de Chaves published a detailed description of the New World for Spanish pilots titled *Espejo de Navagantes*. Volume four of this work and a short biography of Chaves were published in Madrid in 1977 by P. Castañeda, M. Cuesta, and P. Hernandez. From these works, it is evident that Chaves set 5 5/7 degrees of latitude = 100 Spanish leguas, or 10 x 5 5/7 = 57.143° = 1000 leguas. Now 57.143° is approximately equal to 1 radian (180° ÷ π = 57.296°). Thus, Chaves has been interpreted by some as imputing geodetic significance to the Spanish legua, i.e. the radius of a spherical earth is approximately 1000 leguas.

However, 5 5/7° = 100 leguas is merely another way of saying that 1° = 17 1/2 leguas. Further, Chaves also set the legua equal to 4 Italian miles so that 1° = 70 Italian miles = 70 x 1481.5 metres = 103,705 metres and a great circle = 37,334 kilometres, both results about 6.7 % too small. Still, Chaves' estimate was a bit closer to reality than Fernel's measurement.

Petrus Apianus' views as to the circumference of the earth and the length of a degree underwent some minor change between 1520 and 1545. Whereas the map in Figure 8 led to the conclusion that his degree of longitude at the Equator probably was intended to measure 62 1/2 miliaria (Italian or Roman miles), there is no equivocation in his 1545 position, shown in figure 10. Clearly the curcumference of the earth is equal to 5400 German miles, or 21,600 miliaria, from which one degree of longitude at the Equator equals 15 German miles or 60 miliaria This was a mariners' measure, used by Columbus, among others, despite his well known advocacy of Alfragan's 56 2/3 miles per degree. (The German mile here appears equal to the 4 Italian mile league, but there can be no assurance of this. See Figure 17 which indicates otherwise.)

Harisse is our source for Oviedo's views[8] which, stated most simply, were 1 equatorial degree = 17 1/2 leagues. As shown in Table 3, in the absence of further specifications indicating the relationship of the league to other, less variable smaller units, one can only speculate as to the length of Oviedo's degree. It does seem as though the 17 1/2 league degree might be gaining adherants.

The extraordinary variation in the views of the cosmographers, geographers, and navigators considered thus far as to the absolute length of the league or the degree, and the relationship of the two, made it impossible, as a practical matter, to pinpoint the precise location of the Line of Demarcation. This despite many attempts to do so and the seeming simplicity of the specification.

Comment for Figure 9, page 219

The original chart, which reposes in the Biblioteca Apostolica Vaticana, Rome, is considered to be a copy of the then current Spanish padron reál. Ribeiro, "cosmographer and chartmaker" for the Casa de Contratacion was privy to all the geographic and navigational information returned to the Casa by Spanish explorers and seamen. The Casa also encouraged knowledgeable Portugese (like Ribiero) to join its ranks, thus learning many Portugese geographical "secrets". Still, the most valuable such information were the marked up charts and logs used in the Magellan-Caño circumnavigation of the earth, 1519-22. (The original chart shows the track of Magellan's ships (Vitoio and Trinidad). Despite many gaps and distortions in the coastlines of landmasses, the placement of islands, and the somewhat confused treatment of the western Pacific, especially the Chinese mainland and off-shore islands, the overall distribution of the world's lands and seas is much improved. The Mediterranean's length is correct and the Lisbon-Hispañola distance almost so. And despite the hazy treatment of China and the political requirement that the Moluccas be shown in Spains "half of the world", the Lisbon to China westward longitudinal distance is enormously improved at 205-215°.

This outline is based upon a photocopy of the original map provided the author by the Biblioteca Apostolica Vaticana Rome, which also provided permission for its reproduction, and on a photograph of the original map made by Walter W. Roberts. It appeared as page 16-17 of the Times-Life publication "The Explorers" by Richard Humble and the Editors of Time-Life Books, Alexandria, VA 1978.

Ribeiro has chosen to place his compass roses at North and South 15°, 30°, 44°, 55°, and 64.5°. His rhumb lines are essentially E-W, and WNW-ESE, NNW-SSE, NNE-SSW, and ENE-WSW. While he employs 5 formal parallels (and several horizontal lines leading from the rose centers to the appropriate latitude scales), he shows only 3 formal meridians. Very lightly drawn vertical lines appear to have been intended to identify the longitude of the rose centers, but no longitudinal scale is evident.

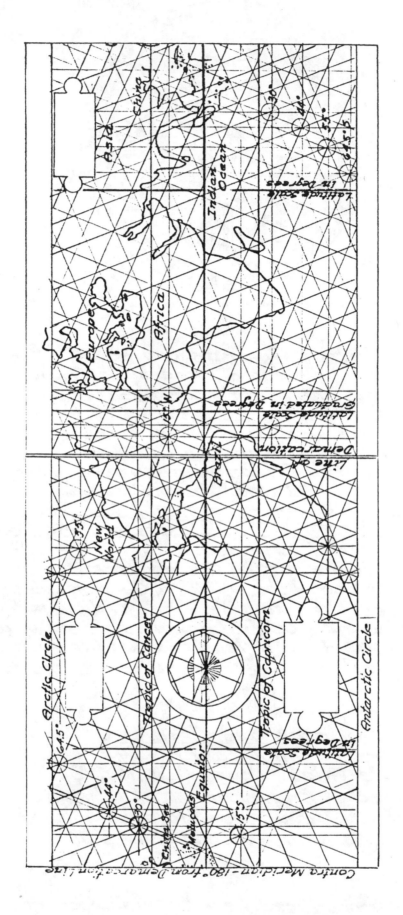

Figure 9. An outline of the Diego (Diogo) Ribeiro Sea Chart of 1529. Comments and credits on page 218.

219

Map to indicate the size of the earth. From: APIANUS, Antverpiæ 1545.
(Orig. size).

Figure 10. Petrus Apianus, professor of mathematics at the celebrated Academy at Ingolstadt from 1527, was one of the more popular cosmographers of the first half of the 16th century. His maps and diagrams were used at several universities as cosmographical manuals. The earliest of several editions in different languages was published at Landshut in 1524, titled "Cosmographicus liber Petri Apiani Mathematica studioso collectus".

The map above, used to illustrate a 1545 edition, expresses a widespread, but not universal view of the size of the earth: the equatorial and polar diameters of the globe = 1718²/₁₁ German miles = 6872⁸/₁₁ Italian miles; the circumference (π × diam.) = 5400 German miles = 21,600 Italian miles. From which a degree of latitude or of longitude at the equator = 15 German miles = 60 Italian miles (the mariners' measure).

Sources and Credits

Map is Figure 64, p. 102a, of A.E. Nordenskiöld's "Facsimile-Atlas", Dover Pubs., N.Y., 1973, first published in English in Stockholm, 1889. For more about Apianus, see Nordenskiöld, pp. 76a, 92a, 93a, 94a, 99a, 100a, 102a and n., 104b, 116a.

Map is reproduced here by permission of the U.S. publishers, Dover Publications, New York.

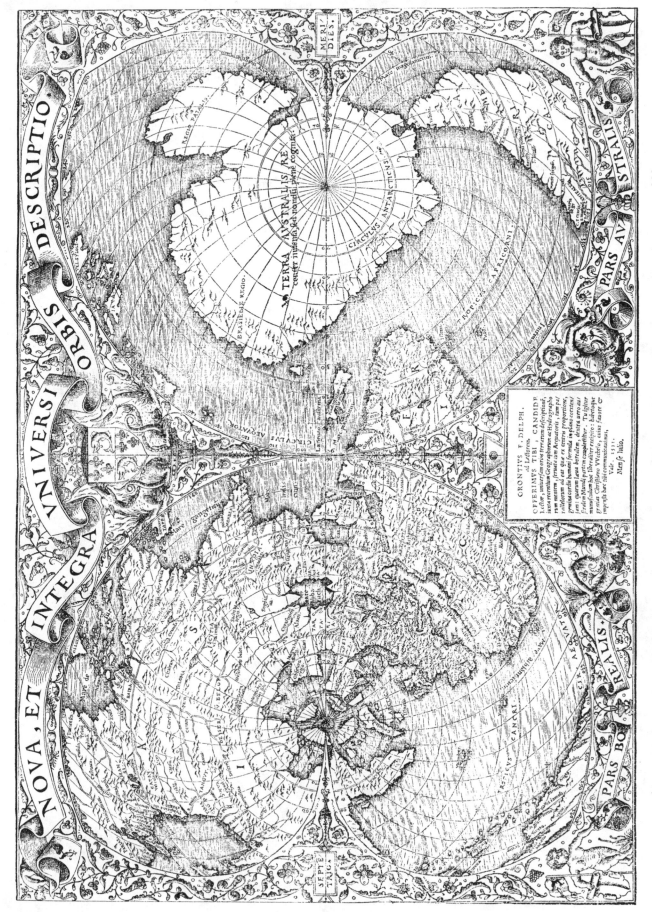

Figure 11. Fineaus' Double Cordiform Map on a projection of Iohannes Werner. Comments and credits on page 222.

Comments and credits for Figure 11, page 221
The poles are the centers of all parallel circles. While Japan is not shown, Mediterranean has a length of 57°, Lisbon to Hispaniola is 81°, and Asia mainland is 129° west of Lisbon.

This map is taken from Nordenskiöld's "Facsimile Atlas.". Nordenskiöld indicates that it served as a model for Mercator's Double Cordiform Map of the World of 1538. (See Figure 14.)

The Werner Projection is one of the "equal area" class, i.e. areas on map are equal to areas on the globe they are intended to represent. While increasing distortion results at longitudes increasingly distant from the mean meridian (in the Fineaus map, 90° longitude), the projection has several attributes. Since it represents a globe of radius " r ", the parallel circles centered at the poles have radii equal to:

$$\pi r(90°- Lat) + 180° \quad and \quad total\ lengths\ of\ 2\pi r\ cos\ latitude$$

i.e., their identical lengths on the globe. Parallel circles are equally spaced (10° of latitude is everywhere the same), meridians are equally spaced on parallel circles, and 10° of longitude bears its proper relationship to 10° latitude, i.e., 1/36 (2πr cos latitude.) (See comment on Figure 14 which employed the same projection.)

This map is reproduced by permission of the U.S. publisher, Dover Publications, New York, NY 1973.

Comments and credits for Figure 12, page 223
Parallels are straight equidistant lines. Meridian spacing on each parallel is obtained by dividing each parallel into the same number of spaces within arbitrarily chosen oval perimeter.

Grynaeus puts Japan 95° west of Canaries, Hispaniola 60° west of Lisbon, the Asian mainland 140° west of Lisbon, and gives the Mediterranean a length of 58°. Nordenskiöld, from whose "Facsimile Atlas" this map is taken, indicates (page 90b) that the major axis is 1.87 times as long as the minor axis.

In this projection degrees of longitude bear proper relationship to degrees of latitude only at (approximately) 25° North or South, being too short between the Equator and 25° North or South and too long between 25° North or South and the appropriate pole.

This map is reproduced by permission of the U.S. publisher, Dover Publications, New York, 1973.

Figure 12. Grynaeus' 1532 Map of the World on a Bordone Oval Projection. Comments and credits on page 222.

Figure 13.
Gerardus Mercator's Double Cordiform Map of 1538
(on the same projection as Orontius Finaeus 1531 map).
Mercator places Japan 103° west of the Canaries; Hispaniola 58° west of Lisbon; the Asian mainland is
131° west of Lisbon; and the Mediterranean has been given a length of 62°

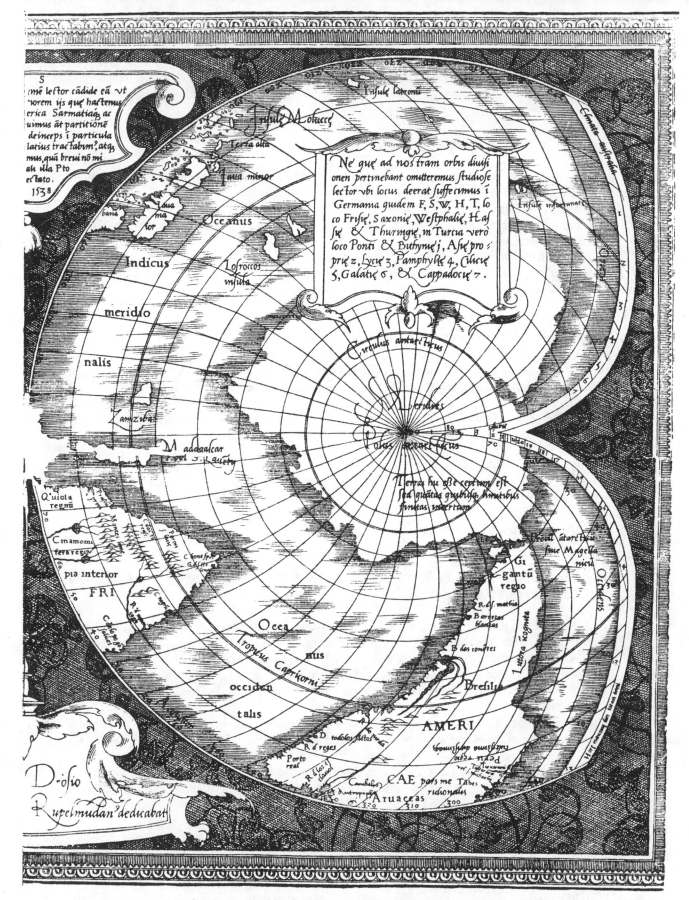

Credits and additional comments for Figure 13, page 224.
Mercator's map of 1538 is taken from Nordenskiöld's "Facsimile Atlas". While it still follows the Columbian concept of the distance westward to the Orient and the Marinus/Ptolemaic length of the Mediterranean, both in error, its depiction of the New World is far superior to any other map up to 1538 and for sometime beyond.

This is reproduced by permission of the U.S. publisher, Dover publications, New York, 1973.

Maps Drawn in the Period 1529 to 1545

Representative of the maps drawn during this period are the double cordiform projection of Orontius Finaeus of 1531 (Figure 11), the Simon Grynaeus map of 1532 on a Bordone oval projection (Figure 12), and the double cordiform projection of Gerardus Mercator of 1538 (Figure 13). None of these maps--despite the evident improvement in the sophistication of presentation--have improved their accuracy in showing the length of the Mediterranean.

With regard to the longitudinal distance from Lisbon to the Asian mainland, the Grynaeus map of 1532 shows it as 140°, and the other two are still at the Columbian 130° (approx.). However, except for the Finaeus, 1531 map, this group has by now achieved a realistic longitudinal distance from Lisbon to Hispaniola.

Exploration and Other Events in The Period 1524-1571

The improvement in the positioning of Hispaniola and, at least in the case of Mercator, of Florida, the east coast of the United States, and the Gulf of Mexico probably reflect the following:

- The great increase in the number of re-supply ships proceeding from Spain to each of the Greater Antilles, to Mexico, and to Panama and Central America.

- The 1524 expedition of Giovanni de Verazzano, in the service of Francis I of France, which explored the U.S. Atlantic coast from Florida to Nova Scotia, and which is credited with the discovery of New York Bay.

- The exploration and mapping of the entire U.S. Coast in 1524-25 by Esteban Gomez, a Portuguese in the Spanish service.

- The establishment by Lucas Vasquez de Ayllon of a short-lived settlement on the South Carolina coast in 1526.

- The ill-fated attempt by Panfilo Narvaez and Alvar Nunez Cabeza de Vaca to colonize Florida in 1528. Narvaez and all but four of the contingent lost their lives, but de Vaca and three others managed to cross Texas, the Rio Grande country of Mexico and New Mexico, part of Arizona and Sonora to Sinaloa, reaching there in 1536. De Vaca's story was the proximate reason for the authorization of the subsequent De Soto and Coronado expeditions.

But other expeditions were reaching out into the Pacific, probing the shorelines of North and South America, exploring the interiors and accomplishing astonishing conquests:

- To confirm Magellan's information on the lengthy route to the Moluccas via the Straits and the Pacific crossing and to maintain a bargaining position with Portugal on the Moluccas, a small fleet under Garcia de Loaysa, with Sebastian del Cano commanding a ship, was sent to Molucca via the Straits of Magellan in 1525.

- In 1527, Alvara de Saavedra was dispatched from Mexico with a relief squadron to aid de Loaysa. Saavedra was supposed to return to Mexico, rather than continue on to Spain around Africa. In several attempts, wind and current forced him back each time and he died in Indonesia.

- Francisco Pizarro, with the backing of the Spanish crown, and in association with Diego de Almagro and Hernando de Luque (vicar of Panama) conquered Peru, in 1531, and Ecuador (Kingdom of Quito), Chile, and part of Bolivia, between 1532 and 1542.

- Francisco de Orellana, in 1541, descended the Napo, in Ecuador, to its junction with the Amazon and the latter for 1700 miles to its mouth. His craft, a crude brigantine built in the rain forests bordering the Napo (by Gonzalo Pizzaro and assigned to de Orellana along with 50 men to get help for the beleagured main force) carried him back to Spain. By this feat, Spain gained a more direct feel for the enormity of the width of equatorial South America.

- Jacques Cartier, in four voyages, explored the Gulf of St. Lawrence, discovered the entrance to the St. Lawrence River and explored that river for 500 miles upstream (as far as modern Montreal). Indians told him of a country called Saguenay, rich in gold, and of a great river flowing to the south, Mississippi. No settlements were established. 1534-1542.

- Francisco de Ulloa explored Mexico's west coast from Rio Fuerte north to the mouth of the Colorado, then south on Baja California's Sea of Cortes coast to Cape San Lucas, then north on the Pacific coast of Baja to about 28° N. He had coasted 2000 miles and shown that Baja California was not an island, but a peninsula. 1539.

- Operating from Mexico, Francisco Vasquez de Coronado and lieutenants (Alarcon, Diaz, Cardenas) explored Sonora, Arizona, New Mexico, Texas, Oklahoma, Kansas, possibly the southeastern tip of Colorado, part of northern Baja California, the lower Colorado River and its California and Arizona banks, and discovered the Grand Canyon, in period 1540-1542.

- Originating in Spain, which they left in April 1538, and "filled out" in Cuba between early June 1538 and early May 1539, Hernando (Hernan) de Soto, Luis de Moscoso, and a large force of Spaniards, Portuguese, and Indians, explored Florida from Tampa Bay northward thru the peninsula and westward thru the panhandle, Georgia, South Carolina, North Carolina, Tennessee, Alabama, Mississippi, Arkansas, and Louisiana, discovering the Mississippi, on 21 May 1541. One year later, to the day, De Soto was dead, having succumbed to fever and exhaustion. His survivors, under Luis de Moscoso, probed deep into Texas, then reversed their tracks, returning to the banks of the Mississippi where they built a small fleet of sailing craft. They then descended the Mississippi to the Gulf of Mexico and coasted Louisiana, Texas, and Tamaulipas (Mexico) to Panuco (Tampico), arriving there in September 1543.

- In a voyage which started at Navidad, on Mexico's Pacific coast, in June 1542, Juan Rodriguez Cabrillo, a Portuguese, and Bartolome Ferrelo, a Levantine, in the service of Viceroy Mendoza of New Spain, explored the Pacific coast of Baja California and California, discovered San Diego Bay, Catalina and San Clemente Islands, went ashore at Ventura, discovered Santa Cruz, Santa Rosa, and San Miguel Islands and reached a point near Fort Ross, California (38° 31' N) on November 14, 1542. Cabrillo died at San Miguel on January 3, 1543 from complications resulting from a fall. Ferrelo took over and on February 18, 1543, started a second voyage from San Miguel reaching the mouth of the Rogue River, in Oregon (42° 30' N), on March 1, 1543. With great difficulty, because of storms, unfavorable winds, and scurvy among most of the surviving crewmen, Ferrelo made it back to Navidad by April 14, 1543. The Cabrillo - Ferrelo expedition plus the Coronado and De Soto expeditions set the stage for further exploratory efforts within the United States, but immediately gave Spain's cosmographers and chart makers vital information on the great east to west length of the land north of Mexico.

- About the time the Cabrillo-Ferrelo voyage was getting underway, Ruy Lopez de Villalobos led an expedition from Mexico to the Philippines which were dutifully renamed after Philip II of Spain. However, colonization was delayed for a generation because of the impossibility of returning across the Pacific against the North Equatorial current and the trade-winds responsible for the current.

- In the period 1565-1571, operating from Mexico, Miguel Lopez de Legaspi colonized the Philippines for Spain. From the very beginning of this program a major effort was mounted to find a practical route back from Manila to Acapulco. In 1565, Urdanetta, de Legaspi's chief pilot and Arellano, one of his captains, independently pioneered a route northward along the coast of Japan and then east across the Pacific with the Japanese current and the westerlies at approximately 42° N, then southward along the North American coast to Mexico. Urdanetta's sailing directions were incorporated in the standard "rutters" for the South Sea issued to Spanish pilots from the 16th to the 18th century.

    With the achievement of a practical return route, colonization of the Philippines proceeded apace and there was established the annual Manila galleon carrying silver from the mines of Peru and Mexico to the Philippines where the ship (usually one, sometimes two) was laden with spices, Oriental silks, and Chinese objets d'art.

    It was colonization of the Philippines and the establishment of regular trade between that land and Peru and Mexico which appears finally to have come to the attention of the map makers of Europe initiating some much needed changes in their world maps.

COSMOGRAPHY, CARTOGRAPHY, AND EXPLORATION IN THE PERIOD 1570-1595

The Abraham Ortelius (Antwerp) Map of 1570

The Ortelius world map, shown in Figure 14, was the first to sharply reduce the length of the Eurasian continent to realistic proportions. Conversely, the longitudinal length from Lisbon westward to the Asian mainland, at about 220°, is only 11° less than the actual length. Despite this improvement, the width of North America is grossly exaggerated while the Pacific is shown commensurately narrower.

The broadening of Mexico and the U.S. can be attributed to inflated itinerary distances reported by the official chroniclers of the Coronado and de Soto expeditions and a misunderstanding by geographers of the precise length of the units used by the explorers.

With regard to distances over the Pacific between the coast of North America and the Asian Islands such as Japan, the Philippines, and the Moluccas, there was another phenomenon at work, tending to shorten considerably the estimates of seasoned pilots and navigators--ocean currents. There was no practical method whereby 16th century seaman could sense the speed of the North Equatorial Current which bore them from Mexico to the Philippines and the Moluccas and that of the Japanese (or North Pacific Current) which brought them back. Thus, unaware, for some time, of their true speed, and being able only to *Text continues on page 230.*

228

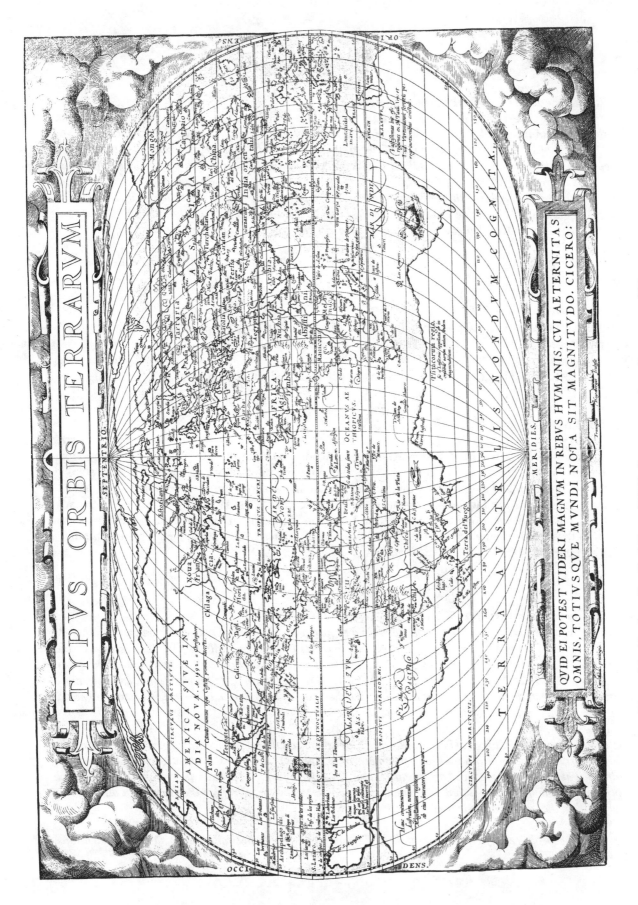

Figure 14. Abraham Ortelius' Map of 1570. Comments and credits on page 230.

Text continued from page 228.
measure the ship's speed relative to the surface of the water, the actual ship's speed and the distance covered were understated.

Returning, now, to the Ortelius map, as might be expected, the Lisbon to Hispaniola distance was almost perfect and the length of the Mediterranean much improved although still about 10° too long.

Rumold Mercator's World Map of 1587

The remarks made for the Ortelius map hold for Rumold Mercator's handsome map shown in Figure 15: the Eurasian continent is still somewhat too long although drastically reduced from the stretching given it by Marinus; the breadth of the United States and South America are grossly exaggerated; and the Pacific Ocean although no longer a strait, as in the maps before 1570, has its width shown significantly less than actual. The longitudinal distance from Lisbon to Hispaniola is shown correctly and that from Lisbon to the Asian mainland is about 11° less than actual. Mercator's Mediterranean is, like Ortelius's improved, but still 10° too long. *Text continues on page 233.*

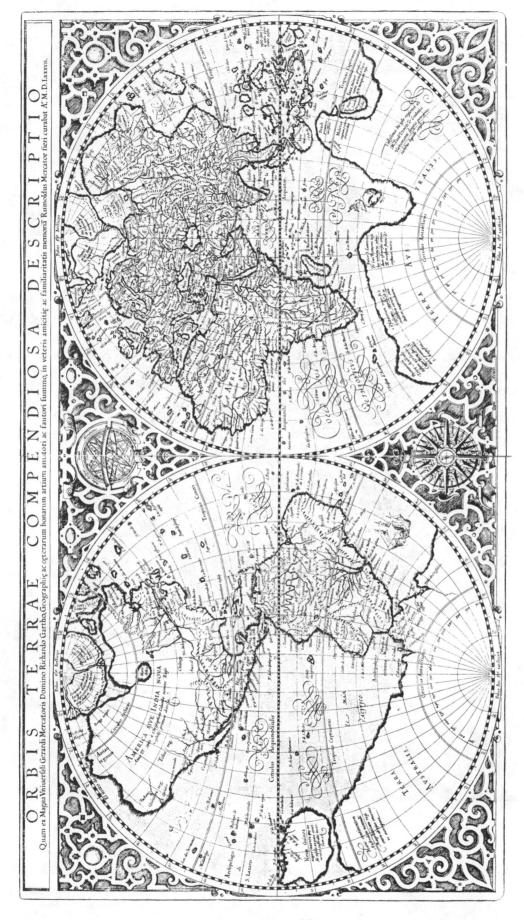

Figure 15. Rumold Mercator's World Map of 1587. Comments and credits on page 230.

Figure 16. De Judaeis' World Map on a Polar Projection. Comments and credits on page 233.

Credit and comments for Figure 16, page 232.
Despite the year, 1593, Japan is placed 152° west of Canaries; Hispaniola and Asian mainland are 66° and 167° west of Lisbon; and the length of the Mediterraneaan is a full 62°. Japan is placed closer to North America than to Asia and the Pacific Ocean north of the Tropic of Cancer has become the Strait of Anian.

Nordenskiöld, from whose "Facsimile Atlas" the De Judaeis map is taken, indicates the map - like Mercator's 1569 map of the polar regions, on which it is based - is drawn on the "equidistant polar" projection, in which degrees of latitude are equal.

This construction was developed by Mercator as a compromise between the stereographic polar projection, which results in degrees of latitude near the equator being considerably longer than those near the poles, and the orthographic projection, which results in just the opposite, However, the equidistant polar projection is best employed for latitudes from about 60° N to the pole, which was the case in the 1569 Mercator polar map. In the De Judaeis map, the relationship between latitude and longitude becomes increasingly distorted in the latitudes from 60° N. southward, the degrees of longitude being 8.5% too long at 50°N., 14% at 40°, 20.1% at 30°, 29.9% at 20°, 41.7% at 10 °, and 57% too long at the equator.

This map is reproduced by permission of the U.S. publishers of the "Facsimile Atlas", Dover Publications, New York 1973.

The Polar Projection of Cornelius de Judaeis (Antwerp) 1593

So far as the cosmographic aspect of de Judaeis' polar projection, Fig. 16, is concerned, it is quite inferior to both the Ortelius and Mercator maps. The Eurasian continent at over 190° is almost as distorted as the maps, pre-1570. This and the grossly broadened "America" reduce the Pacific Ocean to the "Streto de Anian." The length of the Mediterranean is a full 62°, 50% longer than actual. De Judaeis shows the longitudinal distance between Lisbon and Hispaniola at 66°, only 5 1/2° greater than the actual distance. However, because of his reluctance to reduce the length of the Eurasian continent, the longitudinal distance westward from Lisbon to the Asian mainland is still only 167° - and this despite 100 years of exploration in the New World and the Pacific.

De Judaeis, like Mercator and Ortelius, misplaces Quivira and Tiguex (visited by the Coronado expedition of 1539-42), putting them on the U.S. west coast, instead of Kansas and New Mexico, respectively. He appears not to have heard of Magellan's long crossing of the Pacific, nor of the many Spanish crossings since then, including Urdanetta's discovery of the return route from the Philippines to Mexico utilizing the Japanese and North Pacific Currents.

The Barentszoon Chart of the Mediterranean Sea, 1595

One of the better maps of the second half of the 16th century is the Chart of the Mediterranean Sea drawn by Willem Barentszoon and shown in Figure 17. While not a World Map, it does contain valuable cosmographic information. As is usual for sea charts of the day, it contains a myriad of rhumb lines emanating from several strategically placed compass roses, and a scale of latitudes along both vertical borders. It also contains a scale relating degrees of latitude, or of longitude at the Equator, to Italian miliaria, German miles, and Spanish leguas.

From this scale:

$$1 \text{ degree} = 60 \text{ Italian Miliaria}$$
$$= 15 \text{ German Miles}$$
$$= 17 \tfrac{1}{2} \text{ Spanish Leguas}$$

Since the Italian or Roman Miliaria was the most enduring unit of measurement we have encountered in our long search , with a length of approximately 1482 metres, it is evident, then, that Barentszoon considered the degree to have a length of 88,920 metres, and the earth's circumference to be 21,600 Italian miles, 5400 German miles, 6300 Spanish leguas, or 32,000 kilometres.

Also, it is apparent that Barentszoon's Spanish legua equals not 4 Italian miles, or 3.5714 as shown in Table 3 (Reviewer's conjecture #1), but 3.43 Italian miles. At 5079.4 metres, it is about midway between the smallest and the largest Spanish leguas we have encountered (see Table 2, Appendix B). Further, it is evident that Barentzoon did not believe the German mile and the Spanish legua had the same length.

Of greater significance, however, is the fact that despite the passage of over 100 years since Christopher Columbus' first voyage to the New World, we find a competent maker of sea charts employing the same measure, 1 degree equals 60 Italian miles, that Columbus employed.

The Mendana Voyages of 1567 and 1595

Lest one gain the impression that the equivalency, 1 degree = 17 1/2 Spanish leguas, had become universally employed by the second half of the 16th century, Skelton reports[9] that Alvaro de Mendana, who discovered the Solomon Islands in 1567 and the Marquesas group in 1595, employed the equivalency one degree = 20 leguas.

The Mendana voyages, which originated in Callao, Peru, provide another example of how utilization of a strong ocean current - in this case the South Equatorial Current - resulted in a significant underestimate of the ship's speed, and distance covered. Gallego, Mendana's chief pilot, estimated the Solomon Islands landfall to be 1700 leagues (85 degrees) due west of Peru. The true distance is more nearly 121°, equivalent at 10 °S latitude to about 2380 of Gallego's leagues, 13,240 kilometres, or 7150 nautical miles, about 1/3 the earth's circumference - an astonishing feat for a saling craft were it not for the ocean currents. Those currents, driven by the trade winds, bolster the speed of a ship beyond that achievable from wind alone and have been estimated to add 25-40 miles per day to the distance a sailing craft can make from wind alone.

Returning, now , to the Mendana/Gallego equivalency of 20 leagues to the degree[10]. This ratio ultimately became the dominant one throughout Spain as is shown by Table 1, Appendix B. This was at a time when a league is known to have had a length of 6666 2/3 varas castellanas, or 5573 metres. The degree, thus, measured 111,460 metres.

It is not considered likely that Gallego would have employed a ratio between the degree and the league not sanctioned by the Spanish government. However, because the length of the legua has varied so widely with time and place, one cannot conclude that Spain had, in Mendana's time, arrived at an accurate length for the degree - although this is possible.

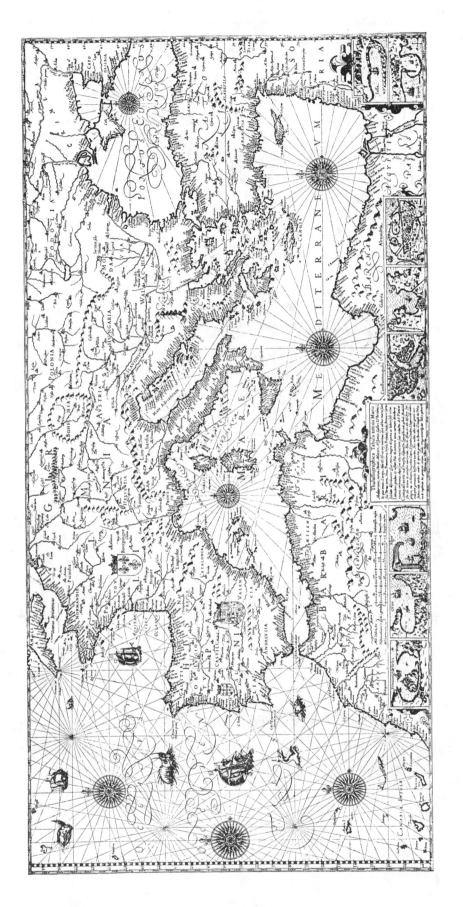

Figure 17: Legend on Chart relates 60 Italian Miles to 17½ Spanish Leguas, 15 German Miles, and 1 degree. Thus, 1 Spanish Legua = 3.42857, say 3.43 Italian Miles, or 5079.4 metres. The Earth's circumference = 21,600 Italian Miles = 6,300 Spanish Leguas = 32,000 Km. This chart taken from A.E.Nordenskiöld's "Facsimile-Atlas", p.39; Dover Publications, N.Y., 1973. It is reproduced here by permission of Dover Publications.

235

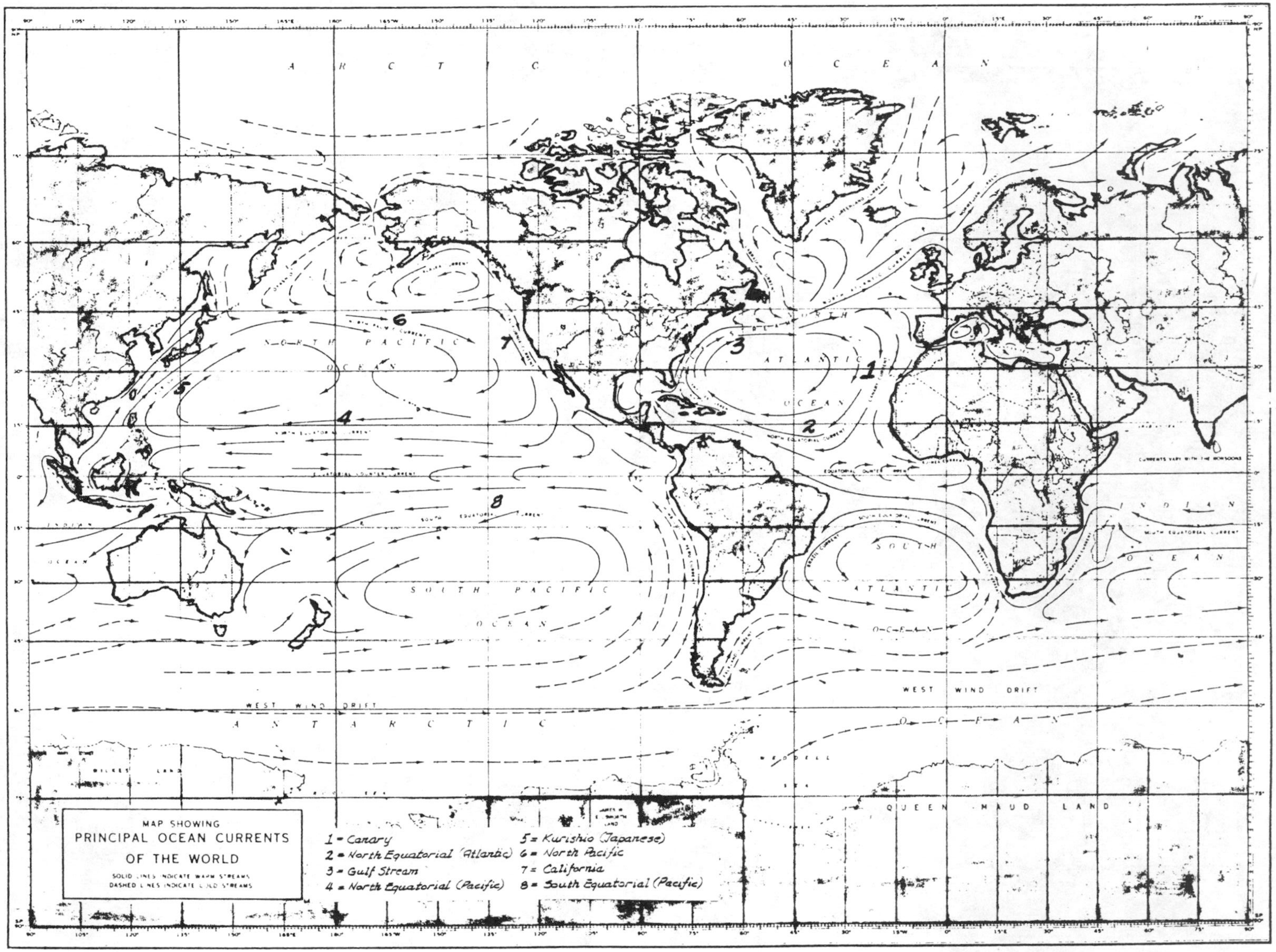

Figure 18. Principal ocean currents of the world. Comments and credits on page 237.

236

Credits and comments for Figure 18, page 236

Columbus pioneered use of the Canary and North Equatorial Currents to reach the New World and the Gulf Stream to return. Saavedra was first to use the North Equatorial Current between Mexico and Moluccas and the Philippines. Urdanetta discovered the Kurishio (Japanese) and North Pacific Currents in pioneering a return route from the Philippines to Mexico. The annual Acapulco - Manila galleons were initiated shortly thereafter (1565). Mendana pioneered use of the South Equatorial Current from Callao, Peru to the Solomons and Philippines (1567).

Ocean currents result from the effects of the wind above the surface of the sea and the effects of temperature and salinity differences within the sea.

"Map Showing Principal Ocean Currents of the World" is taken from the publication "Tidal Currents", of uncertain date, issued by the Environmental Sciences Service Administration.

Extended comments for Table 4, page 238

All lengths and distances are approximate, intended to show whether cartographers had accepted the Marinus concept of an elongated Mediterranean and Eurasian continent with the still further lengthening derived from Marco Polo's travels - a concept accepted in the main by Toscanelli, Behaim and Columbus.

Length of the Mediterranean is taken between the east end of the Strait of Gibraltar and the easternmost shore of the Gulf of Issus.

Distances to Hispaniola are taken as distance to Santo Domingo, approximately 1/4 the length of the island west of the east end.

Distance to the Asian mainland westward from Lisbon is taken to the meridian where 36°N. parallel cuts the China Coast.

It will be noted that - except for the Ribeiro copy of a Spanish padron - every world map followed the "Columbian Cosmography", described above in the first paragraph, until the 1570 Ortelius and the 1587 Rumold Mercator maps. These maps reflect reports of the several Pacific crossings, in both directions, since Magellan's circumnavigation of the globe.

The 1592 De Judaeis Polar Projection is, cosmographically, little improved over the 1506 Contarini Conical Projecttion.

Year	Cartographer	Length, in Degrees of Longitude, of		Degrees of Longitude Westward from Lisbon to		Remarks
		Mediterranean	Eurasian Continent	Hispaniola	Asian Mainland	
	ACTUAL	41.67	129.75	60.46	230.75	See Extended Remarks on other side.
1500	Juan de la Cosa (Fig.1)	42	—	60	—	Sea chart; first to show New World.
1502	Cantino (Portuguese Padron) (Fig.2)	37	—	53.3	—	First to show "Line of Demarcation"
1506	Contarini Conical Projection (Fig.3)	58	197	72	163	First printed map to show continent of America
1507	Waldseemüller (Fig.4)	58	233	48	127	Named New World continent "America"
1508	Ruysch Conical Projection (Fig.5)	55	225	74	135	————
1509	Sylvani Cordiform Projection (Fig.6)	58	Approx. 216	48	Approx. 144	Uncertain of China coast. Shows 1°=62½ Miliaria.
1512	Stobnicza Homeother Projection (Fig.7)	61	231	46	129	First to employ 2 hemispheres and show N. and S. America as one continent.
1520	Apianus Cordiform Projection (Fig.8)	63	229	46	131	Shows length of degree of long. at different lat.
1529	Ribeiro (Spanish Padron) (Fig.9)	42	152	54.6	208	Incorporated results of Magellan circumnavigation of the globe.
1531	Finaeus Double Cordiform Projection (Fig.10)	57	231	81	129	Served as model for 1538 Mercator Map of World.
1532	Grynaeus Oval Projection (Fig.11)	58	220	60	140	Shows Cuba as part of continent north of S. America.
1538	Gerardus Mercator Double Cordiform Projection (Fig.12)	62	229	58	131	Depiction of New World best up to his time.
1540	Ptolomaeus, Basilae, Oval Projection (Fig.13)	59	202	58	158	Unknown cartographer.
1570	Ortelius Oval Projection (Fig.16)	52	140	61	220	First to depart from Columbian cosmography.
1587	Rumold Mercator Stereographic Meridian Projection (Fig.17)	52	140	60	220	Like Ortelius, shows N. and S. America too broad.
1593	De Judaeis Polar Projection	62	193	66	167	Pacific Ocean reduced to "Strait of Anian".

Table 4: Some longitudinal cosmographic criteria applied to world maps and sea charts, 1500-1593. Extended comments on page 237.

CONCLUSION

It should be clear from the foregoing that the cosmographers and cartographers of the 16th century were no more in agreement as to the length of a degree and the circumference of the earth than they were in Columbus' time and before. Furthermore, it is almost impossible to be sure what their positions were because--except for the miliaria, or Italian (Roman) mile, or units defined in terms of the miliaria--several interpretations could be made of every other unit of length employed.

As brought out in earlier chapters, many, but not all, cosmographers elected the 62 1/2 miliaria per degree ratio. Seafaring men, including Columbus, used the 60 miliaria per degree ratio. Occasionally representatives of each group crossed over to the other camp, such as Pietrus Apianus, the cosmographer, who, by 1545 had concluded that 60 Italian miles made the degree. Similarly, the experts at the Badajoz Conference of 1523-24--all seafaring men - while submitting conflicting testimony agreed that 62 1/2 miles made the degree. As late as 1595, Barentszoon said it was 60. Thus, Columbus - who espoused 56 2/3 miles to the degree in cosmographic arguments, but used the 60 mile ratio when at sea--had lots of company.

With regard to the distribution of land and sea in the longitudinal direction, there was just about the same situation. Maps began to reflect the discoveries in the New World from as early as 1500, but, the length of the Eurasian continent and the distance westward from Lisbon to the Asian mainland was shown just about as Columbus had said it was. And, this was so up to about 1570, when the truth about the enormous expanse of the Pacific, after so many crossings, could no longer be kept secret. It is true that the official padrons of Spain, such as the 1529 Ribeiro, began to reapportion the lands and seas more realistically after Magellan's charts and logs had been carefully studied, but this information was priveleged. The padrons were not only not available to cosmographers and cartographers at large, but even Spanish sailing men on official missions were given only the charts pertaining to the region of the proposed voyage and these had to be returned to the government upon their return.

It took 80 years of constant exploration in the Atlantic and Pacific, along both coasts of North and South America, and in the interior of both continents, supported by conquest, colonization and regularized re-supply and trading voyages to develop the information which resulted in the 1570 Ortelius Map of the World. This map, full of gross distortions, is however, recognizable as a world map.

The distortions in this and other world maps which followed (see Table 4 for a comparative assessment of world maps 1500-1593) would not be significantly reduced until the development, in the 17th century, of the telescope by Galileo[11] and a reliable pendulum clock by Christian Huygens. The telescope permitted Galileo to discover and observe the satellites of Jupiter. The pendulum clock permitted their timing and the development of tables predicting precisely when events such as immersion and emersion (of the satellites behind Jupiter's disc) would occur at some reference site (of known longitude and latitude). Such events could be viewed and timed simultaneously at points widely dispersed on the earth's surface. The difference in the times of occurrence locally from that listed in the tables (ephemerides) was an indication of the difference in longitude of the local site from the reference site--1 hour being equivalent to 15 degrees.

The telescope also led to instruments which could determine far more accurately the altitude or zenith distance of heavenly bodies, vital to the determination of latitude, and to vastly improved land surveying instruments. These, coupled with the method of triangulation, permitted-for the first time--the accurate determination of the length of a degree of latitude in all kinds of terrain. Ptolemy's dream, and that of Hipparchus before him, of determining the latitude AND longitude of land points by astronomical observation became a practicality in the late 17th and 18th centuries. The means of determining longitude at sea had to wait still longer.

In the last quarter of the 18th century, over 100 years after Huygens had developed his pendulum clock and almost 300 years after Columbus' first voyage to the New World, the spring wound chronometer--under

239

development for half a century by John Harrison and his son William--was accepted by the British Admiralty for use at sea in His Majesty's Ships[12]. Henceforth, ships equipped with chronometers set at Greenwich time could always determine their longitude by comparing their local time (determined from the heavens and ephemerides) with Greenwich time.

The French Academy (Academie Royal des Sciences), with the encouragement of the minister for home affairs, Jean Baptiste Colbert, and the "Sun King," Louis XIV, took the lead in Europe in developing the instruments and techniques required to determine accurately the latitude and longitude of land points in France, elsewhere in Europe, and in the New World. This led to the re-mapping of France and then the known world. Early in this extensive program mounted by France, the length of a degree of latitude was determined at several points within France, in the Lapland Arctic, and in the equatorial Andes. This led to the conclusion that the earth was oblate, bulging at the Equator and flattened at the Poles.

No less important was the French effort to standardize the systems of measurement throughout the world in one linked to the length of a quadrant of the meridional circumference of the earth. Thus, the metre was established as one-ten millionth of the meridian through Paris stretching from the North Pole to the Equator. The metric system became the legal system of measurement in France in 1799. Since that time its use has spread throughout the world and almost every nation (including the United States) has legalized its use, especially in scientific endeavor.

These developments, however, came long after Columbus' time[13]. The Admiral, for all his great personal attributes of courage, intelligence, imagination, persuasiveness, and leadership, and the skills of seamanship and navigation in which he became adept, was - when it came to cosmography in the 15th century - a man of his time, exposed to a variety of views as to the size of the earth and the distribution of its lands and seas. None of the instruments required for an accurate determination of latitude and longitude on land or at sea were available to him and his contemporaries. If he was wrong in many of his conclusions, he had lots of company - and no one of his time was right. Indeed, it was the impetus provided by the many discoveries and explorations of Columbus and his contemporaries which led to the developments required to begin to depict the world more nearly as it is.

In Appendix A, "European Discoveries and Exploration in the Americas thorugh the Year 1504" (1504 was the year Columbus returned from his 4th and last voyage to the New World), there is presented an abridged treatment of several related topics:

a.   The first Americans

b.   Pre-Columbian contacts with the New World

c.   Christopher Columbus' 4 voyages of exploration and discovery in the New World and those of other contemporary Europeans (through 1504)

d.   An account of the naming of the continent discovered by Columbus.

Appendix B treats, in some depth, on the almost unbelieveable variation in one of the most important units of measurement utilized on land and sea during the 15th through the 18th centuries by Europeans in the Old World and the New- the league (legua, leuga, lieue). This variation was responsible for distorted maps, disputes between countries and much litigation over the specific boundaries of grants made by Old World monarchs to colonists in the New World.

Appendix A

EUROPEAN DISCOVERIES AND EXPLORATION IN THE AMERICAS THROUGH THE YEAR, A.D. 1504

The First Americans

We are all immigrants, or the descendants of immigrants, in America - North, Central and South. The first immigrants came to the "New World" some 30,000 to 50,000 years ago crossing a land bridge from Siberia to Alaska during one of the ice ages. They were neither explorers nor settlers, but hunters - men, women, and children - following the caribou, bison, and horses whose carcasses provided their food, clothing, and shelter.

It was late in the history of human evolution when these hunters and their families first set foot on the American continent. For hundreds of thousands of years, apemen and primitive forms of humanity wandered across Africa, Europe and Asia gradually evolving into Homo sapiens - modern man. Such an evolution did not take place in America. The first Americans found only animals and vegetation, no native men.

At the time, the world was in the grip of the last of the great ice ages. During the past several million years, glaciers from the poles have advanced and retreated many times. In the northern hemisphere, the glaciers actually reached into the temperate zone. What precisely happened to cause the sun's heat received by the earth to diminish is not known. Several hypotheses have been proposed by scientists. The result, however, was that evaporation from the oceans of the world was not matched by runoff from precipitation. And, in the colder latitudes and higher elevations of the world glacial activity was intensified. The net result was a gradual lowering of the seas. Equilibrium between evaporation and ice melt (which reached the sea) was not attained until the level of the seas had been lowered by some 400 to 500 feet. Thus, much new land appeared along the shorelines of the seas. In partial compensation, the immense weight of the glaciers depressed the continental land masses inducing a corresponding rise in the seafloor level.

Bering Strait, the body of water separating Siberia from Alaska is 56 miles wide and about 180 feet deep. During the ice ages (and frequently at other times), with the lowering of the seas, a 1000 mile wide land bridge (dubbed Beringia by some modern scholars) appeared, connecting Siberia and Alaska. In due course, large herds of Siberian bison, caribou, horses, and mammoths seeking new grazing areas found Beringia. Much of the newly exposed land was water logged, but there were many low, rolling hills which were dry and grassy. These provided grazing for the animals. Where the herds went the hunters followed, both groups continually working their way eastward.

Beringia, which included most of the Bering and Chukchi Seas and adjacent land areas, was exposed for two prolonged periods between 34,000 and 30,000 B.C. and 26,000 and 11,000 B.C. During those times, not only was there animal and human traffic across the land bridge, but movement of glaciers along the north American Pacific Coast, in North Central Canada, and the Central Plains (of the U.S.) created ice free corridors which were followed southward, subsequently, by the ancestors of most of the Indians of the New World.

Between 8000 and 5000 B.C., the climate became milder and the oceans filled up. Thus, the land bridge and traffic over it disappeared, although it was still possible to cross on the ice in winter and by boat in the warmer months. Inasmuch as boats began to appear in northern Europe about the seventh millenium B.C., it is reasonable to assume that a knowledge of boats was transferred to the tribes in Siberia not long thereafter and that sporadic migration continued by boat, in both directions, across the Bering Strait until modern times.

At the time that Columbus arrived in American waters, there was an enormous difference between the cultural levels of the various Indian tribal groups. G.H.S. Bushnell has likened this to " . . . a range of mountains, rising from lowlands in Alaska and Tierra del Fuego, through the foothills of the United States and Chile, to the great twin peaks of Mesoamerica and Peru, with lesser heights over the intervening Central and South American areas."[1] In the peak areas, man had reached a high state of culture, but in the lowlands he remained as primitive as his ancestors. Why this disparity? Could it be that the tribes which developed faster and rose to greater cultural levels had benefited from contacts with the highly developed civilizations of Europe, Africa and Asia?

On Pre-Columbian Contacts with the New World

Many fascinating books and articles have been written setting forth various types of evidence of pre-Columbian contacts with the New World. Ivan Van Sertima provides intriguing evidence of Black African contacts, centuries before Columbus, with the Olmecs the first builders of a high civilization in the Americas. Excavations at Tres Zapotes and La Venta have uncovered huge Negroid stone heads dating from 900 B.C. A large number of much smaller figures have been uncovered in Central and South America, many of which, fashioned in clay, gold, copper, and copal, have undeniably Negroid heads. As Van Sertima puts it, "Accidental stylization could not account for the individuality and racial particulars of these heads. Their Negroness could not be explained away . . . Their coloration, fullness of lip, prognathism (jaws projecting beyond upper face), scarification (scratching of face), tattoo markings, beards, kinky hair, generously fleshed noses, and even, in some instances, identifiable coiffures, handkerchiefs, helmets, compound earrings--all these had been skillfully and realistically portrayed . . . " Van Sertima also points out that, around 1927, Leo Weiner, a Harvard linguist studying native languages of Central and South America, found words which suggested an African or Arabic origin [2].

R. A. Jairazbhoy makes a convincing case for the ancient immigration into MesoAmerica and Peru of not only blacks, but Egyptians, Babylonians, Hebrews, and Chinese. Indeed, heads with Semitic features have been found at Tres Zapotes, La Venta, and Vera Cruz, on or near the Gulf of Mexico, and in Peru. Besides various types of artifacts found in the America's, Jairazbhoy finds "evidence" in comparing historical accounts contained in documents such as the "Popol Vuh, the Epic of the Quiche Maya of

Guatemala" with the stories told in bas-reliefs and hieroglyphs associated with the Rameses dynasty of ancient Egypt. Cyrus Gordon has shown many parallels between the "Popol Vuh" and the "Book of Genesis". Jairazbhoy's Chinese influence is related to members of the Shang Dynasty (c. 1121 B.C.) who were forced to flee China when Chao, the Shang ruler, was defeated by Wu-Wang, the leader of the Chou forces. The claim is that some members of the Shang made it all the way across the Pacific, skirting Japan, touching the Aleutians, and ultimately reaching West Mexico, from whence they moved inland[3].

A report of another Chinese contact with America, much later in time, was first brought to the attention of Europeans when, in 1761, the noted French sinologist, Desguignes, published a paper in the *Memoires de l'Academie des Inscriptions et Belles Lettres* reporting he had found in the works of early Chinese historians a statement that in the fifth century of our era, certain travelers of their race had discovered a country which they called Fusang. From the direction and distance as described by them, Desguignes concluded the Chinese had reached Western America, and, in all probability, Mexico.

In 1841, Carl Friedrich Neumann, Professor of Oriental Languages and History at the University of Munich, published (in German) the original narrative of one of the Chinese travelers, a Buddhist monk or missionary by the name of Hoei-shin, who had returned to China from his voyage to the East in 499 A.D. Professor Neumann had found the account in the Year Books or Annals of the Chinese empire, whence it had been entered by Hoei-shin. Neumann's account went considerably beyond Desguignes and, in 1875, was translated into English by one of his students, Charles Leland[4].

Leland's work (or Neumann's) reports the things Hoei-shin saw (and had been told him by natives) that can be interpreted as evidence he had reached Mexico and details of the voyage to that land from China, including the lands touched enroute. It also includes critiques by other experts on Sinology, Meso-America, and North Pacific navigation. The book is scholarly and restrained in its claims, yet convincing in its basic thrust.

Herodotus reports, with some skepticism, the voyage of some intrepid mariners from Tyre and Sidon, who in 600 B.C. had been commissioned by the Pharaoh Necho of Egypt to circumnavigate Africa. Setting sail down the Red Sea, the expedition rounded the southern tip of Africa, noting with some surprise that as they sailed westward the sun was on their right side, to the north. Encountering unfavorable winds and storms, they were forced to go ashore frequently for shelter, water, and food. As a result the voyage took some three years before the mariners returned through the Mediterranean to the mouth of the Nile. This voyage, which appears to currently enjoy a "near consensus" of acceptance, may have included a landfall in Brazil. In 1872, a stone with an unknown inscription was found on a Brazilian plantation. Ladislau Netto, Director of the National Museum in Rio de Janeiro, later pronounced it to be Phoenician and made a translation indicating that Canaanites from Sidon had embarked from Ezion-Geber during the reign of Hiram, passed into the Red Sea and voyaged around "the land of Ham" (Africa). At sea for 2 years, the 10 ship fleet was separated by a storm. The survivors of one ship, 12 men and 3 women, had evidently attempted a settlement near Paraiba, Brazil, and had made the stone inscription to record their saga. In 1968, Cyrus H. Gordon of Brandeis University made a study resulting in the conclusion that Dr. Netto's translation was correct and the event genuine. Others, however, have contested the matter vigorously[5].

Barry Fell's *America B.C.*[6] deals with Libyans, Celtiberians, Phoenicians, Irish, Druids, and others who came to America by ship and left inscriptions and rock structures to mark their presence. Professor Fell, a prominent Harvard marine biologist who has drifted into epigraphy, translates the Bourne (Mass.) stone, known to the Indians long before the Pilgrims landed, from Iberian as "A proclamation of annexation. By this Hanno takes possession." Hanno, it will be recalled, was the Carthaginian navigator who, in the *Periplus*, wrote of a voyage from Carthage in 425 B.C. past the Gates of Hercules and Iberia, past the Canaries, and down the West coast of Africa, for the purpose of establishing or finding sites for new Libyaphoenician towns. Could he have crossed the Atlantic?

Fell reports hundreds of Susquehanna River (Pennsylvania) stones that seem to be Punic grave markers, a Libyan monument in the California desert, Celtic dolmen, and "sacrificial" stones which, in

Ireland or Brittany, would be assigned a rough date of 1200 B.C. Many of Fell's conclusions are not shared by the archeological community.

Tristan Jones, in a most interesting article[7], indicates the coast of Ecuador and Peru was "known to the Ancient Chinese, the Asiatic Indians, the Phoenicians and probably the Persians long, long before either the Norsemen or St. Brendan, Prince Madoc, stray European fishermen, or Columbus stepped on the shores of North America." Jones claims "Chinese artifacts, of great age, have been found on the northern coast of Ecuador." He tells of white bearded men who had been ship-wrecked on the coast of Chile who lived for awhile among primitive Andes Indians.

Jones, a former navigation officer of the British Royal Naval Survey Service, is a veteran of single-handed ocean sailing (345,000 miles in craft under 40 feet in length, including many ocean crossings and 3 circumnavigations of the earth).

Each of the references mentioned makes an effort to describe how these ancient travelers could have reached America--the type of craft utilized and possible routes. The best such treatment is provided by Thor Heyerdahl, who has made two trans - Atlantic and one partial trans-Pacific voyages, the first two in reed hulled sailing craft and the last in an "aboriginal" balsa sailing raft. Heyerdahl is an expert on the subject of the currents and trade winds of the worlds oceans, particularly the Atlantic and Pacific. He points out that "the largest river with its source in Peru is not the Amazon, flowing eastward through Brazil, but the Humboldt Current flowing westward through the Pacific. The mightiest river in North Africa is not the Nile, but the Canary Current with its delta between the Caribbean Islands, emptying African sea water into the Mexican Gulf. The fixed itineraries of these marine rivers span the oceans and form paths between the continents."[8]

Heyerdahl explains oceanography by connecting elements to the failures and successes of the Atlantic and Pacific voyages of the Age of Discovery. He then shows how certain of the earliest claimed contacts with America, preceding by centuries, even millenia, those of the Age of Discovery, could have been distinct possibilites[9].

The foregoing notwithstanding, historians like the late Samuel Eliot Morison and anthropologists like Dr. Marshall McKusick (see reference note 5) say, unequivocally, any contacts with America prior to the Norse at the end of the 11th century A.D. have not been proven and probably never occurred. This includes St. Brendan of Ireland, who is alleged to have sailed the western seas for seven years in search of tropical islands, and, in 570 A.D., to have reached Madeira or the West Indies. Some enthusiasts even claim he reached continental North America. Writers of the 8th and 9th Centuries chronicled his feats at sea, although the best known work, the *Navigation of Saint Brandon* (this alternate spelling of his name appears in the literature) was not written until the 10th century. Crone, who has studied available writings on St. Brendan, concludes he reached the Faroe Islands, about 300 miles due north of Scotland, and one or more of the Outer Hebrides, west of Scotland and north of Ireland, i.e. St. Brendan's voyages were northern journeys, not southern.[10]

It is more certain that Leif Ericson, the son of the Norseman Eric the Red, reached southern Baffin Island and Labrador and quite likely that he also reached Newfoundland in the year 1000 A.D. Whether he continued southward and reached Martha's Vineyard (in the Atlantic about 5-10 miles off the southern coast of Massachusetts) is a matter of some dispute. Crone says "No!" Leif Ericson wintered at a place he called Vinland, where grapevines were seen (?) by a member of his crew, giving rise to later speculation the place may have been Martha's Vineyard[11].

Thorfinn Karlsefini, an Icelander, retraced generally Leif Ericson's voyage in the period 1020-1023, but provided no additional details which would prove precisely how far south he had been.

There were many notable discoveries, explorations, and travels to little known lands between Karlsefini's efforts and Christopher Columbus' discovery of the West Indies in 1492, but these had to do with Africa, Asia, and the islands in the eastern Atlantic. Still, each of these events played an important role in hastening the discovery of America itself.

244

Figure 1. Spain in the time of Christopher Columbus.
Princess Isabela of Castile and Léon married Prince Ferdinand of Aragon in 1469. In 1474, with the death of her brother, Isabela became sovereign of Castile and Léon. In 1479, with the death of his father, Ferdinand became King of Aragon. In 1484, the 'last crusade' to capture the Kingdom of Granada and end Moslem rule in Iberia was launched. Early in 1492, the last stronghold fell. Navarre, in the north, was added the same year.

From "The Life of the Admiral Christopher Columbus by his son Ferdinand", translated and annotated by Benjamin Keen, Rutgers University Press, New Brunswick, NJ. 1959. It is reproduced here by the kind permission of Professor Emeritus Keen, copyright owner.

245

One of the earliest of these was the sensational travels and experiences of the Venetian Polo family-Marco, his father Nicolo, and Uncle Maffeo - in period 1260-1295. (The routes taken in their extensive travels on the Asian continent, and the impact of *The Book of Ser Marco Polo* are described in Chapter 5 of the text.) The publicization of the Polo's travels, including the riches to be obtained by trade with the various Indias (Cathay, Mangi, and Chipangu being only the easternmost), set off a wave of exploration aimed at finding a sea route to India and China by going around Africa.

As it was determined just how huge Africa was, several of the expeditions concentrated on the exploration of parts of Africa itself. From 1427 through 1488 Portugal took the lead among European nations in voyages of discovery and exploration. Diogo de Seville discovered some of the Azores Islands in the period 1427-31 A.D. and eight other expeditions explored Africa's west coast, the last being that of Bartolomeu Dias, who in the period 1487-88 A.D. discovered the Cape of Good Hope and explored the coast eastward from that point about 200 miles to Mossei Bay.

Events Leading up to Columbus' First Voyage

The aggressiveness of Prince Henry the Navigator in initiating this series of explorations was matched by his successors. Thus, it was only natural, when Columbus conceived the idea of a voyage of discovery and exploration to Cathay and Cipanju by sailing westward, instead of rounding the Cape of Good Hope, that he should turn to King John of Portugal to back such a voyage. This was especially appropriate since Portugal owned the Azores and it was Columbus' belief that Cathay and Cipanju lay only about one week's sailing time westward from the Azores. Later, Columbus came to the conclusion that the latitude of the Azores was not as favorable a route for a westward crossing of the Atlantic as that of the Canaries because of the character of the prevailing winds. However, as has been related in the text, Columbus was not able to gain King John's backing for his project. Whether he sought backing, then, from the monarchs of England and France, as some accounts have it, has never been positively established. Sometime in 1484, Columbus went to Spain with his son, Diego.

Through the intercession of Luis de la Cerda, then Count and later Duke of Medina Celi, his host and one of several influential people interested in his project, Columbus was brought to the attention of Queen Isabela, and later her husband, King Ferdinand. As Columbus outlined the project to them, they became intrigued with the possibilities (probabilities to Columbus) of attaining great wealth and power for themselves and their Kingdom of Castile, Leon, and Aragon. Despite their genuine interest, spurred somewhat by the Treaty of Alcocovas (which made the route to the Indies by way of the southern cape of Africa a Portuguese preserve), there were obstacles. Ferdinand and Isabela referred Columbus to various learned commissions in order to obtain both their concurrence that his plan was feasible but also, hopefully, some degree of financial participation in the project. Invariably, their reactions were negative.

In early 1488, Columbus again approached John II of Portugal for sponsorship and, encouraged somewhat, arrived in Lisbon later that year to press his case. In December of the same year, Bartolomeu Dias returned triumphantly to Lisbon, having rounded the Cape of Good Hope. Columbus returned again to Spain, knowing further efforts in Portugal would now be fruitless.

In 1489, Ferdinand and Isabela attacked the fortified city of Granada, the last stronghold of the Moors on the Iberian Peninsula. Victory was not obtained until early 1492. Now Ferdinand and Isabella turned their attentions once more to Columbus' project. But Columbus' high terms for undertaking the voyage was an obstacle. He wanted one-tenth of all the riches that might be found in the Indies, the title of Don Cristobal Colon, and the rank of Admiral of the Ocean and Viceroy and Governor of the Indies. At first, he was flatly refused. However, through the good offices of Luis de Santangel, the Kings Treasurer and a rich man, Isabella and Ferdinand were finally persuaded to authorize the trip and agree to Columbus' terms. The agreement was signed on April 17, 1492. The next three and one-half months were taken up in obtaining the ships, provisions, and crews required for the voyage.

Christopher Columbus' First Voyage of Exploration and Discovery, 1492-1493

After many delays, Christopher Columbus left Palos de Moguer, Spain, on August 3, 1492 with 3 ships and 90 men. The *Santa Maria*, the flagship which carried Columbus, was a decked ship of between 80 and 100 tons. Its master, Juan de la Cosa[12], Columbus' personal staff, and the ship's crew totalled some 50 men. The *Pinta* and *Nina*, the other two ships of the tiny fleet, were caravels of about 60 tons, carrying complements of 20 men each. The master of the Pinta was Martin Alonzo Pinzon. His brother, Vicente Yanez Pinzon, commanded the *Nina*. The two caravels, though smaller than the *Santa Maria*, were faster, more seaworthy, and better suited to the task of exploration which lay ahead. Columbus had wanted a caravel similar to the *Pinta* and *Nina* for his flagship, but had to settle for the *Santa Maria* , which had been designed for Mediterranean service, because of the tremendous demand for ships due to the deportation of all unbaptized Jews from Spain, the deadline for which had been set for August 2nd.

On the fourth day out of Palos, the *Pinta's* rudder came loose from its bearings. After emergency repairs were made, Columbus had his fleet head for the Canary Islands, with the intent of replacing the *Pinta*. This, however, proved to be impossible, so the *Pinta* was repaired and the three ships resumed their journey westward on September 8, some 36 days after having departed Spain. Alternately becalmed or blown off course by variable winds, many days passed with no sight of land. The crews began to grumble, then plot a mutiny against Columbus so as to turn about and return to Spain. Columbus, for his part, cajoled, tried reasoning with his men, and then threatened them with severe punishment and even death if they interfered with his plans. Finally, on the 11th of October, at night, a sailor named Rodrigo de Triana aboard the *Pinta* (which had ranged far ahead of the other two ships) sighted land. Shortly after day break on the 12th of October, Columbus went ashore with selected members of the small fleet and claimed the land for the Catholic sovereigns. He named it San Salvador. While the preponderant number of investigators have concluded, as has Morison, that the island is Watling Island, strong arguments have been advanced for other landfalls and routes within the Bahamas and Cuba.[13] Figure 4a illustrates the Morison version favored by Professor Keen. Figure 4b illustrates the landfalls and routes favored by Arne B. Molander, the Judge - Marden National Geographic team, as well as the Morison route and landfall.

Note:
Figures 3a, 3b, 3c, on pages 249, 250, and 251, and 5a on page 252 are reproductions of illustrations drawn by Bjorn Landstrom for his "Columbus, the Story of Don Cristobal Colon, Admiral of the Ocean Sea..." Macmillan Co., New York, 1966.

Permission to reproduce this and other Bjorn Landström illustrations in this appendix has been graciously provided by the author/illustrator/copyright owner.

Figure 2 . Oil Painting of Columbus by Sarolla. Courtesy of Mariners Museum, Newport News Virginia.

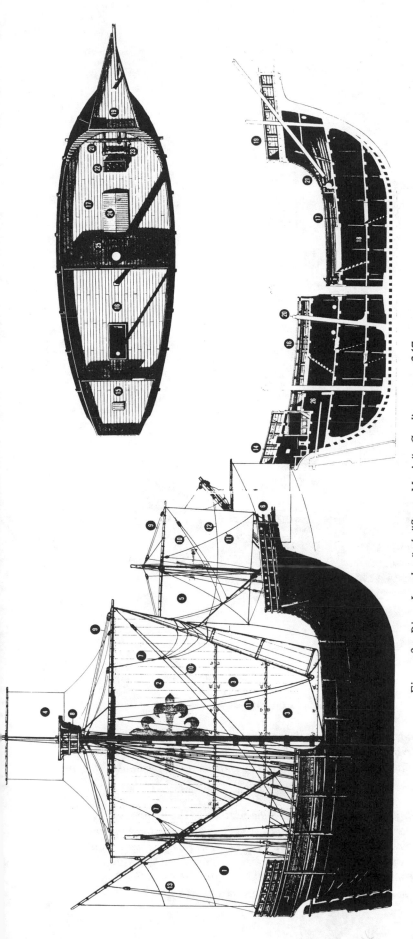

Figure 3a. Bjorn Landström's "Santa Maria". Credit on page 247.

For the past seventy-five years people have been trying to reconstruct the "Santa Maria", the flagship of Columbus' first voyage. In fact, no such attempt can hope to be successful; none of the models or "replicas" we have seen in museums or at international exhibitions can claim to represent anything but, at best, a ship of approximately the same kind and the same size as the celebrated original. We simply do not know what the Santa Maria looked like.

We do know that, unlike the "Pinta" and the "Nina", she was a "nao", a small round-bellied vessel and we know what sails she carried; but beyond this the field is open for speculation. To get a more accurate picture, we would have to analyze and compare all the pictures and descriptions of ships of that period; but there is relatively little contemporary source material, and many questions remain unanswered. The second half of the 15th century saw many changes in ships, especially in the rigging. We do not know whether the "Santa Maria" belonged to the northern type, which among other things had the main shrouds attached with deadeyes and lanyards, and had ratlines in the shrouds, or to the southern type, with tackles to attach the shrouds and a ladder running to the top.

I have chosen the latter type for my reconstruction. I assume she was between 80 and 100 tons, since it is known that the "Nina" was assessed 60 tons, and Las Casas says that the "Santa Maria" was only "slightly" larger. This means that she could be loaded with between 80 and 100 "tons" of wine. A Spanish "tonelada" of wine was equivalent to 213 imperial gallons, almost a ton in weight. By comparing measurements in a mid-15th century Italian shipbuilder's manuscript, I have arrived at these approximate dimensions: keel 56 feet, over-all length 82 feet, beam 28 feet.

Key to the reconstruction: 1. Mizzen. 2. Mainsail. It is generally assumed that the "Santa Maria" had crosses painted on the sails, but there is no evidence to support this view. The cross shown on the mainsail here is the fleur-de-lis cross of the Spanish Order of St. James, but another type of cross is just as likely. 3. Bonnets, to increase or decrease the area of sail. 4. Topsail. 5. Foresail. 6. Spritsail. 7. Martnets, to draw the sail together while the yard is lowered. 8. Top. 9. Topping lifts. 10. Braces. 11. Clewlines. 12. Bowlines. 13. Braces. 14. Captain's cabin. 15. Poop deck. 16. Quarterdeck. There may have been cubicles or small cabins below the quarterdeck for the officers. 17. Main deck. 18. Orlop. 19. Forecastle, and beneath, a platform. 20. Knight with sheave holes. 21. Beam or catena; the anchor cable may have been made fast to this. 22. A metal box with sand, used for cooking. 23. Windlass, used to raise the anchor; alternatively, the anchor may have been raised by a capstan which would probably have been mounted under the quarterdeck. 24. Hatch. 25. Pumps. 26. Compass in binnacle, in front of helmsman.

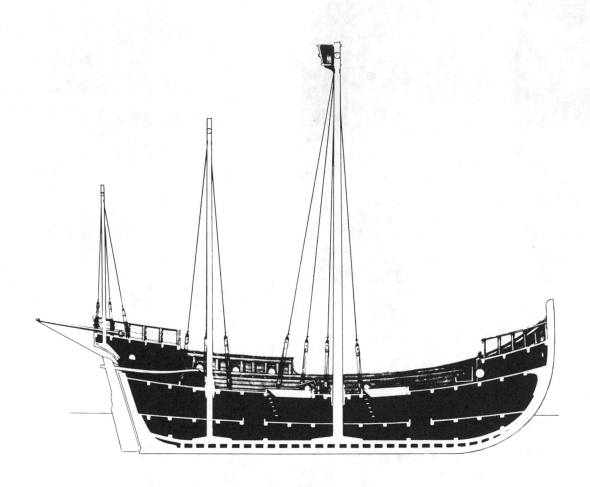

Figure 3b. Bjorn Landström's "Niña". Credit on page 247.

Any reconstruction of a 15th century caravel is bound to be conjectural, for most contemporary drawings of these ships are simplified or obscure. This is how the space on the "Niña" might have been utilized. There were a couple of small berths under the sterncastle, but neither the Niña nor any other contemporary ships had any particular sleeping accommodations for the crew.

Note: The sectional drawing above applies to the "Niña" as she was rigged on the first leg of the maiden voyage, i.e., between Palos and the Canaries. Enroute, the three-masted lateener yawed badly when running with the wind resulting in her rudder coming loose. At Las Palmas, after repairs to the rudder, it was decided to re-rig the "Niña" to match the "Pinta". That ship, of approximately the same size as the "Niña," had square sails hung from the yards on the foremast and mainmast and a lateen hung from the mizzen. To accomplish the re-rigging, the number 2 mast was placed far forward in the bow. (See "Pinta" in figure 3c) and the mainmast probably moved aft a few feet.

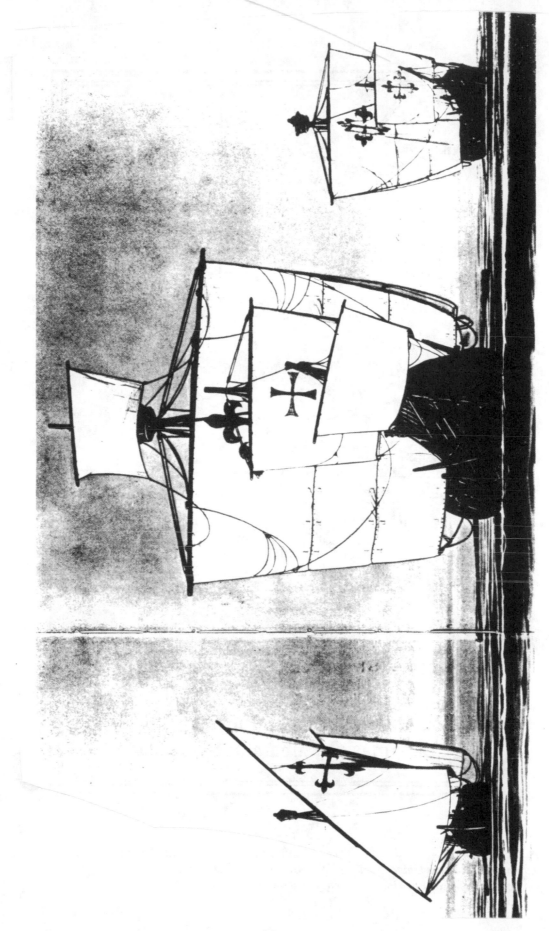

Figure 3c. Bjorn Landström's Niña, Santa María, and Pinta becalmed on Columbus' first voyage. To give steerageway, the men are rowing with long oars, technically known as sweeps. Credit on page 247.

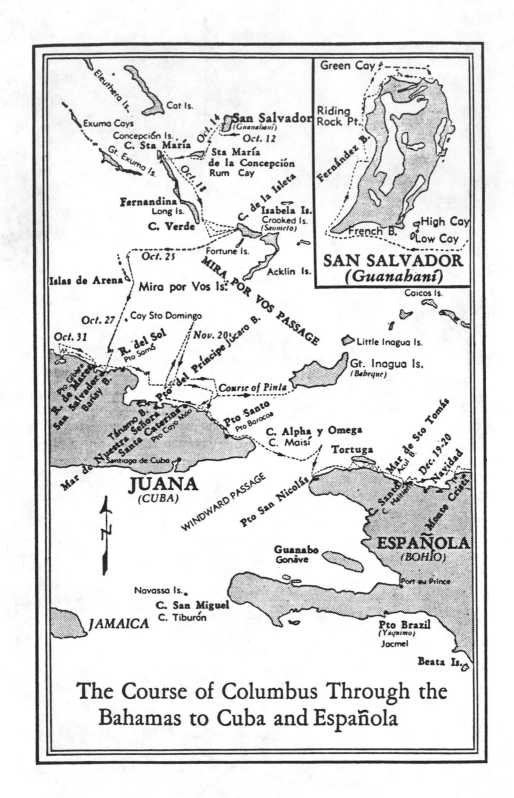

The Course of Columbus Through the
Bahamas to Cuba and Española

Figure 4a
From "The Life of the Admiral Christopher Columbus by his son Ferdinand", translated and annotated by Benjamin Keen, Rutgers University Press, New Brunswick, NJ. It is reproduced here by the kind permission of Professor Emeritus Keen, copyright owner.

252

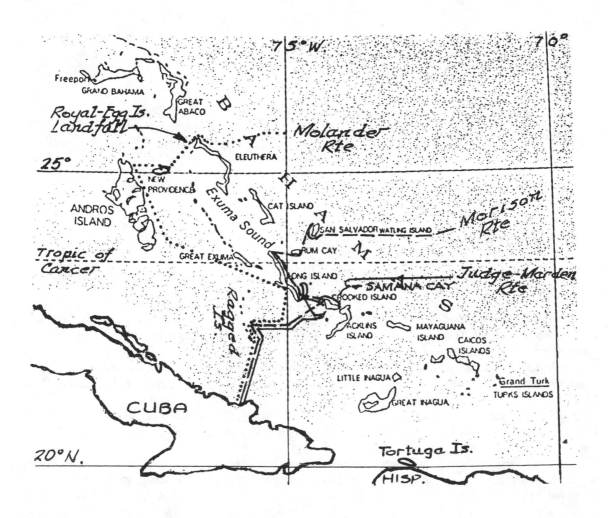

Figure 4b. Approach routes, landfalls, and routes through the Bahamas to Cuba.
These routes were suggested by Arne B. Molander (Northern Route); Samuel Eliot Morison (Central Route; see also figure 4a, this appendix, and figures 1a and 1b, chapter 1); and Joseph Judge and Luis Marden (Southern Route). See also reference note 13, this appendix. This illustration is based on figure 1 of Arne B. Molander's "A New Approach to the Columbus Landfall" from "In the Wake of Columbus, Islands and Controversy," edited by Louis De Vorsey, Jr. and John Parker, Wayne State University Press, Detroit, 1985, and the map supplement of the National Geographic, November 1986, page 566A, Vol. 170, No. 5 - Columbus.

Continued comments from page 247

CORRESPONDING ISLAND NAMES

Columbus	Central Route (Morison)	Northern Route (Molander)	Southern Route (Judge-Marden)
San Salvador	Watlings Island	Egg/Royal Island	Samana Bay
Santa Maria de la Concepcion	Rum Cay	New Providince	Cooked Island
Fernandina	Long Island	Andros	Long Island
Isabela	Crooked - Acklins	Long Island	Fortune Island

Columbus had taken along, as a member of his personal staff, a Christianized Jew, Luis de Torres, to serve as interpreter. De Torres knew Hebrew, Aramaic, and Arabic and it was thought that Arabic, being an Oriental tongue, would be understood at Cipanju and in farthermost India. De Torres was able to converse, after a fashion, with the inhabitants of the island who turned out be very friendly and eager to trade whatever they had for glass beads, hawk's bells, and other trifles. The Indians, as Columbus dubbed them, brought parrots, skeins of woven cotton, darts, and gold pendants normally worn hanging from a hole in their nostrils. Questioned about the source of the gold pendants, they indicated from the south, where a king with much gold lived. They added that to the south and southwest there were many other islands and large countries.

In the course of the next two weeks, Columbus explored San Salvador, an island to the southwest, which he named Santa Maria de la Concepcion, a much larger island with hills, many springs and beautiful meadows and trees which he named Fernandina, after the King of Spain, and a fourth island which he named Isabela after his patron queen. On October 25th, taking advantage of a favorable wind, Columbus sailed south heading for the large country the Indians called Cuba, reaching its northern coast on Sunday, October 28th. Before leaving the Bahamas, Columbus decided to take along an Indian from San Salvador who had appeared unusually intelligent in conversing with de Torres. This man, it was reasoned, should be helpful in conversing with the Indians on Cuba, and so it turned out to be.

Columbus was very impressed with the lushness of the vegetation and the natural beauty of Cuba as well as the friendliness of its inhabitants, who believed the Europeans to be gods. Columbus at first believed Cuba to be Cipanju (Japan) but after several days of exploration of the eastern section of the north coast came to the conclusion that he had found the Asian mainland and that this must be Cathay (China). During their short stay on Cuba, the expedition repaired its ships at a place Columbus named "Rio de Mares," now identified as Puerto Gibaro.

The explorers saw many strange things there, including tobacco leaves, but they found no evidence of any gold, pearls, or spices. These, they were told, could be obtained in abundance in a land to the east called Bohio,later identified as Haiti. Accordingly, Columbus and his fleet left Cuba, heading eastward, on November 13th. They took with them 12 Indians - men, women, and children.

Unfavorable winds, however, forced the ships back to the Cuban coast near a great harbor Columbus named Puerto del Principe (probably modern Jucaro Bay). Columbus tried again on the 19th of November, but after tacking for 3 days and making little progress, the little fleet, less the *Pinta*, which was a better sailer, put into a harbor (not far from Puerto del Principe) which Columbus named Santa Caterina (modern

Puerto Cayo Moa.) Martin Alonzo Pinzon, excited by the stories of much gold on Bohio, had separated from the fleet without Columbus' permission and gone ahead.

At Santa Caterina, Columbus found traces of gold in some gravel at a river where they were taking on fresh water. Columbus was also impressed with the tall pines beautifully suited for ship's masts, planking, and framing - and plentiful enough to build the largest fleet. Continuing eastward along the coast until he reached its eastward end, which he named "Alpha and Omega" (modern Cabo Maisi), he departed Cuba for the third time on December 5th having spent a total of 39 days and explored about 300 miles of Cuba's north coast, reaching Bohio's west coast the next day and entering a harbor he named Puerto San Nicolas. Unable to "gain speech" with the Indians who filled the harbor with their canoes, Columbus followed the coast north and eastward till he came to another harbor which he named Puerto de la Concepcion (modern Moustique Bay). This was opposite (due south) of an island he later named Tortuga. The character of the country here resembling so much that of Spain, on Sunday December 9th, Columbus named this land Española (in English, Hispaniola).

Contact with the Indians started immediately after reaching Pto. de la Concepcion. Despite their shyness and fear, the Indians were gradually emboldened by the calculated kindnesses and gifts of Columbus' men to attempt conversation. With great difficulty, it was learned from them that a great cacique (king) ruled this area, whereupon efforts were redoubled to arrange a meeting between Columbus and the cacique. This finally occurred on December 18th, when the cacique, a young man, and 200 followers boarded the Santa Maria. Gifts were exchanged and it was evident that the Spaniards were held in very high esteem, partially because of their gentle treatment of the Indians and partially because of the imposing appearance of the three ships, their great sails, and their cannon, fired from time to time to impress the Indians.

Again, despite great difficulty, it was learned that the cacique and his followers lived about 15 miles farther eastward, that there was some gold available on Espanola but Babique (Great Inagua Island) was the place it existed in abundance. Columbus decided to go to Babique, but first to explore more of the north shore of Espanola. Thus, shortly after the departure of the cacique, Columbus' two ships (the *Pinta* not having rejoined them) weighed anchor and proceeded slowly eastward. Monday night, December 24th, Columbus having retired, the helmsman, against Columbus' standing orders, turned the tiller over to a boy apprentice seaman. And, in the early hours of Christmas, 1492, the boy allowed the *Santa Maria* to ground on a reef.

Efforts to free the ship with the *Nina's* assistance were unavailing for the ship having grounded at high tide, became more securely inpaled as the tide ebbed. A consequence of this was that the ship's seams started to open, with flooding of the bilge. It was now clear to Columbus that the *Santa Maria* could not be saved. Accordingly, Columbus and the crew of the *Santa Maria* transferred to the *Nina*, taking with them the most vital gear and supplies. At daybreak, Columbus sent two aides ashore to find the cacique and seek his assistance.

So helpful was the cacique (whose name they soon learned was Guacanagari) in bringing ashore all of the supplies that could not be transferred to the *Nina* and in providing secure storage for this, that Columbus began to ponder the wisdom of establishing a small colony at the site. This idea gained strength as the Indians under Guacanagari, and from other places, appeared to have access to substantial sources of gold. When Guacanagari, seeing that gifts of gold pleased Columbus, told him that he would have a great deal of it brought from the Cibao (the gold bearing central zone of Espanola), Columbus' mind was made up. He would build a fort with the timbers and planking of the *Santa Maria* and arm it with her cannon. He had already ascertained that many of the Spaniards were willing to stay behind and man the fort, with the thought of permanently settling in Espanola, while Columbus returned to Castile for reinforcements required in the settling and subjugation of the land. The idea of such a settlement was made palatable to Guacanagari, who feared the warlike and cannabilistic Caribs, by telling him the Spaniards would defend him and his people from the Caribs.

With Guacanagari's assent, the fortress was built and a moat dug around it. Columbus named the fortress "La Navidad" for it was on Christmas morning that the Santa Maria had run aground. He appointed Diego de Arana commander of the fortress and lieutenant to the Viceroy of the Indies (Columbus). Pedro Gutierrez and Rodrigo de Escobedo were to be de Arana's immediate assistants. Thirty-six volunteers, including a physician, tailor, gunner, carpenters, calkers, and others needed for a permanent settlement, together with a great store of trading goods and provisions, arms, and the ship's boat were left with de Arana. Columbus told Guacanagari that he was returning to Castile to obtain jewels of great value for him.

Despite having learned on December 27th that the caravel *Pinta* had been seen at the eastern end of the island, Columbus left Puerto de la Navidad at sunrise on Friday, January 4th, 1493. He believed it most important to advise the Spanish sovereigns of what he had found so that appropriate measures could be taken to secure the discoveries and to stake whatever legal claims were required.

Columbus shaped a northwest course with the ship's boat going ahead in order to avoid the many reefs and shoals near the coast from which he had departed, all the time taking pains to note landmarks which would be useful in finding La Navidad upon his return.

On the morning of January 6th, the *Pinta* was sighted running westward before the wind and shortly thereafter Captain Martin Alonso Pinzon came aboard to make his apologies for having become separated from Columbus' command "against his will". Actually, the lust for gold overcame the thin veneeer of the command relationship and Pinzon had struck out for Babique when the tiny fleet was leaving Cuba. Some Indians aboard had told him much gold was to be found there. Finding no gold there, Pinzon had headed the Pinta southward to Espanola, striking the coast at the mouth of a small river Columbus had named Rio de Gracia, about 45 miles west of La Navidad. Here, by barter with the natives, he obtained much gold. This, he divided with his crew, keeping half for himself. He told Columbus, however, that he had found no gold. While Columbus did not believe the last, he could see no merit in administering anything other than a mild admonition to Pinzon, especially since most of the crew members of the *Pinta* and the *Nina* were Pinzon's townspeople, many being his kinsmen.

Columbus' two ships now continued their progress eastward proceeding very slowly. The wind becoming still more unfavorable, Columbus ordered the ships to anchor near Monte Cristi. Entering a river southwest of that mountain by boat, he found that its sands abounded in gold dust. This river, about 50 miles east of La Navidad, he named Rio de Oro (now known as Rio Yaque de Norte).

It was not until Wednesday, January 16th, that favorable winds permitted the ships to clear Espanola. By now, both ships were leaking badly and required great effort to keep them afloat. The winds were generally fair until February 14th when high winds and heavy seas again separated the two ships. For 3 days, the storm continued without abatement - all hands including Admiral Columbus being sure they would never see, let alone set foot on, land again. Finally, February 17th, the storm abated and land was sighted, which turned out to be the Portguguese island of Santa Maria, the easternmost of the Azores. Unable to anchor there until the next day, Columbus was to endure a precarious 6 days at Santa Maria until the captain of the island was satisfied as to the authenticity of Columbus' mission for the King and Queen of Castile.

On February 24th, a southwesterly wind permitted the Nina to leave Santa Maria. While Columbus' destination was Palos de Moguer (his port of departure), a severe storm which arose on February 27th blew them off course, to the north. Again, it was a miracle the tiny craft made it to land again. As the storm abated on March 4th, Columbus saw he was near the mouth of the Tagus, the river of Lisbon. Entering the estuary, he anchored the Nina at the town of Rastelo, about 27 miles from the residence of King John of Portugal. Columbus immediately dispatched a letter to the King, requesting repairs and provisioning for the *Nina* which would permit him and his crew to return to Palos, Spain. This was granted by the King who, though chagrined that the discoveries had not been made in his name rather than for Ferdinand and Isabela, nonetheless appreciated Columbus' achievement and honored him for it.

Columbus left Portugal March 13th and by noon of Friday March 15th had entered the river Saltes to anchor off Palos.

That evening the *Pinta* also anchored at Palos. She had sailed by the Azores during the storm of February 14-17 and eventually reached Bayona, a Spanish harbor north of the Portuguese border. From there she returned to Palos after an unsatisfactory exchange of letters between Captain Pinzon and the Spanish sovereigns. Pinzon died a few days later, pining away with grief at the manner in which the expedition had worked out for him.

Columbus had dispatched a letter from Rastelo intended for Isabela and Ferdinand, but addressed to his benefactor Luis de Santangel at the Spanish court. He also made several copies of the letter, sending each to other friends at the court, in the hope that at least one would reach the King and Queen. It was not long after his return that Columbus was summoned to their court at Barcelona.

Received by their majesties in full court and seated in their presence, Columbus told his story and exhibited his spoils--gold, cotton, parrots, mysterious plants, curious arms, and the Indians he had brought with him for baptism. There were two results of significance which stemmed from Columbus first voyage. Pope Alexander VI granted bulls to the crowns of Castile and Leon confirming their rights to all lands discovered or to be discovered west of a line of demarcation drawn 100 leagues west of the Azores. The other significant result was the immediate authorization by the Spanish crown of another Columbus voyage of discovery and exploration.

Columbus' Second Voyage of Discovery and Exploration, 1493-1496

The mission of Columbus' second voyage (as drawn up carefully by Ferdinand and Isabella along the lines of, but not identical to, recommendations made by Columbus) was to return to Espanola to relieve the men who had remained there, to augment the number of settlers, and to complete the conquest of that island, as well as of all other lands that had been discovered or should be discovered. Columbus' rank of Admiral and title of Viceroy, with the authority to govern all lands discovered, or to be discovered, were confirmed.

The fleet which left Cadiz on September 25, 1493 consisted of three galleons and fourteen caravels, 1200 men, including 5 missionaires, and animals and materials necessary for colonization. There were no women on this voyage, the decision having been made that subjugation of the island should take place first. The flagship of the fleet was named the *Santa Maria,* like her predecessor. The two other large ships, about 200 tons each, were the *Collina* and *La Gallega.* The *Nina*, which had brought Columbus home on his first voyage, and 11 other caravels made up the remainder of the fleet.

The owner and master of the flagship, and probably second in command of the expedition, was Antonio Torres, brother to the nurse of the Heir Apparent. The Admiral's brother, Giacomo, now known as Don Diego, accompanied him as did the Admiral's boyhood friend, Michele de Cuneo of Savona. The chief surgeon, Diego Alvares Chanca, also chronicled the voyage. Heading the five priests on the expedition was Friar Buil of the Benedictine Order. He had the special responsibility for missionary work among the Indians. A young cartographer, Juan de la Cosa[14] signed on as a seaman. The map of the world which he was to make in 1500 was the first known map to show all the newly discovered islands. Ponce de Leon (who was later to discover and name Florida) was a member of the expedition. It is likely that he was one of the 200 "gentlemen volunteers" who went on the expedition at their own expense to look for gold and adventure. Alonso de Hojeda, of whom we shall hear more, was captain of one of the ships.

Figure 5. Bjorn Landström's 17 ship fleet used by Columbus on his second voyage. Credit on page 259.

Credit for Figure 5, page 258
The drawing of Columbus' 17 ship fleet is taken from Bjorn Landström's "Columbus, the Story of Don Cristobal Colon, Admiral of the Ocean Sea and His Four Voyages Westward to the Indies," Macmillan Co., New York, 1966. All drawings in this book were done by the author, who must be considered one of the greatest living authorities and artists on sailing ships over the millenia.

Permission to reproduce all Landström illustrations in this appendix has been graciously provided by the author/illustrator/copyright owner.

The fleet reached Grand Canary Island on October 2nd and Gomera (on Santa Cruz de Tenerife) on October 5th. There they provisioned for the last time. Twenty-two days after leaving Gomera, at dawn on November 3rd, land was sighted. This turned out to be a large mountainous island which Columbus named Dominica. Not finding a suitable anchorage on the east side of the island, they sailed northward to another island visible from Dominica. Here, Columbus went ashore, naming the island Mariagalante (after the nickname of his flagship), and - with "suitable solemnities" - renewed the possession he had taken in the name of Ferdinand and Isabela, of all the islands and mainland of the Indies on his first voyage.

Leaving Mariagalante on November 4th, the fleet sailed northward to another nearby island which Columbus named Santa Maria de Guadalupe. This was the first island to show signs of being inhabited, so a party was sent ashore. It turned out that this island was inhabited by fierce, man-eating Caribs who, however, disappeared into the forest rather than confront the shore party. The information gained about the Caribs was obtained by a careful examination of the huts they lived in and by interrogating some male and female prisoners of the Caribs, taken in raids on other islands, and being kept to be eaten. These people Columbus freed and took along with him when he left Guadalupe on November 10th, heading for Espanola.

On a northwest bearing, they next sighted islands Columbus named Monserrate (because of its height), Santa Maria la Redonda, Santa Maria de la Antigua, San Martin (later renamed Neustra Senora de las Nieves and still known as Nevis, the birthplace of Alexander Hamilton), San Jorge (St. Kitts), Santa Anastasia (St. Eustatius), and San Cristobal (Saba).

Changing their course to westerly, they sighted an island Columbus named Santa Cruz (St. Croix) on November 15th. Although eager to reach Espanola to the west, Columbus was prevented from doing so by a hard wind blowing from that direction, so he headed north instead and found a whole archipelago of small islands which he named Las Once Mil Virgenes (Virgin Islands). This was on the 16th of November. The next day, with a more favorable wind, the fleet was able to steer west southwest and on November 18th they passed south of a beautiful green island Columbus named Graciosa (Vieques Island). For all of the 19th of November they sailed westward, south of a large island which the Indians called Boriquen and which Columbus named San Juan Bautista (Puerto Rico). Finding a good anchorage off the west coast, they stopped there for two days while fresh water was taken aboard and the men fished, no contact being made with natives.

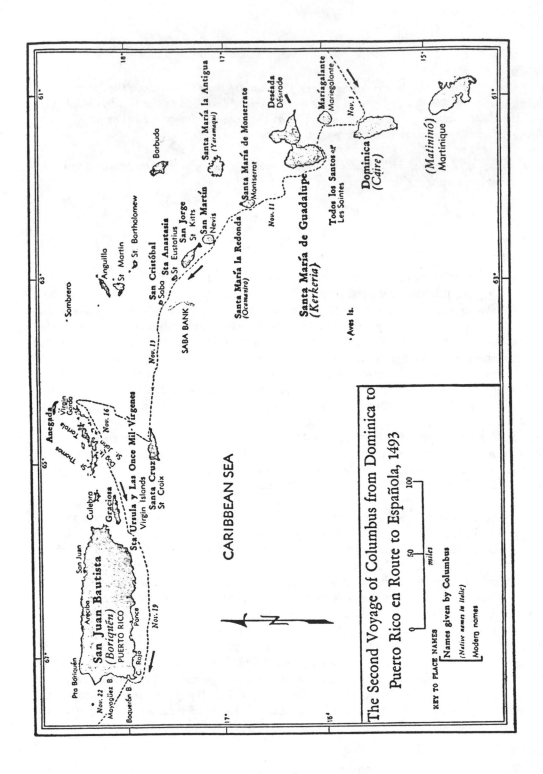

Figure 6. Credit on top of page 265.

260

Figure 6a. "Man eating Caribs Encountered by a shore Party at the Columbus named Island of Santa Maria de Guadalupe, November 9, 10, 1493, on Columbus' 2nd voyage of Exploration and Discovery". This is but one of the imaginative and striking illustrations created by Theodore de Bry (1528-98) for his 14 volume series, "Historia Americae", c. 1594. His engravings were the first to illustrate happenings, native life, flora, and fauna in America with any degree of accuracy and elegance. This illustration is provided courtesy of the Library of Congress.

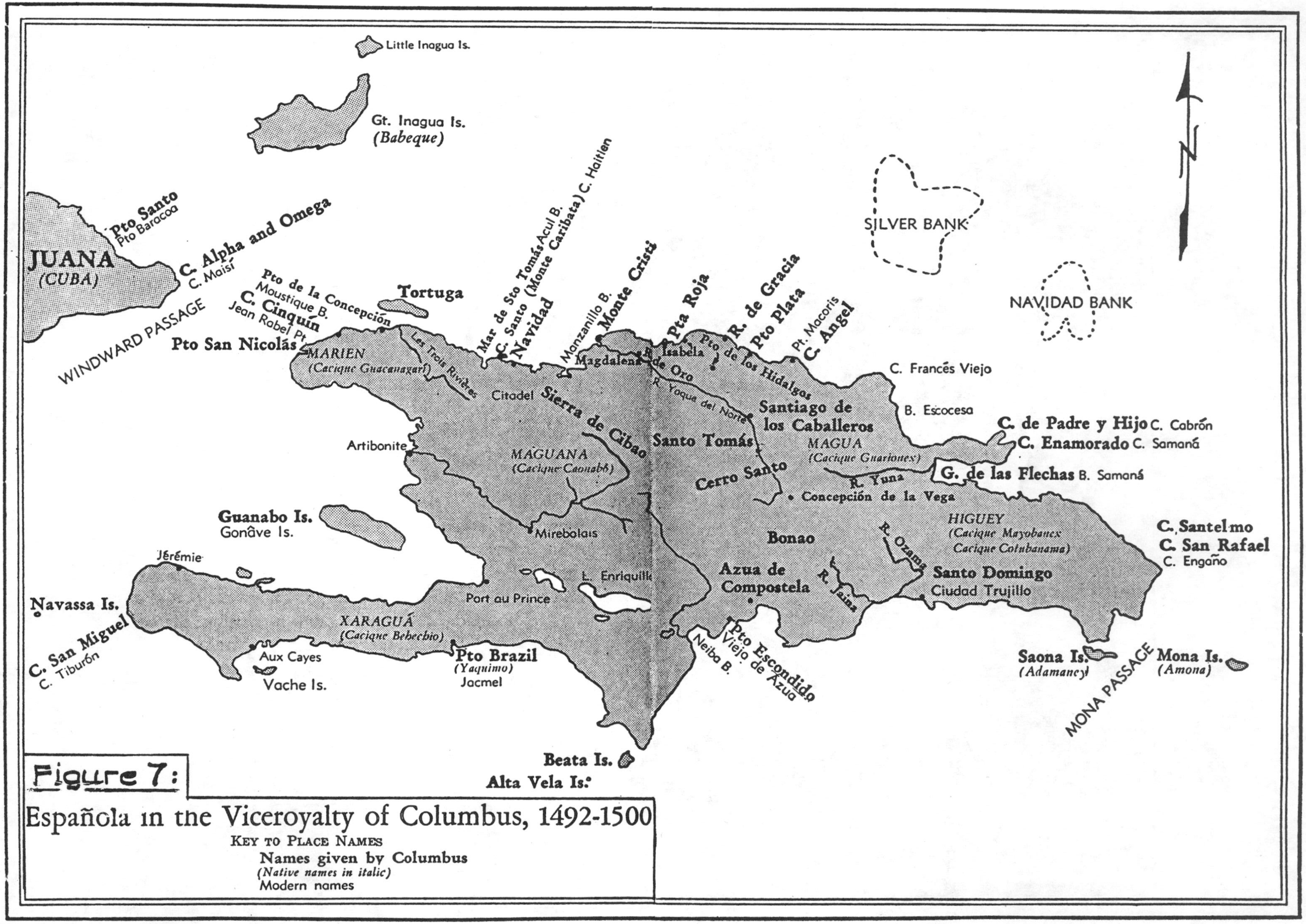

Figure 7:
Española in the Viceroyalty of Columbus, 1492-1500

KEY TO PLACE NAMES

Names given by Columbus
(Native names in italic)
Modern names

Figure 7. See credit on top of page 265

262

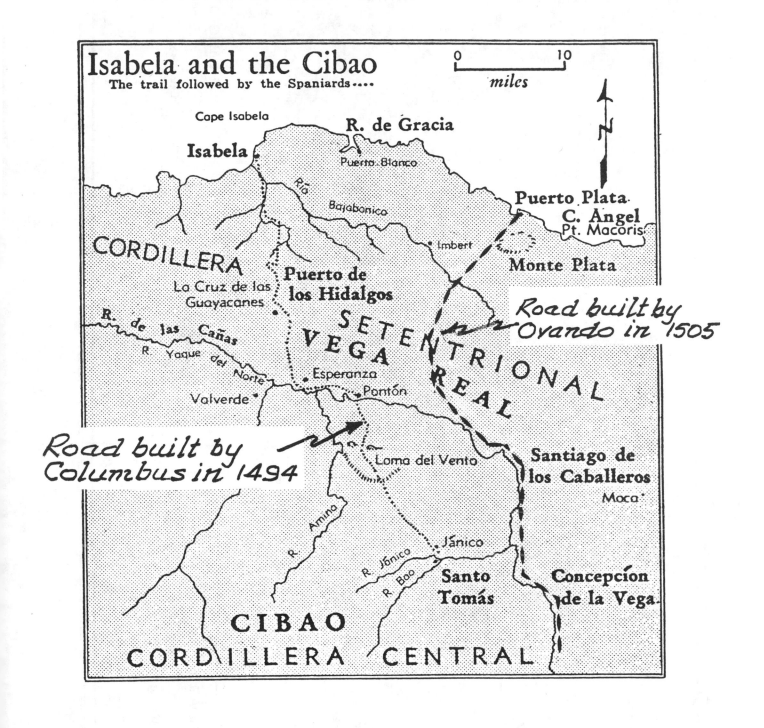

Figure 8. *The earliest roads in America. Credit on top of page 265.*

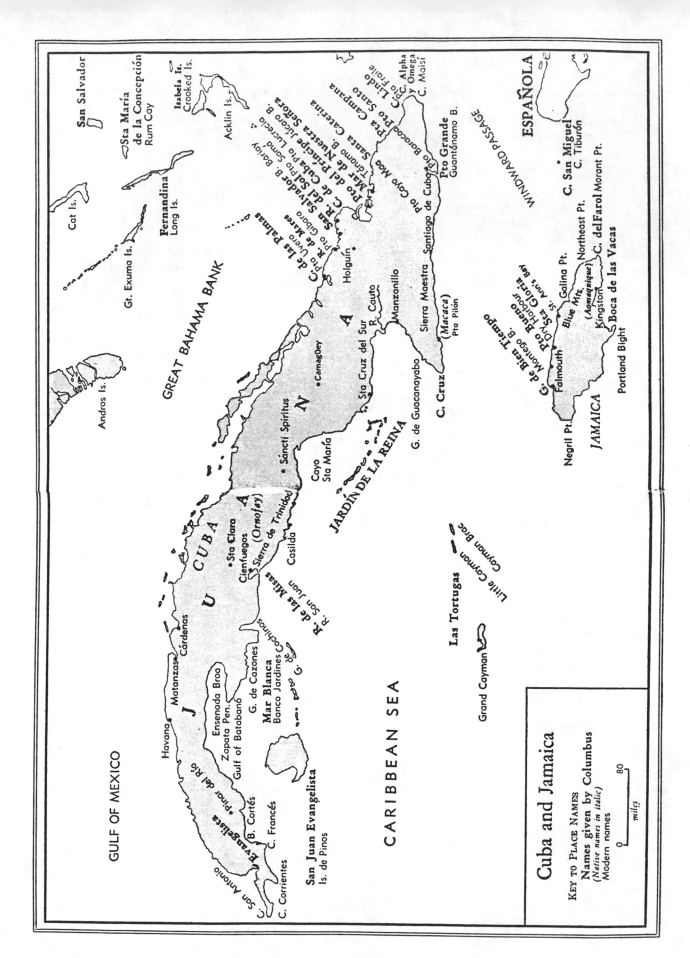

Cuba and Jamaica

KEY TO PLACE NAMES
Names given by Columbus
(*Native names in italic*)
Modern names

0   80
miles

GULF OF MEXICO

San Salvador

Sta María
de la Concepción
Rum Cay

Isabela In.
Crooked Is.

Cat Is.

Acklin Is.

Gt. Exuma Is.

Fernandina
Long Is.

GREAT BAHAMA BANK

Andros Is.

Bariay
San Salvador B.
Pto del Príncipe
R. de Cuba
Mar de Nuestra Señora

Somó
Pto Sama
Lucrecia
Jácaro
Santa Caterina B.
Santa Caterina

Tánamo
Pta Campana
Pto Santo
C. Lindo

C. Alpha
y Omega
C. Maisí

Ptá Fraile

Baracoa

Pto Grande
Guantánamo B.

Coyo Moa

Pto

Santiago de Cuba

C. San Miguel
C. Tiburón

WINDWARD PASSAGE

ESPAÑOLA

de las Palmas
Pta Uvero
C. de Mares
Pto Gibara

Holguín

Sierra Maestra

R. Cauto
Manzanillo

(*Macaca*)
Pto Pilón

G. de Guacanayabo

C. Cruz

• Camagüey

N

A

C U B A

Sta Cruz del Sur

• Sancti Spiritus

Cayo
Sta María

JARDÍN DE LA REINA

Sierra (Ornofay)
• Sta Clara
Cienfuegos
Sierra de Trinidad
Casilda

R. de las Mlas
R. San Juan

Matanzas
Cárdenas

Havana

Ensenada Broa
Zapata Pen.
Gulf of Batabanó
G. de Cazones

Cochinos
Mar Blanca
Banco Jardines

J

Pinar del Río
Evangelista

San Juan Evangelista
Is. de Pinos

B. Cortés
C. Francés
San Antonio
C. Corrientes

CARIBBEAN SEA

Las Tortugas

Little Cayman
Cayman Brac

Grand Cayman

Negril Pt.

de Bien Tiempo
G. de Montego
Pto Bueno
Pto Sta Gloria

Galina Pt.
Blue Mts.
Northeast Pt.
(*Aomaquique*)
S. Ann's Bay
Falmouth
Kingston

C. del Farol Morant Pt.
Boca de las Vacas

JAMAICA

Portland Bight

Figure 9. Credit on top of page 265.

264

Credit for Figures 6, 7, 8, 9,
Illustrations are taken from "The Life of Christopher Columbus By His Son Ferdinand," translated by Benjamin Keen, Rutgers University Press, New Brunswick, N.J., 1959. It is reproduced here by the kind permission of Professor Emeritus Keen, copyright owner.

On November 22nd, the fleet left Puerto Rico on a west northwest bearing passing a small island (Mona Island) where their bearing was changed to northwest. That night they reached Cabo San Rafael at the eastern tip of Espanola's north shore. Continuing westward along the north shore, they recognized the large Bahia de las Flechas where they put ashore a young Indian they had captured on the first voyage and taken to Castile. This lad, dressed in European clothes and speaking some Spanish, promised to induce his people to live at peace with the Europeans.

After sending boat crews ashore at Monte Cristi on November 25th and at other points the next two days with increasingly disturbing findings, the fleet positioned itself off Navidad on Thursday, November 28th and found it burned to the ground.

The true story of what had happened to Navidad and the 39 Spaniards Columbus had left there, will never be known. According to Guacanagari, the Indian cacique with whom Columbus had become quite friendly on the first voyage, soon after Columbus' departure from Navidad, the Spaniards began to quarrel among themselves, each taking as many women and as much gold as he could. This soon led to bloodshed and a split in the Spaniard's forces. One group under Gutierrez and Escobeda had struck out for Cibao, the gold mine country, which was a part of the kingdom of Maguana, leaving de Arana and the others at the fortress. King Caonabo, ruler of Maguana, killed the group which had entered his domain and then marched on Navidad with a strong force. Navidad was defended at this time only by de Arana and ten others who were willing to stay and defend the fort, the others having dispersed to various places on the island. Caonabo surprised the fort by setting fire to it at night, and despite Guacanagari's efforts to aid the Spaniards, the fort's defenders were all killed. Those who had dispersed before the fort was attacked survived for various periods, but all ultimately were killed by Indians not of Guacanagari's following. In the attack on the fort, Caonabo had wounded many of Guacanagari's men, including the cacique, himself.

While this story was corroborated by several other Indians of Guacanagari's following, there existed enough discrepancies to cause many of Columbus' expedition to doubt Guacanagari's veracity and they urged that he and his followers be punished. However, Columbus was not ready for such strong action, as yet. Instead, he searched for a site for a new settlement large enough to house most of the colonists he had brought with him on this second voyage, and more defensible. On December 8th, 1493, he found such a site, about 60 miles east of Navidad, near a large river which emptied into an excellent harbor. Going ashore with all his people, provisions, and equipment, he named the site Isabela, in honor of his Queen. One of the attractions of this site was that it lay astride the shortest and most accessible route from a good harbor to the gold mines of Cibao.

Many among the Spaniards, including Columbus had become ill on the short voyage from Navidad to Isabela and some of the cattle and horses had died. Some of the adventurers who had come on the expedition had expected to find gold on the beaches of Espanola, take all they could, and immediately return to Castile. The enforced labor of helping to build the new settlement at Isabela was not to their liking. Many grumbled and some actually plotted to make Columbus a prisoner, take over the fleet, and return to Castile. The food they had brought from Castile was running out and most had difficulty in adapting to the diet of the Indians. It was in such a setting that in mid-January, 1494, Columbus sent Alonso de Hojeda and 15 men in search of the mines of the Cibao (which Columbus had finally figured out meant stony mountains). And on February 2nd Columbus sent twelve ships of the fleet (the Collina and eleven caravels) under Captain Antonio de Torres back to Castile, carrying many of the sick and discontented. Entrusted to the Captain was a report to Ferdinand and Isabela of all that had happened, of the character of the country, and of the needs there. Soon thereafter, Hojeda returned with good news. There were no mines but there was lots of gold, taken from the many streams of the Cibao by the Indians of that region, who,up to that point at least,had been friendly to Hojeda and his men. Columbus determined to see himself.

On March 12, 1494, Columbus set out with a large force, all the able-bodied men not required to guard the two ships and three caravels in the harbor. He had, however, sent ahead a group to open up the Indian trail, so that it could be utilized by the large force including riders and pack animals. This appears to be the first road built by Europeans in America.

About March 18th, Columbus reached the area which Hojeda had determined was the richest in gold. This place, about 55 miles from Isabela, he named Santo Tomas. He ordered a fort built there which was to dominate the gold country and serve as a refuge, in case of need. Leaving Pedro Margarit in charge, with 56 men who were to build and man the fort and "mine" for gold, on March 21st Columbus returned to Isabela to prepare for his exploration of the southern coast of Cuba. From then until April 24th when he set forth with 3 caravels to explore the coast of Cuba, Columbus reinforced the garrison at Santa Tomas, replacing Margarit with Hojeda, as its first governor, pushed construction of a grist mill and other improvements at Isabela, started work on the upgrading of the road to Santo Tomas, organized a formal pacification of the entire island by a force under Pedro Margarit, and appointed a council to govern Espanola in his absence. His brother Don Diego was given the title of president of the council.

The three caravels Columbus took to Cuba were his favorite, the *Nina*, the *San Juan*, and the *Cardera*. He commanded the *Nina* with Francisco Nino as pilot and Alonso Medel as master. Alonso Perez Roldan was master of the *San Juan* and Cristobal Perez Nino was master of the *Cardera*. When Columbus left Isabela, he was a disappointed man. The facts as they were being determined corresponded neither with his dreams or the reports which he was sending back to the Spanish sovereigns. Espanola was not Cipanju and gold, while available there, was not to be found in the quantities reported by Marco Polo or even Hojeda's first report on Cibao. Nor had he yet been able to fulfill the Sovereigns wishes of winning souls for Christianity. The Indians were willing enough, but Friar Buil was a scrupulous man, unwilling to baptize anyone who had not been taught the elements of Christianity. The interpreters being required elsewhere, there was little instruction available for the Indians, and no baptism. The voyage to Cuba was undertaken to prove once and for all that Cuba was part of the Asian continent, its easternmost cape.

During the period April 24 to September 29, 1494, Columbus explored almost the entire south coast of Cuba from Cape Maisi, which he had named Cabo Alpha y Omega on his first voyage, to Evangelista on the Bay of Cortes; discovered and explored the entire coast line of Jamaica; explored the entire south coast of Espanola; and discovered Mona Island, in the Mona Passage between the east coast of Espanola and the west coast of Puerto Rico ,which he had named San Juan Bautista on November 19th of the previous year. He came within 100 miles of Cape San Antonio, the westernmost point of Cuba, when he decided, on June 12, to end his torturesome exploration of the south coast of Cuba at Evangelista. His food had run out and he and his men were exhausted from their exertions in maneuvering endlessly among the innumerable islands and shoals (frequently going aground and then laboriously freeing their ships). Constantly shifting winds added to their trials. And, at Evangelista, it seemed as though the coast ran on endlessly to the west.

While Columbus hoped that Cuba was a part of the Asiatic mainland, as his calculations indicated, he was not inwardly convinced. Fearing later criticism for not having gone further and obtained more positive proof of Cuba's geographic character, he caused the testimony of the ship's complements to be taken to the effect that Cuba was a peninsula, and it was no use going further. All had to sign their sworn depositions, including Juan de la Cosa who was to show Cuba as an island in his famous map of the world of 1500.

Columbus had intended to visit Puerto Rico and certain other islands inhabited by the fierce, man-eating Carib Indians, before returning to Isabela on Espanola. However, after exploring Espanola's south coast rather meticulously and then discovering Mona Island while en-route to Puerto Rico, he lapsed into a coma on September 25th. Malnutrition and exhaustion, probably compounded by deep disappointment, had taken its toll. His senior aides decided to return to Isabela forthwith and they arrived there September 24th, 1494.

Columbus returned to Isabela still maintaining that Cuba was part of the Asian mainland, this despite having been told by an Indian guide just before reaching Evangelista that Cuba was an island. Despite their sworn depositions to the effect that Cuba was a peninsula, certain of those who made the trip were convinced the Indian was right, especially since signs normally associated with a mainland were lacking. Among those were Juan de la Cosa, the cartographer, Michele de Cuneo (Columbus' friend), and a priest who had made the trip.

If Columbus had found no gold in Cuba and Jamaica, he had at least found many excellent harbors and some fine sites for settlements, among them Puerto Grande (the present U.S. Naval base, Guantanamo) and Santiago de Cuba; Santa Gloria (St. Ann's Bay) and Golfo de Bien Tiempo (Montego Bay) in Jamaica; Puerto Brazil, in Haiti; and Puerto Escondido and Santo Domingo in the present Dominican Republic.

During Columbus' absence of 5 months from Espanola, trouble broke out between the Spaniards and the Indians which was not to be finally settled during Columbus' regime as governor of the Island. Pedro Margarit, whom Columbus had charged with pacifying the island, reducing it to the service of the Catholic Sovereigns, had virtually ignored his instructions. Instead, he bickered with Don Diego and the Council Columbus had set up to govern in his absence. Margarit, insisting that he was the top man, presumably since he was commander of the troops, allowed the soldiers to roam where they willed committing outrages against the Indians, appropriating their goods, their gold, and their women.

The Indians soon responded in kind, murdering Spaniards if they could catch them alone or in small groups, setting fire to huts housing sick men, and in other ways showing their hatred. When the situation appeared to be out of hand, Margarit decided he would go back to Castile, returning on one of the ships which brought Columbus' brother Don Bartolomeo to Isabela in June 1494. Friar Buil and all but one of the priests, deciding the situation was hopeless for gaining souls for Christ, returned on the same ship. They had deserted Columbus and could be expected to describe all of Columbus' weaknesses, failures, and overstatements (or lies) to the Sovereigns with relish. This they did in due course.

Columbus returned to Espanola to find that his brother Bartholomeo had arrived from Castile in June with 3 caravels loaded with the supplies and other necessities he had requested of the Sovereigns in the message carried by Captain Antonio de Torres. This was a god-send to him for he was a very sick man. Between the brothers Bartolomeo and Diego, they provided the care he required to regain his health, but it was to be 5 months before he was well enough to take an interest in affairs on the island and, as a result, learn about what had transpired in his absence and since his return.

By February 1495, Columbus was well enough to undertake pacification measures. In the course of his attempts to bring order to the island, he conducted a military campaign against the more aggressive of the Indian kings and their followers. Many Indians were killed (and some Spaniards) and about 1600 captured. In June 1495, 550 of the choicest specimens were selected and sent back to Seville to be sold as slaves. However, about half died from exposure and malnutrition on the long voyage.

Columbus imposed a tribute of gold[15] (or cotton if no gold was available) on each Indian over 14 years of age, to be paid every three months. Failure to meet the requirement resulted in severe punishment. When the Indians retaliated, whole villages were sentenced to execution or torture. It is estimated that during the period April 1494 to March 1496, when Columbus left Espanola to return to Cadiz, 100,000, or one-third, of the Indian population of Espanola, were killed, committed suicide, or died of malnutrition as slaves or prisoners.

As was to be expected, the Sovereigns had listened to Friar Buil and others and had sent a courtier named Juan Aguado to investigate conditions at the settlement. He arrived in October 1495. Soon thereafter, he reported the deplorable relationship between the Spaniards and the Indians. He also told the

Sovereigns that nearly everyone at Isabela was either ill or disaffected. Anyone who was well enough to do so was inland plundering the country or hunting for gold and slaves. Everyone, he reported, longed to go back to Castile.

Columbus had intended to return to Castile before Aguado's arrival. However, one day in June 1495, an exceptionally violent storm ,one the Indians called "huracan", struck Espanola and, before it abated, caused considerable damage at Isabela including the sinking of all the caravels in the harbor, save the *Nina*. Columbus ordered his carpenters to build a new caravel similar to the Nina and to overhaul, caulk, and repair storm damage to the Nina. This work completed in early March 1496, Columbus and his brother Don Bartolomeo investigated the site of a new settlement at Puerto Plata[16] which they found unsuitable. Then, instructing his brother to investigate a site on the River Ozama, on the south coast of Espanola (the site of modern Santo Domingo), he sailed for Castile on March 10th. Don Bartolomeo was left in charge of Espanola and all of the West Indies Columbus had claimed for the Sovereigns and which the Pope had confirmed by Papal Bull.

After a month at sea, the *Nina* and the *Santa Cruz* put into Guadalupe to take on food, fuel, and water. There, Columbus had a short battle with some Carib women who defended vigorously their village and property. Leaving Guadalupe on April 20, Columbus resumed his voyage back to Castile steering a bit south to avoid the North Atlantic storms. They were successful in this, but ran into the doldrums which extended their voyage to the point that all hands almost perished from starvation. It was not until June 11, 1496, 52 days after leaving Guadalupe and a little over 3 months after leaving Isabela, that they were able to drop anchor at Cadiz.

Columbus must have been a very persuasive person for, in due course when Ferdinand and Isabela were able to receive him, he was able to convince them to provide him with 8 ships. 2 would resupply Espanola with what they needed and 6 would be used to discover the large country with great wealth which--Columbus was now convinced--lay to the south of the islands he had already discovered. It is likely that the sovereigns had, by now, some doubts about Columbus' abilities as an administrator, but none about his loyalty to their Catholic throne, his zealousness in wanting to gain souls for Christ, and his genius as a navigator, sailor, discoverer, explorer, and commander. In addition, they were to a degree, haunted by the realization that if they did not continue their efforts in the New World, another monarch might supplant them, Papal Bull or no!

The Voyages of Exploration of John Cabot, 1497-1498

While the greater part of this appendix is concerned with Columbus' four voyages of exploration and discovery, 1492-1504, it is important in the interest of historical accuracy that mention be made of the voyages and explorations of others during this period. The first such was that of John Cabot.

In May, 1497, five years after Columbus had sailed from Palos on his first voyage of discovery and exploration to the New World, John Cabot left Bristol, England on his first voyage of exploration. He was seeking, like Columbus and several others, a shorter route to India and China by a westward course. Cabot, really Giovanni Caboto, a citizen of Venice, had obtained a patent and financial backing from Henry VII of England for the voyage. Clearing the southern coast of Ireland, the tiny expedition (one ship and 8 men) sailed due west. After about one month at sea, Cabot made his landfall at Labrador, Newfoundland, or Cape Breton Island, Nova Scotia. Going ashore, he claimed whatever he thought he had found for England and returned there, arriving at Bristol around the end of July.

The next year, starting a month earlier, he set out with 5 ships and 300 men, following essentially the same course as the year before. Making his landfall within a couple of hundred miles of his previous landing, but definitely on the mainland of North America (most likely Labrador), he turned southward. Methodically, he explored and mapped the North American coastline noting, in particular, the excellent harbors, the vegetation, the climate, the better sites for possible settlements, and the friendliness of the Indians he had encountered. There is some uncertainty as to whether he got as far south as the Chesapeake Bay or whether the Delaware Bay was the southern limit of his coastal exploration. Again, claims were made for the British Crown and the expedition returned to England without attempting a settlement.

About 10 years later, Cabot's son Sebastian, sailing for the British, sought a northwest passage but did not succeed. He did, however, add considerably to the geographical knowledge of the North American continent.

Vasco De Gama's Voyage to India, 1497-1499

On July 8, 1497, some nine years after Bartolomeo Dias rounded the southern tip of Africa, Vasco de Gama, also in the Portuguese service, left Lisbon with 5 ships. His mission, the same as Dias' original objective, was to find a sea route to India and China. Benefiting from Dias' experience, de Gama followed a southwest course (after leaving the Cape Verde Islands) which brought him close to the Brazilian "Bulge" (without, however, sighting Brazil). There, he altered course to southeasterly and made his first landfall at St. Helena Bay, on the west coast of Africa, about 150 miles north of Capetown. The expedition had sailed about 6700 miles and it had taken them about 4 months to this point.

After a stay long enough to repair his ships, take on wood and water, and rest the crews, de Gama set out on the second leg of the voyage, to round the southern tip of Africa. This done, they proceeded up the east coast to Delagoa Bay at the southern tip of Mozambique. This leg of the voyage was relatively short, some 1300 miles. Still, it was not till early 1498 that de Gama resumed his voyage northward along the African east coast. The third leg of the voyage terminated near Malindi in modern Kenya. This leg was 1700 miles long and the voyagers had travelled a total of 9700 miles and taken about 8 months to do it.

The last leg of the journey, from Kenya to India involved a crossing of the Indian Ocean and a successful landfall at Calicut (modern Khozikode) about 400 miles south of Bombay on India's west coast. This open sea crossing was about 2400 miles long and took about 40 days. The entire voyage, a very successful one at that, had covered a total, one-way distance of over 12,000 miles in 9 months.

De Gama returned to Lisbon in 1499 with a rich cargo and, some say, richer stories. He had discovered a practical, if arduous, sea route to India and Portugal was to profit mightily from this. Moreover, de Gama's successful voyage was to spur Portugal's European neighbors, particularly Spain, to greater efforts in seeking a western route to India.

Columbus' Third Voyage of Discovery and Exploration, 1498-1500

Work proceeded slowly on the preparation for Columbus' third voyage of exploration to the "New World" and it was not until May 30, 1498 that Columbus main fleet of six ships was able to get underway, the *Nina* and the *Santa Cruz* (also known as the *India*) having been sent ahead to Espanola some time earlier. Columbus now decided to send 3 more ships directly to Isabela Nueva[17], while he sailed further south with the remaining three.

The three ships which participated in the exploration were the *Santa Maria de Buia*, Columbus' flagship (somewhat similar to the *Santa Maria*, his flagship on his first voyage); *La Vaquenos*, a 70 ton caravel commanded by Pedro de Terreros; and El Carreo, a much smaller caravel commanded by Hernan Peres.

Columbus proceeded cautiously from Sanlucar de Barrameda (at the mouth of the Guadalquivir) to Porto Santo and Funchal, in the Madeiras, to Gomera and Ferro in the Canary Islands, and Boavista (a leper colony) and Sao Tiago in the Cape Verde Islands. He went ashore at both Boavista and Sao Tiago in unsuccesful attempts to obtain cattle for Espanola.

On July 4th, 35 days after leaving Spain, the three ships left Sao Tiago, taking a southwest course until July 13th when they were becalmed. Columbus estimated they had covered 360 miles since leaving the Verde Islands. The heat had been unbearable. The next 6 days were cloudy and wet, but still hot. All hands, and especially Columbus, suffered greatly from the heat and the salty environment. Sometime during the damp period, the wind returned and Columbus changed his bearing to due west, fearing further calms if he approached any closer to the Equator.

On July 22nd, they saw large flocks of birds flying northwest encouraging them to believe land was near. On July 31st, the Admiral decided to alter course to north by east in the hope of reaching Dominica or one of the Carib islands. They continued on this course till noon when a seaman named Alonso Perez climbed to the lookout station atop the mainmast (called the "Top") and saw land about 45 miles to the west. This was distinguished by three hilltops. So Columbus named the place Trinidad, in honor of the Holy Trinity.

The next day, having sailed slowly all night along the south coast of Trinidad, they dropped anchor in a bay near the mouth of a river. To the south they could see a low lying island which Columbus named Isla Santa. This was however, the mainland of South America (modern Punta Bombeador, Venezuela). It was now August 1st, and 63 days had elapsed since they left Spain. Columbus was now made aware that the food they were carrying for the settlements at Espanola was beginning to spoil. Thus, there was an urgency to their further explorations.

Resuming their course westward on August 2nd, they sailed on till the sea narrowed to a strait, the yellowish waters churning in turmoil. Clearing the strait, they rounded a projection (at the westernmost point on the south coast of Trinidad), which Columbus called Punta del Arenal, and anchored on the lee side. Columbus named the strait Boca de la Sierpe (Serpent's Mouth). He noticed several phenomena here - the enormous tidal range (over 60 feet), the freshness of the sea-almost drinkable (this being the region of the Orinoco Delta), and the mildness of the climate. There was a shortlived contact with the Indians of Trinidad, who, Columbus noted, were well built and paler in color than the natives of Espanola.

Resuming their voyage on August 4th, they headed due north toward a mountainous island Columbus named Isla de Gracia. This, they later determined, was another peninsula of the mainland (Paria). The body of water which they crossed to reach Isla Gracia, Columbus named Golfo de la Ballena, because they had sighted a whale there. This is now known as the Gulf of Paria. The period August 4th to August 13th was spent in exploring the south coast of the Paria Peninsula and in learning what they could from the natives who wore gold on their breasts and pearls around their arms. Questioned about the sources of the gold and pearls, the natives said " . . . further west and north." Thus, on August 13th, Columbus' three ships cleared the Boca del Dragon, the strait between the eastern tip of Paria and Trinidad, and headed northwest by west in the direction of the sources of gold and pearls. They sighted some small islands to the north which Columbus named Los Testigos (The Witnesses) and, shortly thereafter, a large island to the west for which they headed, and which Columbus named Margarita. Sighting again the north coast of Paria, they changed course to southwest until they came to a cape which Columbus named Cabo de Conchas. Unknown to Columbus, the island of Cubagua lay about 30 miles to the west of this cape, and from Cubagua there would be taken (by others) an enormous quantity of pearls or "margarites."

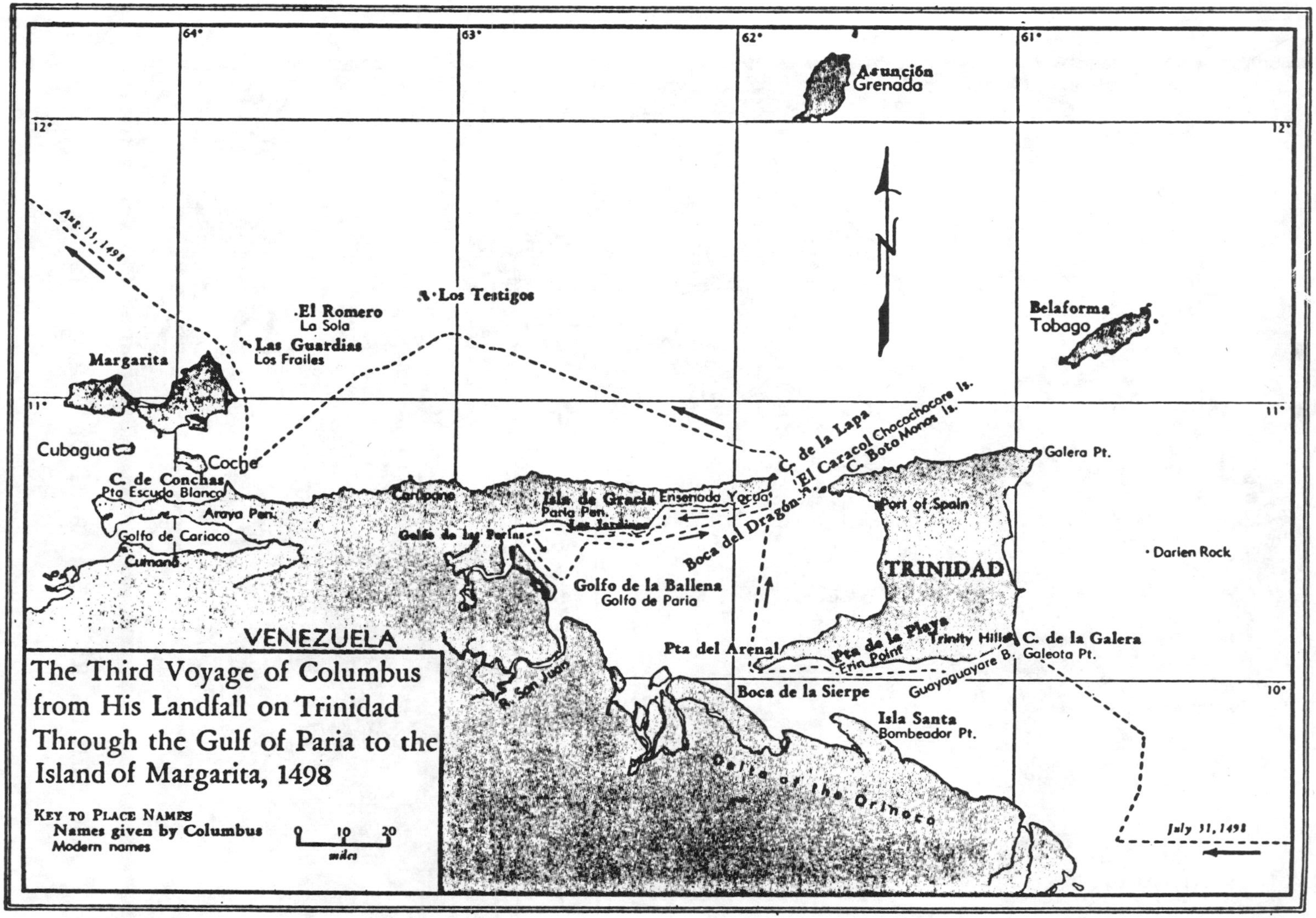

Figure 10. Credit on page 273.

Columbus at the Island of Margarita

Figure 11. Credit and comment on page 273.

He was practically on top of these great pearl fisheries when his illness and his anxiety about the decomposing food supplies made him abandon further exploration on the new continent and head north, around the east coast of Margarita, for Espanola. This was on August 15th. He had spent only 15 days in exploring Trinidad and Venezuela. As he cleared Margarita, Columbus changed course to due northwest. On August 20th, the three ships reached the southern coast of Espanola anchoring between Beata Island and Espanola. Columbus had set his course for the eastern end of the island, figuring currents would move him somewhat westward in the direction of the River Ozama. They actually made their landfall about 100 miles west of Santo Domingo and it took them 10 days to reach the city on the east bank of the Ozama which Don Bartolomeo had founded and named after their father, Domingo (Domenico, in Italian).

Columbus had been away almost 2 1/2 years. To say that things had not gone well in his absence would be an understatement. Many of the settlers had died after prolonged ilnesses and privation, and over 160 of the survivors were suffering from a disease later diagnosed as syphillus. (This was not a disease brought to the New World by the explorers and settlers from Castile, but, to the contrary, appeared to be one shared by many of the Indians on Espanola and the other islands explored. In due course, it was introduced to Europe by returning seamen and disappointed settlers and by those Indians sold as slaves.) Juan Aguado's investigation of Columbus (in late 1495 and early 1496) and his comments gave the impression that Columbus would not be back, thus making Don Bartolomeo's task as Governor much more difficult. As food began to run out and there was no news from home, open revolt flared under the leadership of Francisco Roldan, the alcalde, mayor, or burgomaster, of Santo Domingo. Roldan even incited the Indians against Don Bartolomeo. Still, Don Bartolomeo might have contained the rebellion but for two events--Columbus' return on August 31 and the arrival at Puerto Brazil on September 15, 1499 (a year after Columbus' return) of a fleet of ships under Alonso de Hojeda, Columbus' old lieutenant.

Columbus was a sick man when he returned to Espanola, eager for rest and recuperation. The prospect of becoming involved in what amounted to a civil war dismayed him. Thus, he stayed his brother from stern actions against Roldan while he sought to conciliate him.

Sometime in October 1498, Columbus sent his flagship and El Correo back to Castile with a cargo of Indian slaves, brazil wood, a little gold, and a few samples pearls from Paria. He also sent the Sovereigns a detailed report of the disorder on Espanola in which he asked for priests and a judge capable of dispensing "royal justice." He also sent along a map of the "Earthly Paradise" he had discovered showing how to reach it from Castile and how to get from Margarita to Espanola. He also indicated where he thought gold and pearls could be obtained. Then he asked for instructions.

One of those who received a copy of Columbus' map was Bishop de Fonseca, a man who was in charge of the Sovereigns' overseas interest for 30 years. The Bishop found Columbus to be pretentious and believed that there were several, if not many, qualified Castilians who could undertake voyages of discovery and exploration. One of these was Alonso de Hojeda, who had been with Columbus on his second voyage and had since returned to Spain and become a favorite of the Bishop. Having seen a copy of Columbus' map, he persuaded the Bishop to back him in an expedition to the regions described by Columbus.

Alonso de Hojeda's Voyage to the Spanish Main, 1499-1500

Two ships were fitted out for Hojeda, he was given a copy of Columbus' map and a patent of exploration, and in mid-May 1499 he sailed from Spain "in the Admiral's wake." One of his companions was Juan de la Cosa, the seaman and apprentice pilot on Columbus' 2nd voyage, who was to gather much of the information which went into his famous map of the world to be published in 1500.

Hojeda and his men had their first landfall along the Guiana coast. Then they reached Trinidad, Paria, and Margarita, following which they discovered the rich pearl fisheries of Cubagua. Here, they struck a profitable deal with the natives and then proceeded on westward. Sighting in turn the islands of Bonaire, Curacao, and Aruba, they sailed on and into the Gulf of Venezuela where they saw a village of thatched roof houses built on piles at the water's edge. This, they called Venezuela ("Little Venice"). Continuing westward, they rounded Cabo de la Vela on the Guajira peninsula of Colombia before turning northward toward Espanola.

Hojeda's expedition reached Puerto Brazil on Espanola's south coast in the Xaraguan province on September 15, 1499. There, they promptly started to cut down brazil trees and hunt Indians without Columbus' authorization. Worse, Hojeda soon made contact with the rebels who had been suborned by Roldan and attempted to persuade them to join him in unseating Columbus. This, it seems, would have been an excellent opportunity for Hojeda and Roldan to combine forces in accomplishing this unsavory task. However, this did not occur.

Roldan, who had been "dragging his feet" in reaching an agreement with Columbus, found some common ground with Columbus when Hojeda moved in. He volunteered to capture Hojeda but was unable to do so. He did, however, succeed in making Hojeda's stay so unpleasant and non-productive that Hojeda decided to move on. His little fleet headed for the Bahamas where they captured a number of Indians to be sold as slaves and then sailed for Spain. The expedition arrived home safely in April 1500.

The Hojeda expedition confirmed Columbus' claim that there was a vast continent to the west (whether or not this was Cathay), for they had made their landfall in South America about 200 miles farther south and 400 miles farther east and had travelled westward, along this continent's north shoreline about 500 miles beyond Margarita Island (where Columbus had to turn northward for Espanola). The Hojeda voyage also confirmed that this continent had wealth, friendly Indians, and (at least on the coast) a good climate. Finally, Hojeda had shown that one need not be a Genoese to accomplish voyages of discovery and exploration.

Amerigo Vespucci's Voyage to Brazil, 1499-1500

Sailing from Spain at about the same time as Hojeda (although not part of that expedition, as some have indicated[18]) was a Florentine by the name of Amerigo Vespucci who had become a merchant in Seville. Vespucci, greatly interested in navigation, had obtained a patent to sail to the New World for purposes of discovery. Vespucci's two ships made their landfall in late June 1499 somewhat west (and north) of the mouth of the Amazon (in modern Brazil) and probably no more than 250-300 miles southeast of Hojeda's landfall. However, he continued in the opposite direction to that taken by Hojeda, proceeding south and east along the Brazilian coast.

Crossing the mouth of the Amazon (which must have been an impressive sight as the muddy fresh waters persisted over a stretch of over 200 miles of coastline, Vespucci proceeded almost to Cap Sao Roque, at the north end of Brazil's "Bulge." At this point he had sailed along 1300 miles of Brazil's north coast. Turning, now, in the opposite direction, he came to Guiana and retraced much of Hojeda's route before returning to Spain.

Whereas Columbus maintained (to the end) that the continent he had found was Asia, Hojeda was not so sure, thinking it just might be a new continent. Vespucci had no doubts about it, being convinced the coasts he had explored belonged to a continent hitherto unknown. His conviction in the matter was to lead to the continent being named for him.

Nino and Guerra to the Pearl Coast, 1499

Sometime in early June, 1499, a few weeks after Hojeda and Vespucci had departed on their voyages of discovery and exploration, Pero Alonso Nino (who had accompanied Columbus on his first voyage in 1492-1493) and Cristobal Guerra left Spain bound for the Pearl Coast. They, too, had obtained from Bishop Fonseca a copy of Columbus' map showing how to reach Trinidad, the Gulf of Paria, Margarita Island, the Pearl Coast to the west of Margarita, and Espanola. This was no monumental undertaking. The single ship and complement of forty had but one objective--treasure!

Nino and his companions (for many of those who accompanied him were upper-class Castilians drawn by the prospects of high adventure as well as possible enrichment) were singularly successful. They returned with 76 pounds of precious pearls. This was the first of Spain's ventures to the New World to make real money (on a modest investment) for the participants and their backers. And, it started a stream of similar ventures to the New World.

Pinzon and De Lepe to Brazil, 1499-1500

Vicente Yanez Pinzon, captain of the Nina on Columbus' first voyage in 1492, left Spain in November 1499 on a voyage of discovery and exploration. He touched the coast of Brazil at a point in the neighborhood of 10° S. latitude in January, 1500. (This is approximately where Recife is today.) Turning north, he sailed around the "Bulge" and along Brazil's north coast to the mouth of the Amazon, which he investigated, and thence to the coast of Guiana. Here he turned for home, arriving in Spain in July 1500. He is credited by most historians and geographers as the discoverer of the Amazon River. He had sailed farther south along Brazil's east coast then Vespucci (by about 200 miles) and added to the convincing evidence, by now, that Columbus had discovered a new continent in August 1498.

Behind him by about a month, both in Spain and in Brazil, was Diego de Lepe. He touched Brazil at about 4 ° S. latitude before turning northward for some distance and then returning to Spain.

Pedro Alvares Cabral Claims Brazil for Portugal, 1500

Following the "trails blazed" by Bartolomeo Dias in 1488 and Vasco de Gama in 1497 to 1499, the Portuguese captain Pedro Alvares Cabral, in the service of King Manoel I, set out from Lisbon on March 9, 1500 with a fleet of 16 ships bound for India. On board, as advisers, were Bartolomeo Dias and several veterans of the De Gama voyage. Acting on the consensus view, Cabral steered southwestward after leaving the Cape Verde Islands.

In early April, Cabral's fleet made its landfall and anchored off the coast of Brazil at a point about 18 degrees S. latitude (a little north of modern Caravelos). Going ashore Cabral named the place Santa Cruz. He subsequently explored the area somewhat and then sent one of his ships back to Lisbon with the news of his discovery and claim.

Captain Cabral was, of course, well aware that the Treaty of Tordesillas gave Portugal title to all land east of a line of demarcation 370 leagues west of the westernmost of the Cape Verde Islands. His calculations put him at about 38° to 39 ° W. longitude, somewhat within the zone of legitimacy.

Cabral's claim, the subsequent exploration by Amerigo Vespucci for Portugal, her ownership and occupation of the Azores and the Cape Verde Islands (the westernmost and southernmost of European owned islands in the Atlantic), Papal recognition, the Treaty of Tordesillas and subsequent treaties, but most of all a stream of further explorations coupled with viable settlements, were to hold Brazil for Portugal until the 19th century.

Bobadilla Dispenses "Royal Justice" to Columbus, 1500

Many of the colonists at Espanola, dissatisfied with their lot, had returned to Spain. There, they constituted a vocal and hostile force aimed at unseating Columbus and his brothers. Their stories about Columbus' gross incompetence were confirmed to a degree by the crew members of re-supply and other ships which had visited the island. Certainly the most prominent of these was Alonso de Hojeda who had a chance to talk with Francisco de Bobadilla before he departed from Castile for Espanola in July 1500.

Bobadilla had been appointed by the sovereigns in response to many pleas (including, it will be recalled, that of Columbus himself) to go to Espanola to dispense "royal justice." He was to be governor and judge, with authority to mete out any punishment to Columbus, as well as any others he deemed appropriate. Bobadilla arrived in Espanola on August 23, 1500. He wasted no time in bringing Columbus and his brothers before him. Accusations of severity, injustice, and venality were leveled at the three with such venom that Columbus expected to be put to death. However, Bobadilla put all three in irons and sent them back to Spain. A despatch from Bobadilla to the crown was to follow. Columbus and his brothers arrived in Cadiz on the caravel *La Gorda* on November 20, 1500.

During the course of his imprisonment at Santo Domingo, Columbus became so heartbroken and idignant at the treatment accorded him that he wrote to an old friend, Donna Juanan de Torres, formerly nurse of the infante Don Juan. This letter was read to the Queen before the arrival of Bobadilla's despatch. To the degree possible, confirmation of Columbus' story was sought and obtained from two officials friendly to Columbus. Again there was an "about face." All of Columbus' property, his income from a share of the wealth brought to Spain as a result of his enterprise, and certain of his titles were resotred. Their majesties repudiated Bobadilla's proceedings against Columbus and appointed a new governor, Nicolas de Ovande, to supercede Bobadilla who was to be returned to Spain for trial.

De Ovando left Sanlucar on February 13, 1502, with a fleet of thirty ships to not only supersede Bobadilla, but end, as well, Columbus' tenure as viceroy and governor of that part of the New World he had discovered. Columbus' role as an explorer (if not discoverer) in the New World, was not, however, finished. We will return to his activities after a short treatment of exporation by others initiated (and largely completed) before Columbus' 4th voyage.

Figure 11a. Columbus is clapped in irons by Bobadillo at end of 3rd voyage (de Bry, 1594).
In response to many pleas, the Spanish sovereigns appointed Francisco de Bobadillo their special representative to bring order to Española and mete out punishment to anyone - including Columbus - deserving of it. Shocked at what he found when he arrived at Santo Domingo in August, 1500, he removed Columbus from his office of Governor and imprisoned him and his brothers, Bartolomeo and Diego, pending their return to Spain for trial. In late November, 1500, Columbus arrived back in Cadiz. Hearing Columbus' own account of what had happened, the Sovereigns did an "about face," restoring his impounded property, but not his office of governor, and authorizing still another voyage of exploration for their "Admiral of the Ocean Sea." Bobadillo was to be replaced by Nicolas de Ovando and returned to Spain for trial. He never got there!

From Theodore de Bry's "Historia Americae", coutesy of the Library of Congress.

Bastidas, the First European to See Central America, 1500-1502

Still another expedition sailed from Spain to the New World in October 1500. This one was headed by the notary, Rodrigo de Bastidas, who had accompanied Columbus on his second voyage. To assist in navigation and mapping was Juan de la Cosa who had signed on as a seaman during Columbus' second voyage, had accompanied Hojeda to Venezuela and Espanola in 1499-1500, and had just returned to Spain from the latter expedition. De la Cosa had achieved some recognition internationally as a result of his *Map of the World,* published earlier in the year 1500. Also aboard as a seaman was a lad in his teens who was destined to become the expedition's most famous member, Vasco Nunez de Balboa. The objective of the expedition was exploration of the northern coast of South America.

Reaching the Pearl Coast, they collected--and otherwise acquired--a significant quantity of pearls and gold. However, Bastidas' expedition did not tarry long on the Pearl Coast but pushed on westward. They passed Cabo de la Vela (Colombia) where Hojeda had, the year before, turned north, and went on to explore the Gulf of Uraba (in Colombia, just east of the Panama border) and the Isthmus as far west as Nombre de Dios (about 40 miles east of Cristobal and Colon, at the Atlantic end of the Panama Canal; the site that Columbus was to name Puerto de Bastimentos, in 1503).

Turning northward from Nombre de Dios toward Espanola, they encountered a late summer storm (or hurricane) and lost all their ships while in the Gulf of Xaragua. They managed, however, to save their treasure of pearls and gold. Proceeding by foot to Santo Domingo, Bastidas and his men had some difficulty with Governor Bobadilla who wanted to detain the men and confiscate the treasure on the ground that it had been obtained on Espanola without his permission. The arrival of newly appointed Governor Nicolas de Ovando in April 1502 saved Bastidas and his men from incarceration or worse. Bastidas and most of his men returned to Spain in the mid-summer of 1502 on ships re-supplying Espanola.

It is interesting to note that Bastidas just missed Columbus who had arrived off Santo Domingo on June 29th enroute to Central America and particularly Panama on his fourth and last voyage.

Amerigo Vespucci's 2nd Voyage of Discovery to South America, 1501-1502

Sailing this time for King Manoel of Portugal, Vespucci set out in May 1501 to try again to find the strait, if it existed--which would lead into the China Sea. The successive discoveries had pushed the location of such a strait steadily southward. Vespucci's landfall was about where it had been in 1499. He proceeded eastward past the mouth of the Amazon to Cap Sao Roque, then southward past Pinzon's landfall, past Cabral's, on and on, beyond the Rio de la Plata (which separates Uruguay and Argentina), and far enough down the Argentine coast to experience very cold weather. Precisely how far south he went is a matter of some dispute. He returned to Lisbon on June of 1502 and had found no strait.

Columbus' 4th and Last Voyage, 1502-1504

Columbus was to wait many months before his request to the Sovereigns for authorization for another attempt to reach India, "the land of spices", was approved. When it finally was, on March 14th, 1502, it was clear that the Sovereigns had given the matter much thought for there were severe restrictions as to Columbus' movements in the New World. Among them, he was not to call at Santo Domingo on his voyage westward; only upon his return, if he did not circumnavigate the globe. Also, having made up their minds to approve a new mission, they wanted him to get on with it quickly, or not at all. They were getting a little tired of the trouble "their Admiral" seemed to cause wherever he went.

Columbus' 4th voyage of discovery and exploration started at Cadiz on May 9, 1502. His fleet of 4 caravels consisted of his flagship *La Capitana*, of 70 tons, under Diego Tristan, Captain, with Ambrosyo Sanchez, Master; the *Santiago de Palos* (also called the Bermuda) of 60 tons, under Francisco de Porras, Captain, and Francisco Bermudes, Master; *La Gallago*, of 60 tons, under Pedro de Terreros, Captain, and Juan Quintero, Master; and the *Viscayno*, of 50 tons, under Bartolomeo Fieschi, Captain, and Juan Perez, Master. The flagship also carried the chief pilot of the fleet, Juan Sanchez, the surgeon, Master Bernal, and Columbus' son, Don Fernando Colon. The Santiago de Palos had aboard the chief clerk and accountant of the fleet Diego de Porras, brother of the ship's captain, and Don Bartolomeo Colon, brother of Colombus and former adelantado of Española. Counting the Admiral and all named in the foregoing, as well as seamen, gunners, ship's maintenance personnel, soldiers, "serving gentlemen", ship's boys, and the single priest, Friar Alixandre, there were 143 souls on the expedition.

Stopping at Maspalomas on Grand Canary to take on wood and water, the fleet cleared there on May 25, and passed Ferro (westernmost of the Canaries) the next day. Altering course to west by south, the usual route for Espanola, after 21 days at sea they made land at an island which Columbus named Matinino (which some historians believe to be Martinique, while others believe it to be St. Lucia, about 60 miles north of Martinique). Here they made a brief investigation, while they took on wood and water, and then resumed their voyage, passing to the south of Dominica, St. Croix, Puerto Rico and Mona Island, arriving at the roadstead off Santo Domingo on June 29th.

The Admiral had good reason to put into Espanola for on the crossing, the Santiago had been found to be a bad sailer, heeling excessively in a fresh wind. Desiring to buy or hire another ship to replace the Santiago, but mindful of the Sovereigns' restrictions about stopping at Espanola, Columbus sent Pedro de Terreros ashore to call on Governor Ovando and attempt to obtain the needed replacement. Having noticed the signs of a hurricane brewing, Columbus had de Terreros ask Ovando for permission to enter the harbor on the Ozama. There were, at the time, 28 ships in the Ozama estuary ready to leave for Castile with a varied cargo of gold and slaves. On one of these ships ex-Governor Bobadilla was being returned for trial.

Governor Ovando denied both requests and ignored Columbus' warning about a hurricane. He allowed the fleet to depart. They sailed into the path of the storm and were blown into the Mona Passage where 24 ships were sunk or battered to bits with the loss of all hands, including passenger Bobadilla, Roldan, the mutineer, the cacique Guarionex, and the commander of the fleet, Antonio de Torres, Columbus' friend. In all, 500 lost their lives. 200,000 castellanos of gold (about 2000 pounds) also went to the bottom. Three of the four surviving ships, badly battered, managed to limp back to Santo Domingo. Only one small caravel which had been hired by Columbus', Sovereign appointed, caretaker on Espanola, was able to make it back to Castile. It was carrying 4000 caastellanos (40 pounds of the Admiral's gold). The Admiral's enemies in Castile maintained that he had raised the hurricane by sorcery, while the Admiral's friends spoke of "Divine Justice."

Columbus, for his part, coasted the island, found a protected anchorage, and rode out the storm, albeit not without suffering some damage. After repairs had been made and the men had rested for a few days, they continued their journey westward along the south coast of Espanola, dropping anchor in the harbor at Brazil, in Xaragua, when it appeared another storm was rising.

On July 14th, the expedition quit Espanola heading for the south coast of Jamaica. Alternately losing the wind and then regaining it, they sailed westward and drifted, when becalmed, northward, until Columbus thought he recognized one of the islands of "the Queens' Garden" south of Cuba. Here they anchored, awaiting a favorable wind. With mutiny brewing among the crews, the arrival of an easterly wind came none too soon. Heading westward, on July 30, Columbus sighted and landed on the islet of Guanaja (Bonacca) about 40 miles north of the north coast of Honduras. The next day they anchored off the mouth of a river (on the mainland) which Columbus named Rio de la Posesion (Rio Romano). He named the

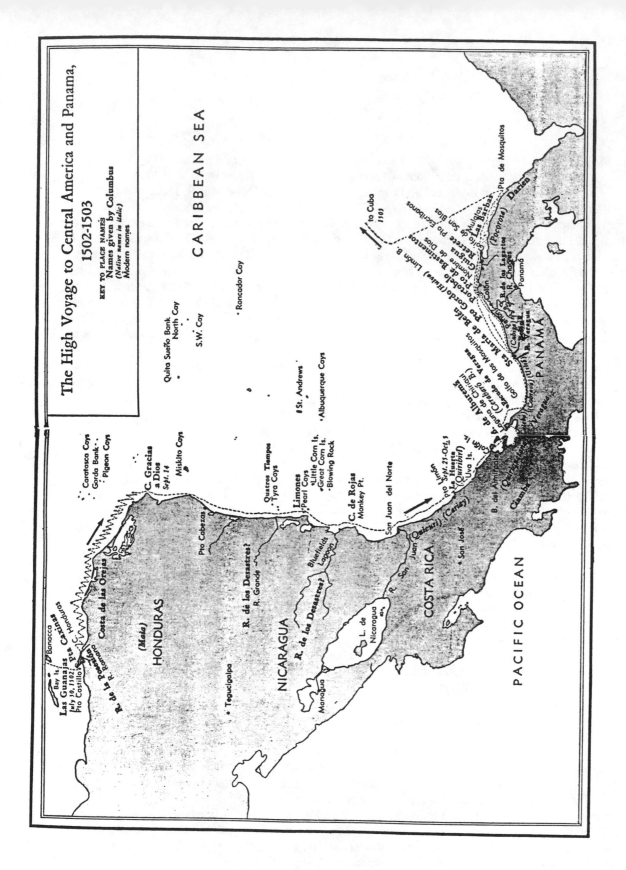

The High Voyage to Central America and Panama, 1502-1503

KEY TO PLACE NAMES
Names given by Columbus
(Native names in italic)
Modern names

CARIBBEAN SEA

PACIFIC OCEAN

HONDURAS

NICARAGUA

COSTA RICA

PANAMA

Bonacca
Bay Is.
Las Guanajas
July 10, 1502;
Pto Castillo
Guanaja
Pta Honduras
C. Honduras
R. de la Posession
Costa de las Orejas
(*Maia*)
Tegucigalpa
Managua
L. de Nicaragua
R. de los Desastres?
R. de los Desastres?
R. Grande
Bluefields Lagoon
R. San Juan (*Quiebri*)
San José
La Huerta (*Quiribri*)
B. del Almirante
Cariarí (*Cariabaro*)
Carambaru
Guaiga (*Cerabaro*)
Gulfo de los
B. de Alburema
Laguna de Chiriquí (*Cerabaró*)
Belén
Sta María de Belén
Pto Gordo (*Huiva*)
Limón B.
pto Sept. 25-Oct. 5
Uva Is.
Veragua
Portobelo (*Huiva*)
Pto de Bastimentos
Nombre de Dios
Retrete pto (*Huiva*)
Guiga
Mulatas
Las Barbas
(*Pocorosa*)
Islas de los Lagartos
R. Chagres
Panamá
R. Belén
Darien
Pta de Mosquitos
Pto de los Escribanos
Is. de San Blas
to Cuba
1503

Carotasco Coys
Gordo Bank
Pigeon Coys
C. Gracias a Dios
Sept. 14
Miskito Coys
Pto Cabezas
Quatros Tiempos
Tyra Coys
Limones
Pearl Coys
Quita Sueño Bank
North Coy
S.W. Coy
Roncador Coy
St. Andrews
Albuquerque Coys
Little Corn Is.
Great Corn Is.
Blowing Rock
C. de Rojas
Monkey Pt.
San Juan del Norte

Figure 12. Credit on page 282.

280

Figure 13. Captain Diego Tristan and boat crew are ambushed by Indians in Panama. (de Bry, 1594)
Sent by Columbus to obtain additional water supplies for La Capitana, Santiago de Palos, and Viscayno before departing from the harbor at Belén, Panama, Captain Tristan and a boat crew of 10 men proceeded far enough up the Rio Belén to insure the water was fresh. While filling their casks, they were attacked by Indians in dozens of canoes (somewhat differently than portrayed by artist de Bry). Tristan and all his men, except a cooper from Seville, Juan de Noya, escaped to carry the news of the tragic ambush.

From Theodore de Bry's "Historia Americae," courtesy of the Library of Congress.

—

harbor Puerto Caxinas (Puerto Castillo). Here there were some fruitful contacts with some friendly Indians. Hearing from one old Indian, whom he took aboard as a guide, of a large rich country lying westward called Maian (Guatemala), Columbus erroneously concluded this must be Cathay. However, since India was now his goal, he knew he had to find a route southward.

Sailing eastward along the Honduran coast looking for the strait which would lead southward, the expedition endured great hardship for wind and current were against them. Alternately sailing by day and anchoring as close to shore as possible for the night, it took them until September 14th to reach Cabo Gracias a Dios where they could turn southward, along the Nicaraguan east coast. Here, they were able to proceed more smartly. On September 16th they sighted the mouth of a large river and anchored there to take on water. While doing this, one of the boats capsized in the breakers and two seamen were drowned. The Admiral called this river Rio de los Desastros (actually Nicaragua's Rio Grande).

Continuing southward, they reached a beautiful island after 3 days. Columbus named this La Huerta (the Garden). It is opposite modern Costa Rica's Puerto Limon. The expedition stayed at La Huerta until October 5th, repairing their ships which were badly worm-eaten and leaking alarmingly. Resuming their voyage along the coast (which was becoming increasingly mountainous and east-west oriented, rather than north-south) they came to a large bay which was entered by a narrow strait. Columbus named the bay Bahia de Alburema (actually Laguna de Chiriqui, in Panama). Here they stayed until October 17th, again repairing their ships and exploring the country. While they found evidence of gold here, the Indians were not interested in trading for what the Spaniards had. They did, however, talk about the golden land of Veragua and of Ciguare on the other side of the high mountain range.

Between October 17th and November 26th, pushing ever eastward, the expedition traversed Veragua (the Indian name for Panama), including the Caribbean harbors at Belen (named by Columbus Santa Maria de Belen), Limon Bay, Portobelo, Nombre de Dios, Puerto Escribanos and a number of Indian villages along the coast or on rivers leading inland. Gold, clearly, was plentiful, but the Indians appeared quite able to do without the presence of the Spaniards.

On December 5th, the expedition turned back because of increasingly severe weather. On December 17th, the weather remaining bad, they put into a harbor the Indians called Huiva, which Columbus named Puerto Gordo. This is probably modern Limon Bay or Manzanilla Bay (at the Atlantic end of the Panama Canal). They remained in this general area until January 3, 1503. Despite several efforts to leave, the heavy seas made this impossible.

On January 3, despite continued poor weather they sailed out of Puerto Gordo heading westward, but again bad weather forced them to seek a safe anchorage. On January 6, the Feast of the Epiphany, Columbus found such an anchorage near a river the Indians called Yebra. After sending boats upstream and finding an Indian village there with more evidence of gold nearby, Columbus decided to establish a settlement there. Because it was on the Feast of the Epiphany that he was forced into the harbor by bad weather, he named the settlement Santa Maria de Belen (Holy Mary of Bethlehem).

By the end of March, 1503, enough huts had been built to house the 80 men Columbus proposed to leave at Belen, under his brother Bartolomeo, while he returned to Spain to obtain supplies and settlers. Accordingly, Columbus and all but the 80 men who were manning the settlement sailed out of Belen on

April 6. The Gallega, which was in the poorest condition to attempt an ocean crossing, was left for the use of the Adelantado, Bartolomeo.

However, this too was not to be. No sooner had the ships disappeared from view than the Indians fell on the tiny settlement. In the fierce fighting which ensued, the Adelantado and six other Spaniards were wounded and one was killed, but the Indians were temporarily beaten off. As the Admiral and his 3 ships were about to leave the bay and enter the open sea, he had second thoughts about the adequacy of his fresh water supply, so he ordered the fleet back to the mouth of the Rio Belen and sent a boat with Captain Diego Tristan and 10 men up the river for water. Tristan and his men witnessed some of the fighting but did not take part in it. Instead, he proceeded far enough upstream to insure the water he was to take on was fresh. While filling his casks, the Indians fell on them from dozens of canoes, killing all but a cooper from Seville, named Juan de Noya. Miraculously, he escaped and reached the settlement with the sad news.

Again, Columbus' brother, the Adelantado, displayed his unusual courage and resourcefulness by evacuating the settlement and setting up an emergency redoubt on the eastern bank of the river. There, he had the survivors build dugout canoes which would be required, with La Gallego's boat, for a quick evacuation to the fleet. The Admiral learned of the situation when after waiting anxiously for Tristan's return, he allowed a boatload of volunteers to approach the sand bar separating the bay from the mouth of the river. From there, one Pedro de Ledesma boldly swam to shore. He soon returned with the news about the destruction of the settlement, the losses in dead and wounded, and the absolute necessity for evacuation of the survivors to the fleet. However, it was not until April 16th that the evacuation was effected and the small fleet left Belen, La Gallego being abandoned there.

Sailing eastward along the coast, they reached Portobelo on April 23rd. There they abandoned the Viscayno whose planking by now was completely riddled by teredo (shipworm). On May 1, 1502, the ships having passed an archipelago which Columbus named Las Barbas (now Las Mulatas) and reached a headland on the Darien coast he named Marmoreo (Marble), they stood to the north with winds and currents easterly. On May 10th, they sighted the islands Columbus had previously named Las Tortugas (now known as Little Cayman and Cayman Brac) about 115 miles northwest of Jamaica. On May 13th, they reached one of the westernmost islands in the Jardin de la Reina, off the south coast of Cuba. Here Columbus decided to wait for more favorable winds to proceed eastward. Spain had been Columbus' destination when he left Belen, prior to turning back. This was clearly not possible now because of the condition of his ships. Without discussion, all hands knew that Espanola had become their destination, now.

On May 20th, Columbus' two ships, barely afloat, hoisted anchors and headed southeastward in an effort to round Cabo Santa Cruz. The unfortunate combination of weak winds of variable direction, geography, and terrible ship conditions so hampered their progress that it was not until June 20th that they were able to round the cape and anchor within sight of an Indian village, a few miles east. There they obtained food from the Indians and did what they could to patch their ships.

Finally, in a race with time, the two ships left Cuba sailing southward. Even the fallback destination of Espanola had now been abandoned. On June 25th the ships were beached at Santa Gloria (St. Ann's Bay), Jamaica. In running the two ships aground, Columbus had them stay as close together as possible. Thus, immediately after the ships grounded, he had them shored and lashed together for stability. He then built housing for the men on the fore and stern castles, since at high tide the decks were awash. Further, having experienced the effect of Spanish greed and lust on Indian society, Columbus restricted his men to the two ships. Specific permission had to be obtained to go ashore and such visits were monitored carefully.

By July 17th, Columbus had established sufficient rapport with the Indians to obtain two native canoes and rowers in which he despatched Diego Mendez and Bartolomeo Fieschi to Espanola for help. Mendes was to obtain a ship from Governor de Ovando and bring it back to St. Ann's Bay to evacuate Columbus and his men. Fieschi was to return to Columbus as soon as both conoes had reached Espanola with the news of Mendez safe arrival there. Mendez was also to insure that Columbus' report to Ferdinand and Isabela was placed into reliable hands at Santo Domingo, for further forwarding to the Sovereigns.

Columbus was to wait almost a year, until June 29, 1504, for the caravel which Mendez had hired in Espanola and guided back to St. Ann's Bay for the evacuation. During this period, he was to endure a mutiny (inspired by the brothers de Porras) which he suppressed with great difficulty, and barely avert an attack by the Indians on the beached ships by a dramatic and daring exploitation of his knowledge of an impending lunar eclipse (see Chapter 1 of text).

Columbus arrived at Sanlucar de Barrameda, after a stop at Santo Domingo to call on Governor de Ovando, on November 7, 1504. The voyage had been tempestous, Columbus was sick with the gout, he was disappointed with the results of the expedition, and he set about to write his will, an enormously complex task because of the wealth and hereditary titles he had earned. The death of Queen Isabela 19 days after his return to Spain deepened his misery.

Columbus died in Vallodolid, Spain on May 20, 1506, still fighting to regain his vice-royalty of the West Indies. Interred at the Convent of Santa Maria de las Cuevas in Seville, his epitaph reads, "To Castile and Leon Columbus Gave a New World."

The Naming of the Continent Discovered By Christopher Columbus

We know that Christopher Columbus was the first European of record to set foot in South America. This occurred in early August, 1498 in Venezuela, during his 3rd voyage to the New World. We also know that Amerigo Vespucci's first land fall on his first voyage to the New World, Brazil, north of the mouth of the Amazon River, occurred in June, 1499. Yet, in 1507, not quite one year after Columbus' death, there was published a little book, *Cosmographiae Introductio,* the work of a professor of geography at the College de Saint Die, in Lorraine, which credited Vespucci with the discovery. The professor, Martin Waldseemuller, had appended a world map showing a new continent in the west and across that continent was the name "America".

Waldseemuller, one of the most important cartographers of the 16th century, had first constructed a large map, 7 1/2 ft. x 4 ft., and a globe. The map appended to "Cosmographiae Introductio" was based on the large map (see outline of this map in Figure 4, Chapter 8).

Waldseemuller, in the book, mentions Ptolemaic tradition and Vespucci's travel accounts as sources and, indeed, sketches of Ptolemy and Vespucci frame two small hemispheres placed above the main map. The American continent is shown for the first time, based on the explorations which had taken place and about which information had been made public, and the name "America" after the Latin version (Americi) of Vespucci's Christian name, is shown prominently across the middle of South America.

When the map and book appeared (1000 were made and distributed widely), there was an immediate outcry from offical Spanish sources and supporters of Columbus' primacy in the New World. Trickery was charged by many, and indeed it would appear that some trickery was involved, but Waldseemuller was not the culprit. Vespucci had written four letters to the Lord of Florence, Lorenzo Piero de Medici, describing his voyages. These had been written in 1500, 1501, and 1502. Unfortunately, "literary pirates" with access to the letters had sensationalized them into two best setllers, *Mundus Novus* which was published in 1504, and *The Four Letters of Amerigo Vespucci*, published in 1505-6. In both publications, it is claimed for Vespucci that he preceded Columbus by a full year. Whether Vespucci had a hand in the deception is not known. Years later, when Waldseemuller became aware of his accidental error, he tried to correct it, but the damage was done. The name stuck.

In 1538, Gerardus Mercator's map applied the name to both North and South America and the geographies which appeared in the next century followed suit. Thus, even in death it seemed that a cruel fate was conspiring against Columbus as so many men had done during his lifetime.

Appendix B

ON THE LEAGUE

Introduction

A prime example of an itinerary unit of measurment which varied widely, if not wildly, at different times and in different places is the league (legua, leghe, leuga, lega, and lieue are the Spanish, Portuguese, Latin, Italian, and French equivalents, respectively).

There appears to be a diversity of opinion as regards its origin and especially its length. Thus, *Webster's New World Dictionary of the American Language* describes the league as " . . . a Gallic mile of Celtic origin; a measure of distance varying in different times and countries; in English speaking countries, it is usually about 3 statute or 3 nautical miles . . . "

The Random House *Dictionary of the English Language* is a bit more succinct and cautious. As they put it, the league is "a unit of distance varying at different times and in different countries; in English speaking countries usually estimated at roughly 3 miles."

Zupko, in *British Weights and Measures*[1], described metrology in use for hundreds of years following the Roman conquest. A sequence follows: ". . . The most frequently used itinerary measures were the furlong or stade (stadium), the mille (mille passus), and the league (leuga) . . . The league had 7500 feet (2200 meters), or 7283 B.I. (British Imperial) feet, or 1500 paces." Kimble[2] confirms this: "Gallic league = 1 1/2 Roman miles" (2223 meters), as does Petrie[3] who says, "The league is a Celtic measure, and is always stated at 1 1/2 Roman miles by ancient writers . . ."

Klein, In *World Measurements*[4], says the Gallic league, or leuga, was 8395 feet, 1.59 statute miles, 2559 metres, or 1.727 Roman miles of 1482 metres, indicating essential agreement with Zupko, Kimble, and Petrie although some growth in the length appears to have taken place. Klein goes on to say that the "old English land league," at 4830 metres, was almost exactly equal to 3 statute miles. The nautical mile was 6080 feet or 1853.2 metres, while the nautical league (British) was 3.104 times the length of the nautical mile. These units are, chronologically, late medieval, perhaps early renaissance. An examination of eight sea charts published between 1608 and 1715 in Amsterdam and London shows that the English nautical league, during this period was taken as 1/20th of a degree of latitude. Two charts dating from 1630 and 1715, respectively, also indicate the English nautical league as equivalent to 3 English nautical miles. A 1777 chart, also providing a plan of Charleston, South Carolina, indicates that one nautical mile equals one minute of latitude, or 1.1 statute miles, while a 1674 map of Jamaica published in London indicates that 70 Miliarium Anglican (English statute) miles make a degree of latitude. All of the foregoing examples are essentially compatible and indicate an effort to couple the length of the league, nautical and land, and the mile, nautical and statute, to the length of some unit of latitude. From Jean Fernel's 1526 - 1528 measurement of the length of a degree (described in Chapter 8) onward, this geodetic unit was being assessed with increasing accuracy.

As to the French league, the "lieue", Funk and Wagnall's French-English dictionary gives for "lieue": "4 kilometres or 2 1/2 English miles." This is equivalent to 2.71 Roman miles.

The Iconographic Encyclopedia of Science, Literature and Art, published in 1851 contained much material from an earlier publication, the Bilderatlas" of Friedrich Arnold Brockhaus. Among this material is a section on "Geography and Planography" containing 44 exceptionally informative maps[5]. Several contain scales along the map borders indicating that 25 lieuen equal 1 degree, or 1 lieue equals 4444.44 metres. The maps were drawn in the early 19th century, when the size of the earth was known rather accurately.

Klein reports on the metrology of France just prior to the installation of the metric system. In this sequence he puts the French league equal to 3000 toises (of 1.949 metres), or 5847.11 metres. Berriman[6] confirms this figure. However, a map titled *Carte de France* [7] depicting geodetic activities of the Royal Academy of Sciences, Paris, in the period 1669-1718, tells a different story. The "Echelle" (scale) of the map says:

> Lieues communes de 25 au degre'
> Lieues marines de 20 au degre'
> Toises de Paris de 57,060 au degre'

The Toise de Paris measured 1.949036 metres and 57,060 toises yielded a degree of 111,212 metres, which when divided by 25 and 20, respectively, indicated that:

> 1 lieue commune = 4448.5 metres
> 1 lieue marine = 5560.6 metres.

A 1768 map of the suburbs of Paris drawn by the noted geographer/geodesist César-Francois Cassiny de Thury confirms and extends the foregoing. By this time four leagues were in use in France. The scale defines these units as follows:

> 1° = 20 grand lieues de France, de 2850 toises (5554.75 metres).
> = 23 3/4 lieues de France, de 2400 toises (4677.7 metres).
> = 25 lieues communes de France, de 2282 toises (4447.7 metres).
> = 28 1/2 petites lieues de France de 2000 toises (3898.07 metres).

Some 20 sea charts published between 1608 and 1780 in Paris, Amsterdam, and London indicate 20 French marine or grande lieues made the degree of latitude. Several of these equate the English and French marine (nautical) leagues.

Another league frequently shown on legends or scales of the late 16th, 17th, and early 18th century sea charts, particularly (but not exclusively) those published by the Amsterdam chartmakers, is the Duytsche Myle, also referred to as the Milaria Germanica, Miliaria Geometrica, and Lieue Allemagne. Of 24 sea charts, published between 1595 and 1734, all set this unit at 15 to the degree of latitude. Ten of these charts also indicate that 20 English and/or French leagues make the degree, while 16 also set the "Spaensche Myle" (lieue d' Espagne, or legua) at 17.5 to the degree. Seven of the 24 relate the German (or Dutch), Spanish, English and French leagues to the degree in the ratios of 15, 17.5, 20, and 20, respectively. Three of the 24 relate the German (or Dutch) and Spanish leagues to the Italian Milaria by setting the 3 units at 15, 17 1/2, and 70 to the degree. Assuming the Italian Miliaria during this period (c.1595) is still the same as the Roman mile defined in chapter 2 as 1481.5 metres, this would assign lengths to the various units mentioned as follows:

1 degree of latitude = 103,705 metres
1 German (Dutch) Myle = 6913.67 metres
1 Spanish League (Legua) = 5926 metres
1 English League = 5185.25 metres
1 French Lieue (Marine) = 5185.25 metres

Two centuries later, a map by Charles-Marie Rigobert Bonne (published in 1780 in Paris) showing several elements of Spain's possessions in what is now the United States and Mexico, carried the following information in its "Echelle":

1 degree of latitude = 17 1/2 lieues d'Espagne
                       = 20 lieues marines
                       = 25 lieues communes de France
                       = 26 1/2 lieues legales de Castile.

By this time, the degree of latitude had been measured several times and found to vary from 110,657.3 metres at the Equator to 111,838.2 metres at the Arctic Circle (later found to be on the high side by 360 metres). With several measurements in France averaging 111,171 metres, it is likely that a degree length closer to this figure than to 103,705 metres was being employed. Using this figure, the league lengths become:

1 lieue Espagne (at 17 1/2 / degree)   = 6352.6 metres
1 lieue marine at 20 / degree   = 5558.55 metres.
1 lieue commune de France (at 25/ degree)   = 4446.8 metres
1 lieue legale de Castile (at 26 1/2 / degree) = 4195.1 metre.

More on the Spanish Legua

The University of Chicago, Spanish-English dictionary gives for "legua": " . . . about 3 miles" (statute) or 4828 m.

The most direct references to the league used by Columbus, the Spanish legua, other than Henry Harisse's remarks contained in Chapter 8, are those of Vignaud, Douglas Phillips-Birt, Morison, Thatcher, and Landstrom.[8] These writers all say 1 league = 4 Roman miles.

Phillips-Birt and Morison, being experienced seamen and naviagtors, feel impelled to add: 1 league = 3.18 nautical miles which introduces a minor incompatibility, for 4 Roman miles equals 5928 meters and 3.18 nautical miles is the equivalent of 5889 metres. The discrepency is 0.5%. However, among the versions of the Roman mile we have encountered was the 1472.5 meter unit of Littre. This length multiplied by 4 gives a 5890 meter league, essentially equivalent to 3.18 nautical miles.

Landstrom, a naval architect, amateur sailor, meticulous researcher and superlative illustrator (his illustrations for his *Columbus* are the most authentic and dramatic of any of the dozens of books on this subject) says:

1 Spanish mile = 8 stadia = 1000 double paces = 1619 yards
4 Spanish miles = 1 league = 6476 yards = 3.2 nautical miles
1 Arabian mile = 2,363 yards

The metric equivalents for the Spanish mile and the league in the foregoing are 1480.4 and 5921.7 m.

It will be recalled from Chapter 8 that the views of Ferrer, Enciso, the experts at the Bajadoz Conference, and Ferdinand Columbus were examined as to the length of a degree and, indirectly, the length of the Spanish legua. Henry Harisse interpreted the views of all except Columbus as according the legua a length of 6152.6 m. This reviewer believes each of the experts, including Ferdinand Columbus, put the league at 4 Roman miles, or 5926-5928 m, although (because of contradictions in the positions of Enciso as well as the experts, less Columbus) a range of lengths from 5081 to 6687 m. are possible for assignment to Enciso's league and a range from 5293 to 6720 for the Bajadoz experts' leagues.

The Employment of the Legua in Exploration Within the United States

Introduction

The Cortes and Pizarro conquests in Mexico and Peru had brought Spain riches undreamed of and hardly imaginable in enormity. While the crown, expedition commanders, and those who helped underwrite the costs of successful expeditions benefited the most, every participant gained. Thus, there was no problem in recruiting personnel for a proposed venture or one underway. During the half-century following Columbus' four voyages of exploration and discovery, Spain became effectively de-populated of young and middle-aged men and women as Hispaniola, Cuba, Puerto Rico, Jamaica, Mexico, Central America, and South America were explored, reduced, and colonized.

Two of the early expeditions of exploration which took place after the conquests of both Mexico and Peru were centered predominantly in what is now the United States. These were the Coronado expedition of 1540-1542 and the De Soto expedition of 1539-1543. Each involved sizeable contingents of men and beasts and each covered over 4000 miles. These were among the largest purely exploratory expeditions ever mounted by any nation. They occurred almost at the same time and were, in a measure, in competition with one another. The objective of each expedition was, of course, gold. Missionaries accompanied each undertaking, but the saving of souls was incompatible with the fighting and looting which was the major occupation of the two forces, if not its overall objective.

A quite different effort was launched in later years by Spanish Jesuits and Franciscans, mainly, plus other European "order priests" who wanted "a piece of the action" in the New World. These priests were zealous, even ardent, in their efforts to evangelize the Indians of America, including those north of Mexico. One outstanding explorer-entrepreneur-missionary was the Jesuit, Father Eusebio Kino, who did his greatest work in the Sonoran desert of northern Mexico and southern Arizona in the period 1684-1711. An account of Father Kino's cosmographical views, including his estimate of the Spanish legua will also be presented.

The Coronado Expedition

On February 23, 1540, a minor Spanish nobleman and aide to the Viceroy of New Spain, Antonio de Mendoza, led a sizeable force of caballeros, foot soldiers, and Indian allies out of Compostella, in Tepic province on the west coast of Mexico. The young leader of this force was Francisco Vasquez de Coronado and his objective was the fabled "Seven Cities of Cibola" somewhere to the north. His mission was typical for Conquistadores in the New World--find it, conquer it, reduce it to the Spanish will, evangelize it--if possible--with the help of the missionaries accompanying the force, and most important of all, secure whatever wealth was there and the source of that wealth.

The story of the expedition, its successes and failures, its battles, and precisely where it went is beyond the scope of our coverage here. Suffice it to say that the expedition traversed modern Sinaloa and Sonora, in Mexico, and Arizona, New Mexico, Texas, Oklahoma, and Kansas. Elements of the expedition

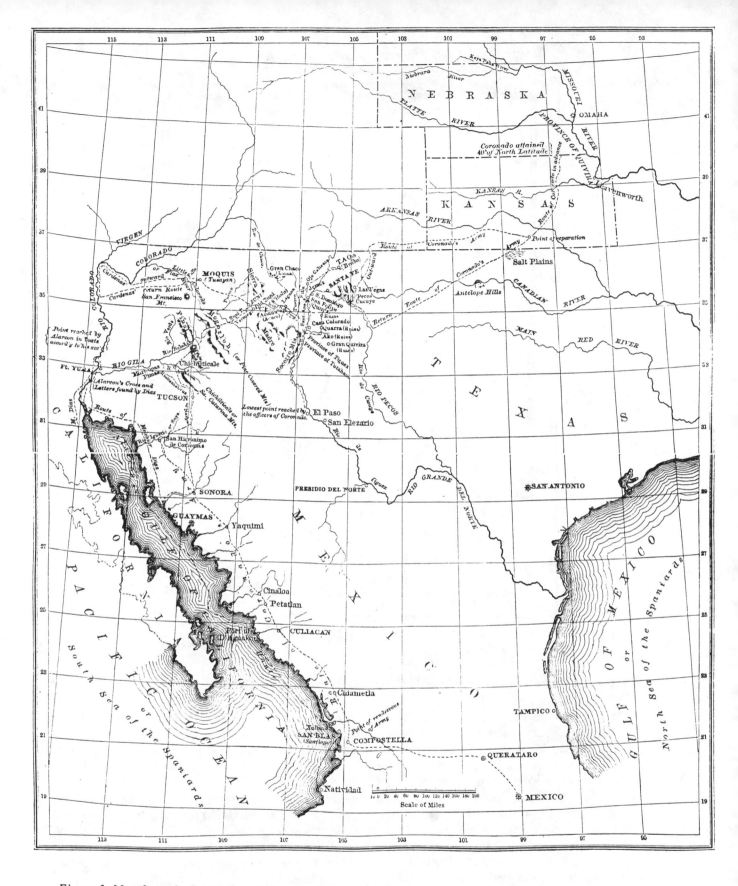

Figure 1. Map from the Smithsonian Report of 1869, "Coronado's March in Search of The Seven Cities of Cibola and Discussion of their Probable Location", by General J.H. Simpson, U.S.A., Washington, 1884.

At several places within the text, author uses the equivalency : 1 league = 3.4 statute miles. The "March" occurred in 1540-42 and its official chronicler, Pedro Castañeda, it must be assumed, used the league, or legua, officially sanctioned by the Viceroy and the Casa de Contratacion.

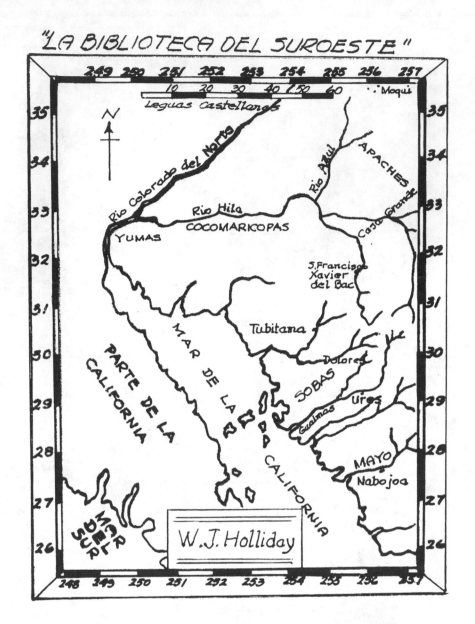

Figure 2. This map appeared on the dustjacket of Herbert Bolton's "Coronado, Knight of the Pueblos and Plains, published jointly by Whittlesey House and University of New Mexico Press. Scaling the Leguas Castellanas (near the upper border) against latitude markings on each vertical border, 6° is approximately equal to 96 leguas castellanas, or 1° = 16 leguas castellanas. The legua of Castellon is shown in Table 1 to have measured 5572.699 metres in 1852. If it had the same length in Coronado's time (c. 1540), then 1° would have measured 16 x 5572.7 = 89,163 metres, or 60.18 Italian (Roman) miles. This is very close to the 60 miles per degree observed by most mariners of the 15th and 16th centuries.

The alert map scanner will note an anamoly- identically sized longitude markings on the top and bottom map borders, some 10 degrees of latitude apart. This is permissable if a scale of leguas were shown for each of the border regions, top and bottom. (A degree of longitude is about 11% longer at 25.5°N. latitude than it is at 35.5°N. latitude.)

In this case, the upper longitude markings bear the more nearly proper length ratios to the latitude markings along each vertical border than do the lower markings.

crossed the Colorado River into California and Baja California. Others established contact with the Hopis of northern Arizona and learned about the Grand Canyon, which, subsequently, they were the first Europeans to visit. One detachment explored the Rio Grande from the vicinity of modern Albuquerque as far south as El Paso. A seaborne support unit under Hernando de Alarcon was the first to enter the mouth of the Colorado River and explore it as far north as Yuma. The route of the expedition, including all the probes of subordinate elements is illustrated in Figure 1.

General J. H. Simpson, U.S. Army (Retired), who during his active years had explored or otherwise seen service in much of the country Coronado "visited" in the U.S., undertook to write an account of *Coronado's March in Search of 'the Seven Cities of Cibola' and Discussion of Their Probable Location.* This was subsequently published by the Smithsonian Institution in their report for 1869.

General Simpson, who had been an Army Engineer, researched the *Relations of Pedro de Castaneda,* the official chronicler of the expedition in order to obtain the details he needed for his report. The report, as a result, is studded with distances and lengths stated in leagues and converted by Simpson to miles. The leagues are Spanish leguas and the miles are statute. General Simpson uses the equivalency 1 league = 3.4 statute miles (3.69 Roman or Italian miles, or 5742 metres). An old hand at estimating distances covered by mounted and foot soldiers on the march, Simpson does not, however, describe how he attained his equivalency.

Some credence is lent to the Simpson equivalency by a copy of a known map of unknown origin depicting the region surrounding the lower Colorado River and the Sea of Cortez (Gulf of California). The date of its production was circa 1700. A reproduction of the map is shown in Figure 2. From the information on the map, it is clear that 1° = about 16 leguas. Since the legua is identified as the "Legua castellana," which is shown in Table 1 to have a length of 5572.699 metres, the ratio between legua and statute mile is 5572.7 ÷ 1609.3 = 3.46. The length of a degree of latitude becomes 16 x 5572.7 = 89,163 metres, or 60.18 Italian miles. This is so close to the marine standard of 60 Italian miles/degree that one concludes this was the intent of the cartographer.

It is somewhat surprising that as late as 1700 cartographers would be using the equivalency 1°= 60 Italian miles, or its equivalent, 16 leguas castellanas. Still, some people "never seem to get the word," as we have seen in reviewing world maps of the 16th century.

The De Soto Expedition

Another Spanish expedition, even larger than the Coronado expedition, and in the United States for a longer period, May 1539 to September 1543, originated in Spain, departing Sanlucar in April 1538. Hernando (Hernan) de Soto, its commander and newly appointed governor of Cuba and captain-general and adelantado of Florida, had gone to Panama at the age of 19 and served under Pedrarias in the reduction of that country, Costa Rica, Nicaragua, Honduras, and El Salvador. Later, he had served under Pizarro in the conquest of Peru and had become quite wealthy from his share of captured Incan treasure.

As the title of its commander might suggest, this expedition was expected to colonize, as well as conquer and reduce, the land called Florida. This was conceived for the purposes of de Soto's contract as extending from 25° N to 37° N, and from the Atlantic coast indefinitely westward.

The fleet of 10 ships, bound from Spain to Cuba via Gomera, in the Canaries, carried 1000 souls and 350 horses. Of the 1000, 600 were horsemen and foot soldiers, 250 crewmen, and the remainder priests, friars, artisans, and some families. At sea for 8 weeks, the expedition, during the period June 1538 to May 1539 when it was in Cuba, spent its time in replacing its losses in livestock while at sea and in recruiting some additional personnel, particularly Indians who would serve as interpreters, bearers, livestock tenders, and warriors, if required.

"ANCIENT" LEGUAS STILL IN USE IN SPAIN, AS OF 1852, WHEN METRIC SYSTEM WAS MANDATED BY ROYAL DECREE.

PROVINCE	LENGTH OF VARA IN METRES	LENGTH OF LEGUA IN VARAS	VARAS CASTELLANAS	KILOMETRES	NUMBER OF LEGUAS PER DEGREE OF 111,111.11 METRES
ANTIGUO SISTEMA (REAL)	0.835905	$6666\frac{2}{3}$	$6666\frac{2}{3}$	5.572699	19.938
ÁLAVA	DO	DO	DO	DO	DO
ALBACETE	0.837				
ALICANTE	0.912	6091.61	6646.15	5.55565	20
ALMERÍA	0.833	6689.91	$6666\frac{2}{3}$	5.572699	19.938
ÁVILA	0.835905				
BADAJOZ	DO	$6666\frac{2}{3}$	$6666\frac{2}{3}$	5.572699	19.938
BALEARES	USED CANA OF 0.782 M.				
BARCELONA	USED CANA OF 1.55 M.				
BURGOS	0.835905	$6666\frac{2}{3}$	$6666\frac{2}{3}$	5.572699	19.938
CÁCERES	DO				
CÁDIZ	DO	$6666\frac{2}{3}$	$6666\frac{2}{3}$	5.572699	19.938
CANARIAS	0.842				
CASTELLÓN	0.906	6150.88	$6666\frac{2}{3}$	5.572699	19.938
CIUDAD-REAL	0.839	7970.49	8000	6.68724	16.615
CÓRDOBA	0.835905	$6666\frac{2}{3}$	$6666\frac{2}{3}$	5.572699	19.938
CORUÑA	0.843	6610.56	DO	DO	DO
CUENCA	0.835905				
GERONA	USED CANA OF 1.559 M.	USED HORA DE CAMINO DE (3.761572 KILOMETRES) AS 4500 VARAS CASTELLANAS ITINERARY MEASURE.			
GRANADA	0.835905	$6666\frac{2}{3}$	$6666\frac{2}{3}$	5.572699	19.938
GUADALAJARA	DO				
GUIPÚZCOA	0.837				
HUELVA	0.835905	$6666\frac{2}{3}$	$6666\frac{2}{3}$	5.572699	19.938
HUESCA	0.772	8000	7388.4	6.176	18
JAÉN	0.839				
LEÓN	0.835905				
LÉRIDA	USED MEDIA CANA OF 0.778M.				
LOGROÑO	0.837	6657.94	$6666\frac{2}{3}$	5.572699	19.938
LUGO	0.855				
MADRID	0.843	6610.56	$6666\frac{2}{3}$	5.572699	19.938
MÁLAGA	0.835905	$6666\frac{2}{3}$	DO	DO	DO
MURCIA	DO	DO	DO	DO	DO
NAVARRA	0.785	7000	6573.71	5.495	20.22
ORENSE	0.835905				
OVIEDO	DO	$6666\frac{2}{3}$	$6666\frac{2}{3}$	5.572699	19.938
PALENCIA	DO				
PONTEVEDRA	DO				
SALAMANCA	DO	$6666\frac{2}{3}$	$6666\frac{2}{3}$	5.572699	19.938
SANTANDER	DO	DO	DO	DO	DO
SEGOVIA	0.837				
SEVILLA	0.835905	$6666\frac{2}{3}$	$6666\frac{2}{3}$	5.572699	19.938
SORIA	DO				
TARRAGONA	USED MEDIA CANA OF 0.78M.	USED HORA DE CAMINO DE (4.457881 KILOMETRES) AS 5333 VARAS CASTELLANAS ITINERARY MEASURE.			
TERUEL	0.768	7256.12	$6666\frac{2}{3}$	5.572699	19.938
TOLEDO	0.837	6657.94	DO	DO	DO
VALENCIA	0.906	6663.46	7222.22	6.037092	18.4
VALLADOLID	0.835905	$6666\frac{2}{3}$	$6666\frac{2}{3}$	5.572699	19.938
VIZCAYA	DO	DO	DO	DO	DO
ZAMORA	DO				
ZARAGOZA	0.772	7218.52	$6666\frac{2}{3}$	5.572699	19.938

Table 1

As with the Coronado expedition, a recounting of the experiences of the de Soto expedition, its many battles, its losses in personnel, its epic achievements in exploring almost the entire U.S. South, and its precise itinerary are beyond the scope of this coverage. Suffice it to say that the expedition explored virtually all of Florida, much of Georgia and South Carolina, Alabama, Mississippi, Arkansas, and Louisiana, and parts of North Carolina, Tennessee, and Texas. It's sea components, including the evacuation fleet built in Louisiana on the banks of the Mississippi (after the death of its commander, de Soto) explored and coasted the entire U.S. coast from Tampa Bay to Mexico and the Mexican shore to Panuco (Tampico).

Figure 3 illustrates the foregoing, including the Indian villages visited, some of which were the sites of bitter battles and hair-raising close calls. Not clearly visible (because of problems of scale and reproduction of Figure 3) are the major U.S. rivers crossed and explored, including the Suwannee, Apalachicola, Flint, Chattahoochee, the various branches of the Altamaha, the Savannah, Tennessee, Coosa, Alabama, Tombigbee, upper Yazoo, Mississippi, White, Arkansas, Ouachita, Red, Sabine, Neches, Trinity, and a branch of the Brazos.

The most authentic, detailed and balanced account of this remarkable expedition is that of the Report of the U.S. De Soto commission, published by the Government Printing Office in 1939, on the 400th Anniversray of the landing at Tampa Bay, Florida. The members of the Commission were particularly circumspect in determining precisely where the various elements of the expedition went and in estimating the distances traversed between the many points of interest described in the original (or early) chronicles of the expedition. As can be expected, they had to cope with the problem of converting estimates of distances made in Spanish or Portuguese leagues (by four chroniclers with different ideas as to what a league was) to statute miles.

The following is a direct quotation from p. 104 of the Report:

The time taken in passing from point to point is sometimes stated in days and then we have to estimate the possible rate of travel during those days under the given conditions.

Less frequently distances are in leagues, and although Garcilaso (de la Vega) is so much in need of watching, it happens that he supplies us sometimes with very useful information of this kind . . . The league, we are informed, originated in Gaul where it was equivalent to 1.4 statute miles. In England, in the Middle Ages, the league was nearly 3 statute miles, the common league of France was 2.764 statute miles, the French posting league 2.422 statute miles, the Flanders league 3.9, the Portuguese league 3.84, the Spanish league 4.214, and the Spanish judicial league 2.634 statute miles. This last has been the one usually assumed by the most competent students of the movements of Spanish explorers in the New World. It might be supposed that the Portuguese chronicler, the Gentleman of Elvas, would have employed the standard usual in his own country. Tests of the actual usage of our chroniclers . . . show clearly, however, that the league they had in mind fell between two and a half and three miles, and even so they are more apt to over-estimate distances than the reverse . . . An allowance of 2.6 miles per league has been found most satisfactory . . .

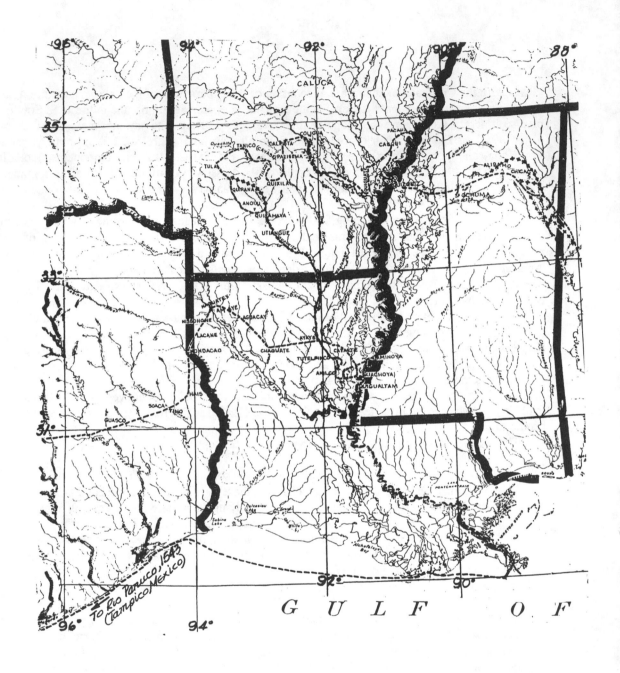

Figure 3, part 1. The route of the De Soto expedition of 1539-1543.

Scale
1" = approx. 78.5 Stat.Mi.
  = "   30.3 Leguas
(1 Legua = 2.6 Stat.Mi.)

Figure 3, part 2. The route of the De Soto expedition of 1539-1543.
Credit and comments on page 296.

Credit and comments for Figure 3, page 285
This map and the Statute Mile to Spanish Legua equivalency are taken from the "Report of the Fact Finding Committee of the U.S. De Soto Expedition Commission," Washington D.C., 1939. Hernando De Doto discovered the Mississippi on May 21, 1541 and died of a fever exactly one year later at the Indian village of Guachoya on the Louisiana side of the river. His weighted body was consigned to the great river he discovered. His surviviors, under Luis de Moscoso, reached Panuco (Tampico), Mexico on September 10, 1543, having descended the Mississippi and coasted Louisiana, Texas, and Tamaulipas State, Mexico, in brigantines built on the banks of the Mississippi, at Guachoya.

Garcilaso's metrological information, put into tabular form, and with useful equivalencies follows:

League Identification	Stat. Mi.	League In Roman Mi	Metres
Gallic	1.4	1.52	2253
English (Middle Ages)	3	3.26	4828
French (Common)	2.764	3.0	4448
French (Posting)	2.422	2.63	3898
Flanders	3.9	4.24	6276
Portuguese	3.84	4.17	6180
Spanish	4.214	4.58	6782
Spanish (Judicial)	2.634	2.86	4239

Father Kino in the Sonoran Desert

The Reverend Charles Polzer, S.J., has written a fact-filled booklet[9] on the work of the revered missionary-explorer Eusebio Francisco Kino (nee Kuhn), S.J., among the Pima, Papago, and Yuma Indians of the Sonoran Desert of northern Mexico and southern Arizona. Father Polzer's book contains a map drawn by Padre Kino, circa 1710, showing the areas he explored and the missions he established. This map, reproduced as Figure 4, has a single distance scale in terms of "leguas castellanas." The left border of the map is graduated into degrees of latitude. By scaling between the leguas and degrees, 130 leguas just about match 9° of latitude, or 14.44 leguas = 1 degree.

This figure, however, is suspect, for, as noted on Figure 4, the latitude divisions shown on the map are somewhat distorted from actual conditions. Two well known landmarks, Cabo San Lucas and the juncture of the Gila and Colorado Rivers, are shown on Kino's map as having a difference in latitude of about 11° 38', instead of the more realistic 9° 52'. The distance along any suitable meridian between these parallels is about 170 of Padre Kino's "leagues castellanas." Thus, the more likely relationship is 17.23 leguas per degree.

By Father Kino's time, the length of a degree of arc along a meridian of the earth's surface had become reasonably well known. In 1669, under the auspices of the French Academie Royale des Sciences, Jean Picard carried out a measurement of a degree of arc along the meridian of Paris northward to the environs of Picardy. The result--57,060 toises (111,212 metres)--was promptly publicized by the Academy. From 1669 on, the French and other European countries regularly carried out similar measurements, refining Picard's calculations. Figure 5 tends to confirm that Kino was aware of Picard's work. Thus, the legua castellana of Figure 4 would figure out to be 111,212 metres ÷ 17.23 (leguas/degree), or 6455 metres.

296

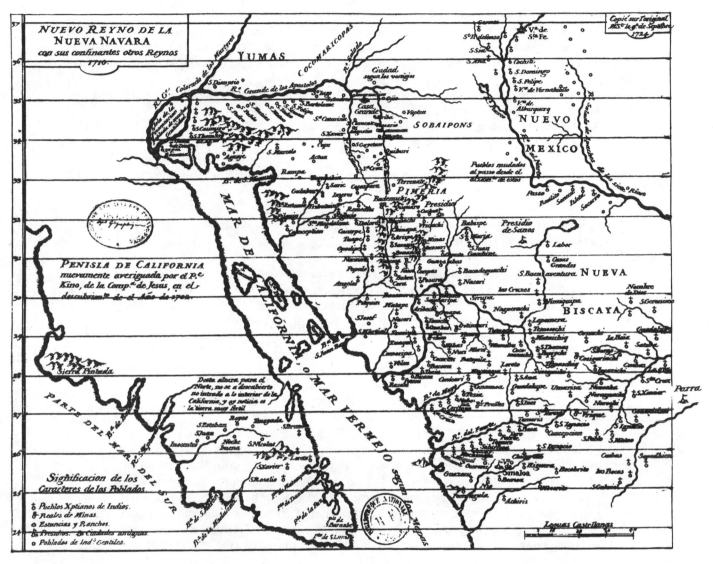

Figure 4. 1710 Kino Map - D'Anville Collecvtion, Bibliotheque Nationale, Paris, & Rev. Charles Polzer, S.J. See page 15, "A Kino Guide," by Rev. Charles Polzer, S.J., Southwestern Mission Research Center, Tucson, 1976. Comparing latitude notation on left border with scale at lower right, 130 leguas castellanas is about equal to 9° of latitude, or 1° ≈ 14.4 leguas castellanas.

See comment on page 298.

Comment on Figure 4, page 297.

Map is somewhat distorted. The juncture of the Gila River (Rio Grande de los Apostoles) and the Colorado River is shown at about 35° 11', about 2° 28' north of its true location. Cabo San Lucas (southernmost point in Penisla de California, or Baja California) is shown at about 23° 33', about 0° 42' north of its true position.

If latitudes are corrected, scaling Kino's Leguas Castellanos against the new grid would yield the relationship 1° = 17.23 Leguas Castellanas.

Fr. Kino did his own surveying. Considering the distances involved and the harshness of the terrain, the resulting map is remarkably faithful to actual conditions.

Table 1, however, lists the legua of Castellon as 5572.699 metres, while showing three other leguas which bracket our figure for Father Kino.

The Employment of the Legua in Surveying and Land Measurment within Mexico and, Particularly Mexican California

Francois D. Uzes has treated the titled subject rather exhaustively in Appendix N of his *Chaining the Land, A History of Surveying in California* [10]. Uzes quotes a number of authorities in this part of his book. It is useful to examine elements of these quotations pertinent to our theme. These follow immediately, after which a synthesis of the whole, with comment, is provided.

LENGTH OF VARA
1854 Annual Report,
Commissioner of the General Land Office

In a report of the 14th November, 1851, from the surveyor general of California, it is stated that all the grants, etc, of lots or lands in California or that of Mexico, refer to the "vara" of Mexico as the measure of length; that, by common consent in California, that measure is considered as exactly equivalent to thirty-three American inches. That officer then enclosed to us copy of a document he had obtained as being an extract of a treaty made by the Mexican government, from which it would seem that another length is given to the "vara," and by J.H. Alexander's (of Baltimore) Dictionary of Weights and Measures, the Mexican vara is stated to be equal to 0.92741 of the American yard.

This office, however, has sanctioned the recognition, in California, of the Mexican vara, as being equivalent to thirty-three American inches.

The Mexican vara is the unit of all the measures of length, the pattern and size of which are taken from the Castilian vara of the mark of Burgos, and is the legal vara used in the Mexican republic. Fifty Mexican varas make a measure which is called cordel, which instrument is used in measuring lands.

The legal league contains 100 cordels, of 5,000 varas, which is found by multiplying by 100 the 50 varas contained in a cordel. The league is divided into two halves and four quarters, this being the only division made of it. Half a league contains 2,500 varas, and a quarter of a league 1,250 varas. Anciently, the Mexican league was divided into three miles, the mile into a thousand paces of Solomon, and one of these paces into five-thirds of a Mexican vara; consequently, the league had 3,000 paces of Solomon. This division is recognised in legal affairs, but has been a very long time in

Map made by Kino in 1702

Figure 5. This map, in French, uses the lieue and sets 20 such units = 1 degree of latitude. Assuming that Kino knew the French Academy had determined the length of a degree to be about 57,060 toises de Paris (or close to 111,212 metres), this would put the lieue = 2853 toises, or about 5561 metres. This reproduction is, by permission, taken from "Early Arizona: Prehistory to Civil War," by Jay J. Wagoner, Tucson: University of Arizona Press, 1975. Photograph courtesy of Donald B. Sayner collection.

disuse-the same as the pace of Solomon, which in those days was called vara, and was used for measuring lands. The mark was equivalent to two varas and seven-eighths-that is, eight marks containing twenty-three varas-and was used for measuring lands.

The following excerpts are from an article titled, *Weights and Measures of Provincial California*, by J. N. Bowman, in California Historical Society Quarterly, Vol. XXX, No. 4, December, 1951.

LEAGUE

The juridical league was a measure both of distances and of areas. It was standardized at 5,000 varas, so its length depended on the length of the vara. The Mexican Law of 1857, establishing the metric system, gave the length of the leagues as 4.19 kilometers or 2.6035 miles (22). In the Alta in October 1857, the length of the league in the province is given as 4,635 yards or 2.633 miles-evidently taken from Lee's Tables of 1849; and in 1861 Deputy Surveyor General Healy testified that the Spanish league was "2 5/8 miles and a fraction of a chain over," or 2.625 miles.[57] Based on the 33-inch vara, it was 2.604 miles, while, on the 33.372-inch vara, it was 2.633 miles. For practical purposes in Alta California 2.6 miles can be used.

MILLA

The milla or mile was an old Spanish unit of lineal measurement and was somewhat in general use in provincial California. In the plan for colonization by foreigners in Upper and Lower California, as proposed by the California Junta de Fomento, 1825-27, provision is made that there shall be assigned, between allotments, "lots of a square milla, or of 1666 2/3 varas, which is the same thing . . . " The milla occurs occasionally in Spanish testimony in the private land-grant cases; it was used in a land grant in the lower San Joaquin Valley, and J. J. Vioget used it in making his plat, now lost, and in his testimony regarding his survey of New Helvetia in 1840-1841[58].

The Mexican milla was one third of a league[59], and this also was the unit used by Vioget in his plat and in his testimony in the Sutter private landgrant case mentioned above[60]. This makes the milla about 0.87 of a mile.

The "Spanish Weights" in the Fitch documents states that the Spanish league is divided into three miles, but without indicating whether this was for the long or the short league[25].

PULGADA

The pulgada or inch was the most used subdivision of the vara, which contained 36 pulgadas; the length of the pulgada, therefore, depended upon the length of the vara.

Where the 32.99206-inch vara was in use, the pulgada was 0.9164 of an inch, but the 33-inch vara made it 0.9166, and the 33.372-inch vara made it 0.927, of an inch. In any of these cases, the pulgada was a little less than an American inch.

In California the pulgada was rarely used in mission and governmental records . . .

SITIO

Sitio was quite generally used to designate the square league. See "Square League" below.

SQUARE LEAGUE

The league, or square league, or sitio, or sitio de ganado mayor was the unit of the Alta California ranchos, as determined by the Law of 1824 and the Regulations of 1828. This stock rancho unit varies with the length of the vara. With the 33-inch vara it contained 4,340.28 acres, while with the 33.372-inch vara it was 4,438.68 acres, or a difference of 98.40 acres for each league. The former figure is quite close to the league as established by the Mexican Law of 1857 at 1755.61 hectares or 4,388.11 acres[22]. For reasons mentioned below under "Vara," the smaller of these figures can be taken as the one used in Alta California, and the latter figure in patenting land grants confirmed after 1855.

The seeming discrepancy is not as important in practice as it is in mathematics. Even though the standard alcalde varas were used as a basis in land measurement, many factors entered to affect adversely the seeming exactness at this point. The cordel, or what is now called the surveyor's chain, was made of sisal, hemp, hair, or occasionally rawhide. The cordel was to be "well twisted and stretched," but the measuring was done by holding the vara measure and the cordel in the hands; also no allowance was made for the natural shrinking and stretching of the cordel. It was pulled along the ground by cord-bearers, usually on horseback when measuring the land, and on foot when measuring house lots. The marks on the ground, indicating cordel lengths, were made by the men on horseback, who struck the ground with sticks tied to the ends of the cordel. It was pulled in the general direction of a selected landmark but not necessarily in a straight line. Wide streams, swampy places, precipitous areas were not measured at all, but were estimated as being of a certain extent agreed upon by all the parties concerned, the number of leagues measured being expressed in terms of "poco mas o menos", a little more or less . . . . .

VARA

The vara was the basic unit of lineal measure in Spain and in the Spanish colonies. On it was based the length of the league and the area of the sitio or square league, and it also determined the length of the milla.

In Alta California it was the most used of all the units, followed closely by the fanega. The vara was used in determining the length of the cordel in land measurement and in measuring the size of churches and buildings.

The original vara, as mentioned earlier, was cut on the outer wall of the cathedral of Burgos at an unknown date, and so became known as the vara de Burgos or vara de Castilla. In 1721 a standard unit of wood was made of the vara de Burgos, and later one was made of metal. A metal unit was also sent to Mexico in the eighteenth century.

The length of the vara de Burgos has been placed at 32.90957 or 32.99206 inches. In Spain, however, variations from this standard became quite general locally and among the provinces, not only in Spain but also in the new world. In time its length ran from 31.496 inches in Colombia to 44.092 inches in Argentina; and in Portugal, which also used the vara as a unit, it was 48.28 inches (12,22-23,30, 63).

In Mexico the generally accepted length was quite close to that of Burgos. With one exception which gives the length as 35.74 inches (63), there is general agreement among metrologists that the length was about 33 inches, but usually a little less . . . .

On May 15, 1857, Mexico established the metric system, to be effective on January 1, 1861, and the publication of the law contained conversion tables. The vara before 1838 was given as equivalent to 837.31 mm or 32.96489 inches, but in that year it was made equivalent to 838 mm or 32.99206 inches[22]. Hall, in his Mexican Laws, gave the vara as equal to 837 mm[70]. The International Congress of Berlin in 1863 gave the length of the Mexican vara as 836.695 mm or 32.9507 inches, the Spanish vara as 835.905 mm or 32.8775 inches, and in the other South American states the vara as somewhat larger.[71]. In 1891 the Bureau of American Republics made the vara of the South American states range from 33 inches to 38.874 inches[20]. In 1906, Chief Justice Fuller in his statement of the case of Ainsa v. U.S., gave the length of the vara 32.9927 inches for Arizona[72].

Synthesis and Comment on Quotations from Uze's *Chaining the Land* . . ., Appendix N.

The system of measurement employed in Mexico, Mexican California, and probably in most of Spanish America during the 19th century for surveying, land measurement and related purposes utilized the following relationships:

1 vara (yard)	= 36 pulgades (inches)	
50 varas	= 1 cordel	
100 cordels	= 1 judicial league	Thus:
5000 varas	= 1 judicial league,	and since
1 judicial league	= 3 millas,	
1 milla	= 1666 2/3 varas	

While the vara varied somewhat throughout Mexico and Mexican California, its length was pretty close to 33 U.S. inches. Throughout Spanish America, the vara varied from a low of 31.496 inches in Colombia to a high of 44.092 inches in Argentina. The statement is made that " . . . The original vara . . was cut on the outer wall of the cathedral of Burgos at an unknown date, and so became known as the 'vara de Burgos' or 'vara de Castilla'". This standard was introduced into Mexico in the early 18th century, but it is likely that a crude approximation of the standard was introduced much earlier.

Based on a 33 inch vara, the judicial league of 5000 varas measured 2.604 statute miles. This checks reasonably well with Garcilaso Vega's estimate of the judicial league as given in the De Soto Commission Report of 1939, described earlier. That there were other leagues, and in particular other Spanish leguas, has been shown by the greater part of Appendix B, thus far.

It is noteworthy that the Spanish judicial league was rated at 5000 varas castellanas and the milla 1666 2/3 varas. Table 1 shows the official ancient legua (antiguo sistema real) to be 6666 2/3 varas castellanas, exactly 4/3 of the judicial league. It is logical to assume that the longer league was composed of 4 millas to 3 for the shorter league. In either case, the milla at about 1393 metres was shorter than Columbus' Roman or Italian mile.

The Legua in Spain

On more than one occasion reference has been made to the various leguas in use in Spain prior to the insitution of the metric system in that country in 1852. It would be useful, now, to summarize the situation in Spain as regards the length of the various leguas in use throughout the country.

Before the institution of the metric system, there was a nationwide, law-mandated system of weights and measures commonly referred to as "Medidas y pesas legales de Castilla." In this system, the "legua de 6666 2/3 varas" measured 5,572.600 metres, one vara equalling 0.835905 metres. However, of the 49 provinces reported on in "Equivalencias of 1886[11], only 16 used the officially mandated vara and legua. 8 provinces used varas with lengths other than 0.835905 metres, but employed the official legua, defining its length in terms of the official vara, i.e., 6666 2/3 varas castellanas. Thus, only 24 of the 49 provinces employed the official legua of 5.572699 kilometres, 19.938 to the degree.

5 provinces employed leguas with lengths other than 5.572699 kilometres. Navarra had the shortest legua at 5.495 kilometres, 20.22 of these leguas making the degree, while Cuidad Real had the longest at 6.68724 kilometres, 16.615 to the degree. Only one province, Alicante, employed a legua which appears to have been defined after the degree had been determined, in France, to be 111,111.11 metres, 1/360 of 40,000,000 metres. In Alicante, the legua was defined as "20 al grado," thus measuring 5.55555 kilometres. This was probably the intent when the national legua of 6666 2/3 varas, measuring 5.572699 kilometres, was set--the degree, at that time, being considered equivalent to 111,453.98 metres.

The fact that as late as 1852 leguas were in use in Spain whose relationship to the degree were 16.615, 18, 18.4, 19.938, 20, and/or 20.22, makes it a likely conclusion that prior to vigorous efforts to standardize the legua, there was even a greater disparity in the length of this itinerary unit of measurement throughout the country. It is significant that 18 provinces employed no itinerary unit at all, while 2 used the "hora de camino" instead of the legua. Gerona defined its measure as 4500 varas castellanas, or 3.761572 kilometres, while Tarragona used the hora de camino de 5333 varas castellanas, or 4.457881 kilometres.

Table 1 presented a summary of the itinerary units in use in Spain prior to the introduction of the metric system, as taken from "Equivalencias of 1886."

Table 2 summarizes the leagues, leugas, leguas, leghes, and lieues found in the literature which have been mentioned in this appendix. The German and Dutch myle (meil) is not included for lack of sufficient specific data on its length.

Clearly, much of the variation resulted from efforts to relate the unit to the length of a degree of latitude during an era in which geodesists gradually approached the true length of a degree of latitude (and found that its length varied somewhat with latitude). Origin is, of course, also responsible for much of the variation since the league was not solely Gallic in origin but Greek/Roman and Babylonain/Persian, as its similarity to the schoenus and parasang and its relationship to the Roman mile and Greek stade demonstrate.

Identification and Authority	Length of League in			
	Metres	Roman Miles of 1480 m.	Statute Miles of 1609 m.	Nautical Mi. of 1852 m.
Roman-British Leuga: Zupko's "British Wghts. & Meas."	2200	1.49	1.37	1.19
Garcilaso's Gallic Leuga: De Soto Commission Report	2253	1.52	1.4	1.22
Gallic Leuga: Klein's "World of Measurements"	2559	1.73	1.59	1.38
French Posting League: Garcilaso/De Soto Comm. Rept.	3898	2.633	2.422	2.105
Funk & Wagnall's French-English Dictionary: the Lieue; Padre Kino's Legua used in Mexico and Arizona (this appendix); and Colombian Judicial legua based on 5000 varas of 31.496 ins. (this appendix).	4000	2.71	2.5	2.17
Mexican California Judicial Legua based on 5000 varas castellanas of 33 ins. (this appendix)	4191	2.83	2.604	2.263
Garcilaso's Spanish Judicial Legua: De Soto Comm. Report	4239	2.864	2.634	2.289
Brockhaus' French Lieue (see text) and Garcilaso's French Common League	4444	3	2.76	2.40
Univ. of Chicago Spanish-English Dictionary: Legua			About 3	
Klein's and Garcilaso's Middle Ages English League	4830	3.26	3	2.6
Random House Dictionary: "In English speaking countries"			"Roughly 3"	
Webster's Dictionary: "In English speaking countries..."			"About" 3 or	3
Barentszoon's 17½ leguas=60 Italian miles=1 degree: Chapter 8 and Vignaud's Portuguese Leghe: "Toscanelli and Columbus"(Opt.1)	5074.3	3.4286	3.153	2.740
Vignaud's Portuguese Leghe (Opt.2):17½ leghes=62½ Italian miles=1 degree	5285.7	3.571	3.284	2.854
Vespucci's Portuguese Leghe of 16 2/3 to the degree (Opt.1): Harisse's "Diplomatic History of America	5327	3.599	3.310	2.876

Table 2, part 1. An array of leagues, leguas, leghes, and lieues, from Roman times to 1852.

Description				
Navarra's Legua of 7000 varas of 0.785 metres: "Equivalencias..." and Table 6d.	5495	3.713	3.414	2.967
Vespucci's Portuguese Leghe (Opt.2): Harisse's "Diplomatic History"...	5549.	3.750	3.449	2.997
Alicante's "legua de 20 al grado: "Equivalencids.." and Table 1.	5555.6	3.754	3.452	3.000
The legua of 24 Spanish provinces and the national standard; "Equivalencias" and Table 1.	5572.7	3.765	3.463	3.009
Klein's British Nautical League	5586	3.77	3.47	3.02
Argentine Judicial legua based on 5000 varas of 44.092 ins. (this appendix).	5600	3.78	3.48	3.02
Pre-metric French league; Berriman and Klein	5847	3.95	3.63	3.16
The league used by Columbus: Vignaud, Morison, Phillips-Birt, Thatcher.	5890-5920	3.98-4.0	3.66-3.68	3.18-3.20
Valencia's legua of 7222.22 varas Castellanas of 0.835905 metre; "Equivalencias.." and Table 1.	6037.09	4.079	3.751	3.260
Harisse's Spanish legua, used by Ferrer, Enciso, and "Spanish Experts "at Badajoz Conference of 1523-24: "Diplomatic History".	6153	4.157	3.823	3.322
Huesca's legua de 8000 varas of 0.772 metre: "Equivalencias".	6176	4.173	3.838	3.335
Garcilaso's Portuguese Leghe: De Soto Comm. Report.	6179.88	4.176	3.84	3.337
Garcilaso's Flanders League: De Soto Comm. Report	6275.1	4.24	3.9	3.388
Padre Kino's Map (Figure 10) adjusted to correct latitudes gives the relationship 1 degree = 17.73 leguas castellanas. See text for other conditions.	6455	4.36	4.012	3.485
The legua of Ciudad-Real, 8000 varas castellanas of 0.835905 metre, was largest listed in "Equivalencias.	6687.24	4.518	4.156	3.611
Garcilaso's Spanish Legua: De Soto Comm. Report	6780	4.581	4.214	3.980

Table 2, part 2. An array of leagues, leguas, leghes, and lieues, from Roman times to 1852.

REFERENCE NOTES

Chapter 1

1. P. 63 *Admiral of the Ocean Sea, A Life of Christopher Columbus,"* by Samuel Eliot Morison; Little, Brown and Company, Boston, 1942.

2. *Columbus, the Story of Don Cristobal Colon, Admiral of the Ocean Sea, and his Four Voyages Westward to the Indies,* p. 191; MacMillan Co., New York, 1966. This is unquestionably one of the best illustrated of the many books about Columbus. Landstrom's coverage of naval architecture, navigation, and cartography are highly professional.

3. The Indies was a term used in Columbus' day in Europe, North Africa, and the Middle East to designate a rather large and poorly known area. It included any part or all of Greater India (the Indian Continent proper); Middle India (Ethiopia and Somalia); and Lesser India (Southeast Asia, including Indonesia; South China, known as Mangi; North China, called Cathay; and Cipanju (Japan). See Appendix A, and in particular reference note 13 for a list of sites, other than Watling Island, which have been suggested for San Salvador.

4. The Columbina, now found in the Seville Library, is an aggregation of Christopher Columbus' books and manuscripts collected by Ferdinand Columbus. It contained copies of the great explorer's correspondence, his logs, a map he had used on his first voyage, all his works in manuscript, and many books by medieval and ancient authors with notations by Columbus in the margins.

5. It is presumed that all of Toscanelli's letters were written in Latin. Ferdinand's remark was made when describing the first letter to Columbus, and its attachments. He may have felt it redundant to make the same remark about Toscanelli's second letter to Columbus. None of Columbus' letters to Toscanelli have ever been located.

6. Henry Vignaud was born in New Orleans, Louisiana in 1830, and educated there. His initial vocation was journalism, but he got caught up in the Civil War and served briefly as a Captain in the 6th Louisiana Regiment. When Union forces captured New Orleans in 1862, he was made prisoner. Shortly thereafter, he escaped and reached Paris, never returning to the United States. In France, he served briefly in the Confederate diplomatic service and in the Roumanian legation at Paris, but by 1872 was serving the United States as a translator. In 1875, He was appointed second secretary of the U.S. legation in Paris and in 1885, first secretary. For thirty four years, he was the indispensible member of the Paris mission, retiring in 1909 at the age of 78. He was 71 when *La Lettre et La Carte de Toscanelli* was published. His special interest in Columbus grew out of his close association with Henry Harrisse, another French-American scholar and

biographer of Columbus, and with the Peruvian scholar Manuel Gonzales de La Rosa. Gonzalez, in fact, was the one who first concluded that the famous Toscanelli letter of 1474 was not genuine and convinced Vignaud in the matter.While, to this point, we have employed Vignaud as "le bête noir," we shall lean heavily on his meticulous scholarship in what follows, for this unusual man had a broad versatility of interest, powers of intense application and mastery of details, and - evidently - a boundless capacity for work.

7.   See p. 7, *Admiral of the Ocean Sea.*

8.   Ovieda, *La Historia general de Las Indias,* Seville, 1535; Las Casas, *Historia de las Indias*, 1563, but not printed until 1875, in Madrid; Gomera, "Historia de las Indias," 1553; Garcilaso de la Vega, *Primera parte de los commentarios Reales,* Lisbon, 1609.

9.   See pp. 65-71, *Latin America, An Historical Survey*, Rev. J. F. Bannon, S. J. , reprinted by Glencoe Press, Encino, Cal., 1977.

10. The "repartimiento" was a system whereby Indians were apportioned to official enterprises or to colonists. It differed but little from slavery and appears to have been installed during the Bobadilla regime, but with royal sanction. This system had been used previously by the Spaniards in Valencia, in Majorca, and in the Canaries, thus it was not new, nor strange. However, if it worked after a fashion in these places, it was because of the effective control exercised by the Crown. In Espanola, much farther away from Crown control, grave abuses set in. Even after Ovando was required to inaugurate a new policy whereby the Indians were to be considered free men, vassals of the Crown of Castile, there was no significant improvement in the lot of the Indians. As vassals, they were subject to tribute, and as Christians, tithes - neither of which they understood or consented to willingly.

11. The "encomienda" was a semi-feudal institution whereby a group of Indians was "commended" to a Spaniard in reward for meritorious service (by the Spaniard). The "encomendero" was to collect the royal tribute due from "his" Indians. While the relationship was supposed to be that of guardian-ward, it soon degenerated into master-servant when the Indians failed to pay the tribute and the obligation was translated into service. Resistance to forced service was met with violence and cruelty to which the Indians in desperation, responded with rebellion, resulting in decimation of their ranks. See pp. 69-71, Bannon's "Latin America . . ."

12. Chios was a Greek island off the coast of Moslem Turkey (near Smyrna) and the purpose of the Chios voyage was partly trade and partly relief for the islanders who were threatened by the Turks.

13. Noli was a Savonese port on the Riviera, about 20 miles northeast of Nice.

14. On none of his voyages, while employed by Genoese interests, does it appear that Columbus was a member of the ship's complement. However, it is likely that Columbus being an agent of the firm which owned or leased the ships, was assigned - and probably sought - part time and contingency functions relating to navigation and ship handling.

15. During the Crusades, Venice and Genoa provided the transport services for the armies of Europe and, as a result, became maritime powers. Its seamen became skillfull navigators and ship

handlers. Genoese, particularly, roamed the Mediterranean and occasionally ventured out past the Straits of Gibraltar seeking new markets and likely spots to colonize. About 1270, Lancelot Malocello rediscovered the Fortunate Islands. In 1291, Doria and de Vivaldo, seeking a water route to India, sailed southward along the West African coast deep into the "Sea of Gloom," as the Moslems called it (probably close to the Tropic of Cancer). Genoese are believed to have discovered the Madeiras and several of the Azores. In 1317, Diniz, King of Portugal contracted with Emmanuel Pesagno (Manuel Pessanha), a Genoese sea captain, to create a Portuguese navy. Pesagno, dubbed Lord High Admiral of Lusitania, was to maintain a minimum cadre of 20 Genoese mariners in Portugal to protect its coast and train its sailors. They were to serve as pilots, captains, and instructors in the Portuguese fleet. When not so employed they were to serve as Portuguese agents in Flanders or Florence. The Portuguese-Genoese relationship was to endure for well over 150 years. When Columbus swam ashore near Cape St. Vincent, the relationship was still strong, but beginning to wane.

16.   The destination was El Mina, a trading post and fort set up on the Guinea Gulf by King John of Portugal. It was at El Mina that Columbus first saw gold being panned or sluiced from streams by black slaves. The impression it left was profound. The search for gold (and pearls when he could find them) was to become the centerpiece of his (and Spain's) exploratory objectives in the New World. And, the utility and hardiness of blacks in its production was to lead to their enslavement.

17.   *Tratactus de Sphaera* (Treatise on the Sphere) by John of Holywood (1265-1321), self-named "Sacrobosco," was based on Ptolemy and the Arab astronomers. Its simplicity of presentation made the work highly popular. Twenty-five printed editions appeared before 1500.

18.   A "great circle" is a circle on the surface of a sphere, the plane of which passes through the center of the sphere. In the case of the earth, meridians and the equator are examples. In the case of the celestial sphere, celestial meridians, the equinoctial (or celestial equator), and the ecliptic (apparent path of the sun) are examples.
       A "small circle" is a circle on the surface of a sphere, the plane of which does not pass through the center of the sphere. In the case of the earth, parallels are examples. In the case of the celestial sphere, diurnal circles (those which describe the apparent daily path of a celestial body) are examples.

19.   *The Teaching of Navigation in Spain and Portugal in the Time of Cabrillo,* an article in the publication *The Cabrillo Era and His Voyage of Discovery,* Cabrillo Historical Association, San Diego, 1982. Admiral Teixeira da Moto's assessment of Columbus' knowledge of celestial navigation falls into the mold established by most Portuguese reviewers. This, unhappily, is suspiciously chauvinistic on the subject of relative Portuguese/Spanish achievements during the Age of Discovery.

20.   To assist navigators in determining their latitude, they were provided with tables known as 'Rules of the Leagues' which gave the number of leagues it was necessary to sail on various bearings in order to achieve a change in latitude of one degree. Given the length of a degree of longitude at various latitudes, the navigator was able to approximate the change in longitude achieved over each course or bearing.

21.   Small ships, particularly but not exclusively, are subject to various types of acceleration induced

by wind and wave, i.e. surge, heave, side-slip, roll, pitch, and yaw. These accelerations may occur simultaneously or in succession although a ship under way need not necessarily experience violent or even significant manifestations of each type of acceleration on any particular tack, course, or bearing. The net effect is to make a small ship's deck, a highly unsatisfactory platform for reading the altitude of heavenly bodies with any degree of precision.

22.   The traditional gesture referred to as "the pilot's blessing," or "taking the north in order to mark it" consisted of raising the arm with flattened palm between the eyes, pointing at the North Star, and bringing the palm straight down on the compass card to see if the needle varied from true north." Understandably, this is hardly a precise method, but in most cases it seems to have worked reasonably well. See Morison's *Admiral of the Ocean Sea,* pp. 203-204.

23.   The earth completes 365.2422 rotations about the sun in one year. In this period, it makes 366.2422 rotations about the stars. Hence, the (apparent) angular velocity of the stars is 366.2422 divided by 365.2422 x 360° = 360.9857°/day = 15.041°/hour. However, the apparent velocity of Polaris must be reduced by the speed of Columbus' ship which was making about 107 statute miles/day, equivalent to 1.75° of longitude/day (or 0.073°/hour) at latitude 28° N, making Polaris' apparent velocity 14.97°/hour. We have used 15°/hour, at latitude 28° N.

24.   p. 184, *Admiral of the Ocean Sea . . .*

25.   p. 655, ibid.

26.   III. Ptolemaeus Romae, 1490, titled *Secunda Europe Tabula* (see A. E. Nordenskiöld's *Facsimile-Atlas,* Dover pubs., New York, 1973) shows the entrance to the harbor of Lisbon at 40.53°, north latitude, instead of 38.68°. Vignaud (see p. 194 of his "Toscanelli and Columbus, The Letter and Chart of Toscanelli," Books for Libraries Press, Freeport, N.Y., reprinted 1971) confirms that in the last quarter of the 15th century Lisbon was placed on the 40th or 41st parallel. Portugal"s "Sacred Promontary", Prince Henry's school for pilots and navigators at Cape St. Vincent is shown on III. Ptolemaeus Romae, 1490, at about 38.2° N. latitude instead of almost precisely 37° N. XV. Ptolemaeus Romae, 1490, titled *Quarta Africae Tabula* shows the Canary Islands lying between 10 and 15° north latitude, instead of from 27.6 to 29.2° north. These are not isolated instances. The determination of latitude, ashore, by the best astronomers and geographers of the late 15th century was frequently in significant error.

27.   See Al-Zarqali (1029-1088 A.D.), p. 109, *Muslim Contribution to Geography,* by Nafis Ahmad, Ashraf Press, Lahore, Pakistan, 1972 and Roderick Webster, *The Astrolabe, Some Notes on its History, Construction and Use,* pp. 1 and 2, Paul MacAlister & Associates, Lake Bluff, Ill., 1974.

28.   P. 186, *Admiral of the Ocean Sea, . . .* However, Hanbury Brown, on p. 57 of his *Man and the Stars,* Oxford University Press, 1978, reports that " . . Amerigo Vespucci made quite a reasonable estimate of his longitude in 1499 while off the coast of South America by observing a conjunction of the moon and Mars . . ." and comparing the local time with the predicted time of the event shown in an almanac prepared for Ferrara, Italy. This was a special application of the "method of lunar distances", or "lunars", developed by Regiomontanus in 1475, data for which were included in his ephemerides intended for use during the next 30 years. In general, the angular separation between the moon and a star, or the moon and the sun, is measured and time

310

and date noted. Reference is made to an almanac prepared for a standard place (Greenwich, Ferrara, etc.) to determine the time the same angular separation occurs at that place. The difference in the predicted and observed times represents the difference in longitude, 1 hour equalling 15°. Giorgio Abetti's *History of Astronomy*, p. 54, Henry Schuman, New York, 1952, tends to confirm Brown as to the "method of lunar distances" being employed by Vespucci. However, until the 18th century the general method could not be employed for two reasons: first, astronomers could not predict the angular separation of moon and star with adequate precision; and there was no instrument with which this angular distance could be measured with sufficient accuracy from the deck of a moving ship.

29. See J.L.E. Dreyer's *A History of Astronomy from Thales to Kepler*, pp. 289-290, Dover Pubs., New York, 1953, first published in 1906 under the title "History of the Planetary Systems from Thales to Kepler," Cambridge University Press.

30. pp. 478-479, *Admiral of the Ocean Sea*, . . .

31. Among the few survivng maps of the Columbian era drawn to a large enough scale to permit a reasonable identification of the vicinity of Nurnberg are the "Ptolemaeus Ulmae, 1482" (see map XXIX, of Nordenskiöld's "Facsimile-Atlas," Dover Publications, 1973) and "V. Ptolemaeus Romae, 1490, Quarta. Europe, Tabula (see map V of Nordenskiöld's Atlas).

32. See p. 655, Morison's *Admiral of the Ocean Sea*, . . .

33. P. 329, *The Random House Dictionary of the English Language*, New York, 1973.

34. P. 327, *The Mapmakers*, by John Noble Wilford, Alfred A. Knopf, New York, 1981.

35. See pp. 64-66. *Admiral of the Ocena Sea...*

36. Pp. 57 and 66, *North America from Earliest Discovery to First Settlements...*, David B. Quinn, Harper and Row, New York, 1977.

37. *The Cartography of the First Voyage,* is an Appendix to *The Journal of Christopher Columbus,* translated by Cecil Jane, Clarkson N. Potter, New York, 1960; pp. 217 through 219.

38. Lucius Annaeaus Seneca (4 B.C. - 65 A.D.), philosopher and writer, composed 9 tragedies. "Medea," one of these, contains the famous prophesy which impressed Columbus so much. "An age will come after many years when the Ocean will loose the chains of things, and a huge land lie revealed; when Tiphus (Jason's pilot) will disclose new worlds and Thule no more be the ultimate." Seneca became a tutor to Nero and continued in close association with him after his succession. However, in 65 A.D. he fell out of favor, was accused of conspiracy, and forced to take his own life.

39. Pierre d'Ailly (d. 1422), Cardinal of Cambrai, France, composed a comprehensive world geography, around the year 1410, which came to be known by one of its treatises *Imago Mundi..* D"Ailly followed Marinus of Tyre in making Eurasia long and the Atlantic (the only ocean known to lie between the Orient and Europe) narrow. While his work was written before Ptolemy's *Geography* was rediscovered by Western Europe, his position was not altered by

subsequent reading of the Alexandrian sage, as is attested by later tracts added to *Imago Mundi*. D'Ailly's work was not printed until sometime between 1480 and 1483. Columbus read and re-read them and became a fervent disciple of d'Ailly's thrust (if not all of his sometimes contradictory statements) starting around 1485 and continuing past at least the first two of his voyages of discovery and exploration.

Chapter 2

1. While Abraham, "the father of all nations" is supposed to have visited and lived briefly in Egypt (uncertain date, possibly 1850 B.C.), it was during the reign of the Hyksos that the Jewish pre - sence in Egypt began. First the Biblical Joseph, then his father Jacob (re-named Israel) and his twelve brothers and their families came to live in Egypt (in the Delta region). They multiplied greatly during the several century span leading to their Exodus from Egypt, led by Moses, in 1290 B.C. Again, following the fall of the Temple, in 587 B.C., many Jews settled in Egypt.

2. Rome was pretty much drained, itself, after the 3 Punic wars, spanning 118 years, with the Carthaginians. During the 2nd Punic war, Hannibal - in the period 218-208 B.C. - crossed the Alps and handed the Romans a series of defeats before being defeated himself, on Carthaginian soil, in 201 B.C.

3. A. E. Berriman, *Historical Metrology*, E. P. Dutton, New York, 1953; p. 1 and 117-120, describes measurements of the platform, or base, of the columnar structure built by Pericles (c. 438 B.C.) and known as the Parthenon. In terms of Attic feet of 0.30864 m, the platform measured 100 x 225, and had an area of 1 myriad (10,000) square cubits.

4. See p. 54, *Islamische Masse und Gewichte, Ungerechnet ins Metrische System*, by Walther Hinz; E. J. Brill, Leiden, 1955, and Table 4, Chapter Four, herein.

5. This section is a rationalization of information (frequently appearing to be in conflict) taken from the following sources:

- A. E. Berriman's *Historical Metrology*, Chapters I, V, VI, and IX.
- Hallock and Wade's *Evolution of Weights and Measures*, pp. 26-40, the MacMillan Co., New York, 1906.
- H. Arthur Klein's *World of Measurements*, pp. 53-72, Simon and Schuster, New York, 1974.
- W. M. Flinders Petrie's *Inductive Metrology*, Chapters IV, V, VI, and X, Hargrove Saunders, London, 1877. In particular, see spread-sheet "Synoptic Tables" on pp. 142-143, or condensed version, Table 4 of this chapter.

6. Greek colonization of Southern Italy, Sicily, North Africa, Southern France and Spain dates from c. 700 B.C., about when Rome was founded. The Pelasgic or Ancient Greek metrologic units were in use by the Greek colonizers and, hence, were implanted in southern Italy from which they spread northward.

7. References for section on cubit are the same as for previous section on the foot.

8. See pp. 123, 124, Berriman's *Historical Metrology*. "Tela" means cloth and "merc" probably

means merchandise or trade. The variation in size of the braccio increased during the Medieval/Renaissance period, for lengths reported for Italy alone, range from 0.3476 m to 0.7443m. The brachium (braccio, pic) is treated in some depth in Chapter 4, under "Toscanelli's Degree."

9.   Ibid.

10. Exodus 12, 37: "The Israelites set out from Rameses for Succoth, about six hundred thousand men on foot, not counting the children."

11. This section is a synthesis of information found in Berriman's *Historical Metrology*, pp. 1-3, 73, 74, 116-133; Hallock and Wade's *Evolution of Weights and Measures*, pp. 25-38; Klein's *World of Measurements*, pp. 55-61 and 68-72; and Petrie's "Inductive Metrology," pp. 81-102, 142 and 143, and 144-149.

12. Not to be confused with Petrie's Babylonian foot of 315.4 mm which he found in Egypt, Greece, Italy (and Roman colonies), but not in Assyria, Persia, Syria, and Asia Minor!

13. See also *Pre-Columbian Contacts in the New World*, Appendix A.

14. The Assyrian "Great U" of Oppert is evidently a form of the "ancient Babylonian double cubit" (which Lehmann found to be the precise length of a seconds pendulum). This unit had an ascribed length range of 990-996 mm. Petrie's Assyrian "Great U" measured 1.014 m.

15. Klein, *World of Measurements*, p. 68, quotes Leonardo da Vinci (1452-1519) on the views of Vitruvius Pollio, an architect-engineer of the 1st century B.C.

     "Vitruvius declares that nature has thus arranged the measurements of man: four fingers make one palm and four palms make one foot; six palms make one cubit; four cubits make once a man's height, four cubits make a pace, and twenty four palms make a man's height."

One is immediately suspicious about these 6 foot Romans until one recalls that 6 Roman feet equals 1.778 metres, or 5 feet 10 inches.

16. Pp. 17 and 29, *Historical Metrology*, by A. E. Berriman.

17. Pp. 69, 70, *Historical Metrology* and p. 209, "Mathematics in the Time of the Pharaohs," by Richard J. Gillings, Dover Pubs., New York, 1982 (first published by MIT Press, Cambridge, Mass., in 1972).

18. Pp. 208-209, *Mathematics in the Time of the Pharaohs*.

19. Ibid.

20. J. B. J. Delambre in his classic *Histoire de l'Astronomie du Moyen Age*, Paris, 1819 (see Johnson reprint, N.Y., 1965) cites Christman (fl. Frankfurt, c.1590) as a proponent of the 6 grain school and Jean Fernel (fl. Amiens c. 1528) as a user of the 4 grains per digit relationship.

Barley grain available domestically (in the U.S.) fits both definitions, i.e. 6 grain widths or 4 grain lengths equals approximately 3/4 inch.

21. Gillings' *Mathematics in the Time of the Pharaohs*, Figure 22.1, p.220.

22. See Hallock and Wade, *The Evolution of Weights and Measures*, p. 6.

23. See pp. 126-128, 140, 142, *Inductive Metrology*. Petrie indicates the ancient hasta had a length of about 0.4525 to 0.4668 metre. On this basis, a yôjana would measure 7240 to 7470 m.

24. See footnote on pp. 218-219 of the Hamilton/Falconer translation of *The Geography of Strabo*, Henry G. Bohn, London, 1857. Major James Rennell, 1742-1830, was a military surveyor, cartographer, and historian. Besides preparing several atlases on parts of the Indian continent, he authored the *Geographic System of Herodotus*, 1800, and "Western Asia", 1809-11.

25. P. 25, including footnote 3, Hallock and Wade, *Evolution of Weights and Measures*.

26. See page xi, Preface to Vol. III of the Hamilton/Falconer translation of *The Geography of Strabo"*.

27. See p. 23, *Evolution of Weights and Measures*, Hallock and Wade.

28. Same as footnote 21.

29. See footnote 17, *The Ancient Measurements of the Earth*, by Aubrey Diller, p. 9 of February, 1949 issue of ISIS.

30. See pp. 7 and 8, ibid.

31. The geodetic appearance of the Greek Olympic foot and the system based on it has, however, been noted by several metrologists including Jomard, in 1812, Watson, in 1915, and Berriman, 1953.

32. See p. 351, *The Great Pyramid, Its Secrets and Mysteries Revealed*, by Piazzi Smyth, Bell Publishing Co., New York, 1978; originally published in Great Britain in 1880 as "Our Inheritance in the Great Pyramid."
   Smyth claims the Sacred cubit was used in the dimensioning of the Great Pyramid of Gizeh and that the base length, expressed in Sacred cubits, equals 365.24, the number of days in a tropical year, and the height of the pyramid (147.8 m) multiplied by 10^9 equals the mean distance of the earth from the sun in its yearly elliptical orbit. (A pyramid height of 149.7 m would have accorded better with the currently accepted value of 93,003,000 statute miles for the astronomical unit.) Smyth also claims the designers of the Great Pyramid had a knowledge of π (pi) as well as a remarkable sense of the distribution of land and water areas on earth. The latter is evidenced by the placement of the huge structure on a meridian (31°10' 35" E) and the "general" parallel, 30° N (actually 29° 58' 51"), on which there is more land surface and less water than on any other meridian or parallel.

33. Lehmann argues that the theory of the pendulum must have been known to the early Babylonians, who doubtless derived it from the use of the plumb line in their building operations. See Lehmann, p. 89 *Uber das Babylonishe metrishe System und dessen Verbreitung,* Verh. der Physikalischen Gesellschaft zu Berlin (Berlin, 1889), vol. viii, pp. 88-101; also in abstract, pp. 167-168, vol. lxi, Nature (London, 1889).

34. Because the earth is not a sphere, but an approximate ellipsoid of revolution with a polar radius of 6,356,912 metres and an equatorial radius of 6,378,388 metres, the force of the earth's gravity is greater at the poles than at the equator. If the earth did not rotate about its own axis, and if there were not other complicating factors at play, this ellipticity would be sufficient to cause the acceleration of gravity at the poles to be 6.63 cm/sec^2 greater than at the equator (and about 4.35 cm/sec^2 greater than at latitude 31° N or S).

One of the complicating factors is the earth's rotation. This induces a centrifugal force (in any object) which is directed outward, i.e., perpendicular to and away from the earth's rotational axis. This force, which is greatest at the equator and zero at the poles, tends to offset the earth's gravitational force. The acceleration of this force has a maximum value (at the equator) of 3.392 cm/sec.^{2} (At latitude 31°, its value is 2.907 cm/sec.^{2}) The effect of this centrifugal force on the difference between the earth's gravitational force of attraction at the poles and at any other point on the earth's surface is to intensify or increase this difference. Thus, were there no additional complicating factors, the total or net difference in the earth's gravitational force of attraction (or acceleration of gravity) at the poles and at the equator would be about 10 cm/sec.^{2}

A second complicating factor present is the non-spherical distribution of mass of the earth which compensates for about half of the difference. Thus, the overall difference in the gravitational force at the poles and at the equator - at sea level - is 5.1805 cm/sec^2, the numbers being 983.2329 cm/sec^2 at the poles and 978.0524 cm/sec^2 at the equator. At 31° N latitude and approximate sea level (as in southern Babylonia), the net acceleration of gravity would be about 979.8 cm/sec.^{2}

The period (time required for one complete oscillation) of a simple pendulum is given by the relationship:

$$T = 2\pi \sqrt{L/g},$$ where   T is the period in seconds
     L = length of the pendulum in centimeters, and
     g = acceleration of gravity, in cm/sec^2

For a "seconds pendulum," the period or time required for a complete oscillation is 2 seconds, and the time for one-half a complete oscillation, t, is one second. Thus,

$$t = T/2 = \pi \sqrt{L/g} \text{ and } L = (t/\pi)^2 g, \text{ or}$$

L (at 31° N latitude, and sea-level altitude) = (1 sec/π)^2 x 979.8 cm/sec^2 = 99.28 cm = 992.8 mm.

The "double cubit" of the Sumerians was 10 palms, or 990-996 mm, bracketing the length of the "seconds pendulum".

35.   P. 12, *Evolution of Weights and Measures.*

36.   Hallock and Wade call this unit "digit", but Berriman has a unit called "digit", one-half the size of Hallock and Wade's unit, which is much closer to the Egyptian, Ancient Greek and Roman digit of approximately 0.0185 m. It will be recalled that the Ancient Greek "knuckle" was equal to 2 digits or fingers. The term "digit" may have been an imperfect translation by the Rev. W. Shaw-Caldecott. For this reason, we call the unit equal to 1/3 palm, or 2 of Berriman's "digits", the "thumb" or "knuckle", either of which could have been intended and both of which appear more rational than "digit".

37.   Le Pere, *Memoires sur l'Egypte pendant les Campagnes du General Buonaparte,* vol. ii, pp. 32, 279. An account in English of Le Pere's work is provided in Berriman, p. 74.

38.   See pp. 50 and 64, *Inductive Metrology.*

39.   The account of the Rhind Mathematical Papyrus is taken from the following sources:

      a) *The Rhind Papyrus,* by James R. Newman which is Chapter 2, Volume One of the four volume work, *The World of Mathematics,* copyrighted by Dr. Newman in 1956 and published by Simon and Schuster, New York, the same year;

      b) *The Nature of Mathematics,* by Philip E. B. Jourdain, which is Chapter 1, Volume One, of *The World of Mathematics;*

      c) *Egyptian Mathematics and Astronomy,* which is Chapter IV of O. Neugebauer's, *The Exact Sciences in Antiquity,* Dover Publications, New York, 1969; and

      d) *Ancient Egyptian Measures and Weights,* which is Chapter VI of A. E. Berriman's, *Historical Metrology.*

40.   Not to be confused with the Theban prince of the same name who expelled the foreign Hyksos from Egypt and established the 18th Dynasty, the first of the New Kingdom.

41.   Herodotus said, "The warrior class in Egypt had certain special privileges . . . each man had 12 arurae of land assigned to him free of tax (the arura is a square of a hundred Egyptian cubits, the Egyptian cubit being the same as the Samian). All the warriors enjoyed this privilege together, but there were other advantages that came to each in rotation." See Berriman's *History of Metrology,* p. 70.

    Samian, of course, refers to the ancient Ionian island of Samos, a center of Greek culture off the Aegean coast of modern Asian Turkey. The Rev. Canon Rawlinson, from whose translation of Herodotus' account of his travels in Egypt the foregoing quotation is taken, makes the point that Herodotus had written his book in Greek and for Greeks. At the time of Herodotus' travels Greece was involved in a continuing series of wars with Persia which had conquered Egypt, almost all of Asian Turkey, and much of Thrace. The units of linear measure in use in Samos, as in much of the rest of Asian Ionia were the same as the Persian. To avoid offending the sensibilities of his Greek audience, Herodotus referred to the Samian cubit when he really meant the Persian cubit, equivalent to 20.68 British inches, or 0.525 metres. (See Piazzi Smyth's *The Great Pyramid,* pp. 339-342.)

42. Pp. 207-210 and 219, 220, *Mathematics in the Time of the Pharaohs*

43. Pp. 154-158, *The Measurement of Length: Yesterday, Today, and Tomorrow,* By Lewis Barnard, Jr., contained in "Report of the 50th National Conference on Weights and Measures, 1965, "published as National Bureau of Standards Miscellaneous Publication No. 272, April 1, 1966.

Chapter 3

1. This would be vigorously contested by Piazzi Smyth, Royal Astronomer of Scotland, who in the years 1864-1865 conducted a series of investigations at the site of the Great Pyramid of Gizeh near Cairo, Egypt. Among his startling conclusions: the designers of the Great Pyramid knew the dimensions of the earth, they had a knowledge of π (the ratio of the circumference of a circle to its diameter), they had determined rather accurately the mean distance of the earth from the Sun, and they sited the Great Pyramid at the intersection of the meridian and parallel, each of which crosses more land and less sea than the other.See "General Summation" (pp. 597-602) and Plate II of Piazzi Smyth's *The Great Pyramid... .* See also ref. note 32, Chapter 2, herein.

2. See pp. 31-84, E. H. Bunbury's *A History of Ancient Geography,* first published in London in 1879. The second edition, published in 1883, was re-published in 1959 by Dover Publications, New York.

3. See Aubrey Diller's *The Ancient Measurements of the Earth,* ISIS, February, 1949.

4. J. L. E. Dreyer,*A History of Astronomy from Thales to Kepler* (New York, Dover Pubs., 1953, originally published in the U.K. in 1906 under the title *History of the Planetary Systems from Thales to Kepler*), pp. 108-122. Revisions to the original edition by W.H. Stahl, N.Y. University, were included in the Dover edition.

5. Earth is taken as a sphere with an Equatorial or Meridional circumference of 40,000 km. (40,000,000 metres). The original basis for the establishment of the metre was as the 10,000,000th part of a meridian quadrant arc. When set as the national standard in France, in 1795, after several arc measurements in various parts of the world had been carried out, the length of the meter was put at 443.44 lines, of which 144 equalled the pied. In 1798, this was changed to 443.296 lines to allow for flattening of the earth at the poles. The line (ligne, in French) thus became equal to 2.2558 mm. The metre has remained as defined even though more recent measurements of the earth show the equatorial circumference to be 40,062,810 m. and the meridional circumference 39,929,335 (the average of these two being 39,996,122.4 m.). Considering the earth a sphere with a circumference of 40,000 km. introduces an error of 0.157% in longitudinal measurements and 0.177% in meridional. One degree, therefore, is being taken as 40,000 km. ÷ 360 = 111.11 km.

6. See H. C. Hamilton and W. Falconer, *The Geography of Strabo, Literally Translated,* Henry G. Bohn, London, 1857, pp. 56, 218, and 219; and, for greater detail, Pascal F. J. Gossellin's *Geographie des Grec analysée,* Paris, 1790.

7. See pp. 173-174, Dreyer/Stahl's, *A History of Astronomy*.

8. A gnomon, in its simplest form was merely a vertical pole of known height. The ratio of the length of the gnomon's shadow, at any time during daylight hours, to the height of the gnomon is equal to the tangent of the sun's zenith distance (90°- altitude) at that time. At true noon, the sun's zenith distance is a minimum for that day. North of the Tropic of Cancer, at noon on the day of the summer solstice the zenith distance of the sun is a minimum for the year (the sun is highest in the heavens; its altitude is a maximum). At any site north of the Equator at noon on the day of the winter solstice, the sun's zenith distance is the greatest of the zenith distances taken at noon for any day of the year. On the days of the vernal and autumnal equinocti, the sun's zenith distance at noon is equal to the latitude of the site (for any place on earth). Without a knowledge of trigonometric functions (and these would not be invented by Hipparchus for another half-century), the gnomon required augmentation by some device which would enable direct reading of the sun's zenith distance or its altitude. Figure 2c illustrates such a device.

9. Strabo, Greek geographer and historian, born about 63 B.C., travelled widely, settled in Rome about 29 B.C., where he died about 19 A.D. His two great works were "Historical Memoirs," in 46 books, and "Geographica," in 17 books. The latter is our most important source on ancient geography up to the time of Ptolemy, and, in particular, on Eratosthenes and Posidonius. Strabo's information on the Earth measurements of Eratosthenes and Posidonius was obtained from Book I of Cleomedes' "cyclice theory meteoron", considered a good summary of Stoic astronomy. Cleomedes, believed to be a Lysimachian, was a disciple but not a contemporary of Posidonius. In Book II, which includes explanations of the Moon's phases and of eclipses and some data on the planets, Cleomedes exhibits an understanding of atmospheric refraction not possessed by Posidonius.

    For good coverage on the great scientific figures of the ancient world see George Sarton's *A History of Science*, Harvard University Press, Cambridge, Mass, 1959; T. L. Heath's *Greek Astronomy*, E. P. Dutton, New York, 1932; M. R. Cohen and J. E. Drabkin "A Source Book in Greek Science", McGraw-Hill, New York, 1948; and (in French) J. P. J. Delambre "Histoire de l'Astronomie Ancienne", V. Courcier, Paris, 1817.

10. See Delambre's *Histoire de l'Astronomie Ancienne*, p. 89

11. See Heath's *Greek Astronomy*, pp. 109-112.

12. In any such "double measurements of the sun" taken at noon at two sites on a common meridian, it goes without saying that the intent was to make both observations simultaneously, or one year apart. It is most likely that Eratosthenes, a practical and prudent astronomer, as well as geographer and geometer, "read the sun" at Alexandria over a period of at least several days before and after each solstice, summer and winter, the readings being taken several times before, during and after the sun reached culmination. For several days in the time frames bracketing each of the solstices, the variation in the sun's altitude, or its zenith distance, at noon is less than one minue per day (see Table 1, Chapter 1), so that the requirement for simultaneous readings at the two sites on solsticial observations is not absolute. However, the daily change in the sun's altitude at noon increases markedly between the solstices and the equinocti making simultaneous readings at the two sites a necessity, where precise tables of the sun's declination are unavailable, and desirable under any circumstances.

With regard to Delambre's highlighting of the gnomon's basic deficiency (yielding a shorter shadow by some 15'), so long as both observation sites are north (or south) of the sun at noon, the error is eliminated (for we are dealing with a difference in zenith distance of the sun, taken at each site). If one observing site is north and one south of the sun at noon, the error is increased to 30 minutes.

Eratosthenes ignored the effects of parallax and refraction-parallax properly so, since considering the sun's rays at Syene and Alexandria parallel introduces an error of less than 0.02 second. He was probably unaware of atmospheric refraction which introduced an error of 8 seconds in the difference in apparent zenith distances observed at the two sites during the summer measurement and 19 seconds during the winter measurement. It is questionable whether his instrument could show differences in shadows much smaller than a degree.

Finally, there is some question as to whether Eratosthenes ever visited Syene. The point is moot. He certainly could have sent an assistant, properly instructed, to Syene to do what he was doing at Alexandria - obtain the maximum altitude of the sun during the period of the summer solstice and the least of the maximum altitudes observed each day for the general period of the winter solstice.

13. For Strabo's views, see the Hamilton/Falconer ... *Geography of Strabo* ... pp.. x and xii, Preface to Vol. III. For St. Isidore's, see George H. T. Kimble, *Geography in the Middle Ages* (London, Methuen & Co., 1938), pp. 23, 24 and E. G. R. Taylor, *Some Notes on Early Ideas of the Form and Size of the Earth, Geographic Journal* (London), Jan., 1935, on pp. 66 and 68.

   The Roman mile is one of the very few units of linear measurement which has appeared to exhibit a resistance to change over long spans of time and over wide areas (essentially those which came under Roman dominance during the period 264 B.C. to A.D. 117). Still, the literature contains estimates of this important unit, linking ancient, medieval, and Renaissance measures, which vary from 1472.5 to 1488 metres. Unless otherwise indicated, we shall use the Roman mile of 1000 paces = 5000 feet of 0.2963 metre = 1481.5 metres.

14. See Dreyer's ... *History of Astronomy* ..., p. 175.

15. Ibid., footnote 4.

16. Aubrey Diller, *The Ancient Measurements of the Earth*, ISIS, Feb., 1949, on pp. 6-9.

17. Kimble, *Geography in the Middle Ages*, pp. 8, 9 and Taylor's "Notes," p. 66.

18. Book XVII, Chapter I, Paragraph 24, *Geography of Strabo.*

19. E. H. Bunbury, in his *History of Ancient Geography* Volume I, Chapter VI, Note C (pp. 209-210), cites (for support) Ukerts *Geographie de Griechen und Römer*, Vol. i, pt. ii,(pp. 51-72); Leake *On the Stade as a Linear Measure*, first published in 1839 in the Journal of the Geographical Society, vol. ix; and Dr. Smith's *Dictionary of Greek and Roman Antiquities*, p. 893. Bunbury's masterful work in two volumes was first published in Great Britain some time during the mid-19th century. The second edition, published in 1883, has been reproduced and was re-published in 1959 by Dover Publications, New York.

20. Gaius Plinius Secundus, called Pliny the Elder to distinguish him from his nephew, born 23 A.D. and died in 79 A.D. (when Vesuvius erupted), was a naturalist, writer, and Roman Navy

Admiral. He is most famous for his *Natural History*, in 37 books.

21. Polybius, Greek historian (?198 B.C. - ?117 B.C.), who because of the Roman conquest, first of Macedonia, and later of Greece, spent much time in Rome and other parts of the Empire where he made some powerful friends. He wrote a history of Rome in 40 books, of which only the first five remain. For specific reference see pp. xi and xii, preface to vol. III, Hamilton/Falconer translation of *Geography of Strabo*.

22. *Posidonius and the Circumference of the Earth*, by I. E. Drabkin, ISIS, 1943, p. 510.

23. George Sarton's *A History of Science*, Harvard University Press, Cambridge, 1959, pp. 99-116; Pascal F. J. Gosselin, *Geographie des Grecs analysée ou les systèmes d'Eratosthene, de Strabon, et de Ptolemee compares entre eux.*, Paris, 1790 (one of Sarton's sources); and Hamilton and Falconer's *Geography of Strabo* Vol. i. pp 1-98, 101-147, 161-203, 239, 255-256, 332, 457-462, 487; Vol. ii, pp. 70, 76,195, 240, 243-248, 261, 271; Vol. iii, pp. 44, 70-79, 84, 124, 130, 149-156, 183-192, 208, 220, 276, 284, 293 - and additional references to be cited.

24. Eratosthenes accepted the word of Pytheas of Massila (in Gaul), Greek navigator in the time of Alexander the Great who claimed to have visited Britain and Thule and coasted Europe from Gades (Cadiz) on the Atlantic to the Tanais (Don R. on the Black Sea). Strabo would not accept Pytheas word on Thule and left it off his own map which, except for this and certain other offshore islands, was almost identical to Eratosthenes' (Figure 2d). When Strabo's *Geography* was first translated into Latin in the 15th century, Thule was assumed to be Iceland, the assumption of Hamilton and Falconer in their own translation: see Vol. i, pp. 99, 100, 157, 173, 299. However, A. E. Nordenskiöld says mariners in Ptolemy's time (2nd century A.D.) considered Thule to mean southwestern Norway. See A. E. Nordenskiöld's *Facsimile-Atlas*, p. 34, footnote 1.

25. There are two solsticial days--summer and winter--when the sun is, respectively, farther north and south than at any other time of the year. This is the most important, but not the only factor, influencing the length of the day. There are two other factors, although minor, which deserve mention. The following explanation is provided by Dr. David Dearborn of the Lawrence Livermore Laboratories, Univ. of Calif. (and former Director, Steward Observatory, Univ. of Ariz.):

> 1) At the time of the solstices almost all of the sun's apparent motion among the stars is to the east, while during the equinocti the sun is moving both east and north (or south). Thus, the true sun moves faster than the mean sun near the solstices, and slower than the mean sun near the equinocti. At the equator, where the declination of the sun has no effect on the length of the day, this tends to make the days of the solstices (both of them) shorter than the length of the days of the equinocti.

> 2) Because the earth's orbit is an ellipse (not a circle), the earth does not move the same distance around the sun each day. Near perihelion (when the earth is nearest the sun), the earth moves further around the sun than at any other time. At aphelion, its orbital speed is the slowest. Perihelion occurs near the time of the winter solstice. This causes the apparent sun to move even faster to the east at that time (early January). Thus, the length of the day, as seen at the equator, is shortest somewhere between the winter

solstice (21 Dec.) and perihelion (2 Jan.), actually 23 December. Consistent with the foregoing, the longest day at the equator would be found between aphelion (6 July) and the autumnal equinox (23 Sept.), actually 17 September.

The effects described in the foregoing are not restricted to the equator and can cause variations in the length of a day at most latitudes of a fraction of a minute.

26.   See, for instance, Note 8, p. 3, vol. II, *History of Ancient Geography.*

27.   P. 176 and footnote 5, p. 177, *A History of Astronomy . . .* See also pp. 98-101, *Larousse Encyclopedia of Astronomy,* published in the United States by Prometheus Press, New York, 1959, for a good discussion of atmospheric refraction. See also reference note 31 , hereunder, for Drabkin's comment on the altitude of Canopus at Rhodes.

28.   See the Hamilton/Falconer translation *The Geography of Strabo,* Book II, Chapter II, Paragraph 2.

29.   Same reference as 28 except Book II, Chapter V, Paragraph 24.

30.   *The Ancient Measurements of the Earth* by Aubrey Diller, ISIS, February ,1949, p. 7.

31.   *Posidonius and the Circumference of the Earth,* by I. E. Drabkin, ISIS, 1943, pp. 509-512. In footnote 5, p. 510, Drabkin also comments: "Even if allowance is made for the effect of refraction, the star Canopus in Posidonius' time attained an elevation of more than 1° above the horizon in the city of Rhodes (and more on the rest of the island). Actually, it was visible for about 2 1/2 hours.

32.   Marinus of Tyre, great Phoenician geographer, known principally through Ptolemy's *Geography* which the latter admits is based, with certain corrections, upon the work of the former. No precise data on the dates and places of birth and death exist. E. H. Bunbury, *A History of Ancient Geography,* Vol. II, p. 519, concludes " . . . he was very nearly contemporary with the geographer of Alexandria (Ptolemy)." (1st-2nd cent. A.D.)

33.   P. 144 (¶2, Chapter III, Book II), of the Hamilton/Falconer translation of *The Geography of Strabo,* London, 1854.

34.   P. 154 (¶6, Chapter III, Book II), same as above.

Chapter 4

1.   R. V. Tooley, *Tooley's Dictionary of Mapmakers* , Alan R. Liss, New York, 1979; p. 102.

2.   J. P. J. Delambre, *Histoire de l'Astronomie Ancienne* , V. Courcier, Paris, 1817; pp. 310-312.

3.   Ibid, p. 312.

4. See pp. 37 38, *Facsimile-Atlas*, by A. E. Nordenskiöld, Dover Publications, New York, 1973, first published in Sweden in 1889.

5. See Figure 5 of Chapter 3.

6. See Figure 2d of Chapter 3.

7. See pp. 8, 9, *Geography in the Middle Ages*, by George H. T. Kimble, Methuen & Co., London, 1938. See also *Some Notes on Early Ideas of the Form and Size of the Earth*, by E. G. R. Taylor, Geographical Journal (London), Jan., 1935. Professor Taylor, in concurring, indicates that 500 feet = 400 palmipes (short cubits). In Table 5, Chapter 2 (herein), the palmipes (one palm + one pes, 5/4 pes) is listed as having a length of 0.37037 metre.

8. See p. 17, A. E. Berriman's *Historical Metrology*, E. P. Dutton & Co., New York, 1953. Berriman says "... in Egyptian metrology there was a linear unit, of 20 digits, called remen and a distance of 500 remen became the stade used for itinerary measurements by the Greeks and by the Romans." Since the remen equalled 5/4 Roman feet, or about 0.3704m, this would make the stade 625 Roman feet=185.2 metres, and 10 such stades, he shows, equalled a "Geodetic Mile", approximately equal to one minute of latitude (or the modern Nautical Mile). This is shown in Table 5, chapter 2.

9. See Kimble, pp. 8, 9.

10. See Nordenskiöld, p. 4b.

11. J. P. J. Delambre *Historie de l'Astronomie Ancienne* , V. Courcier, Paris, 1817; p. 304.

12. J. L. E. Dreyer's *History of Astronomy . . .* , p. 178, note 1.

13. Ibid, pp. 242, 243, " Author gives as the source of the 7.6 mile per yojan equivalency: Colebrooke, *Notes and Illustrations to the Algebra of Brahmagupta*, p. xxxviii., "Essays", p. 467.

14. P. 116, *Muslim Contribution to Geography*, by Nafis Ahmad, Ashraf Press, Lahore, Pakistan, 1972.

15. P. 201, *A History of Science*, by George Sarton, Harvard Univ. Press, Cambridge, 1959.

16. See Table 1, Chapter 2

17. P. 57, "Islamische Masse Und Gewichte," by Walther Hinz, E. J. Brill, Leiden, 1955.

18. Pp. 40, 41, Berriman's *Historical Metrology.*

19. Pp. 8, 80, 88, 96, 116, *Muslim Contribution to Geography.*

20. P. 372, Delambre's *Histoire de l'Astronomie Ancienne.*

21.   Ibid, p. 374.

22.   See p. 389, *History of Ancient Geography,* by J. Oliver Thomson, Biblo and Tanner, New York, 1965. See also p. 678, Vol. 2, *History of Ancient Geography,* by E. H. Bunbury, Dover Pubs., New York, 1959.

23.   See Kimble, pp. 23, 24 and Taylor, pp. 66 and 68.

24.   See footnote 15, p. 8, Aubrey Diller's, *The Ancient Measurements of the Earth,* ISIS, February 1949, in which he discusses Hans Von Mzik's Erdmessung, *Grad, Meil und Stadion, nach den altermenischen Quellen* (Studien zur arm. Gesch. VI) (1933), 90-113. See also Taylor, p. 66.

25.   Neither in footnote 15 nor elsewhere in his article, does Diller explain the cryptic designator "by air (asparez)"; nor apparently, does von Mzik, nor Ananias, himself. What is reasonably certain is that Ananias is describing a system of measurement in Armenia, an area south of the Caucasus, lying in present-day USSR, northeastern Turkey, and northwestern Iran. Over the centuries, this area was over-run by the Assyrians, Medes, Persians, Macedonians, Romans, and others. It would appear that in Ananias' time, the two dominant influences were Hellenic and Persian and the "asparez (by air)" designator refers to the Hellenic influence.

26.   A. E. Kennelly *Vestiges of Pre-Metric Weights and Measures,* p. 166, J. J. Little, and Ives Co., N.Y., 1928.

27.   Same reference as 24, footnote 17.

28.   W. M. Flinders Petrie, *Inductive Metrology,* pp. 134 and 142-143, does, however, identify an ancient Pythic foot of 0.240-0.245 metres which gained more widespread use in its double form, 0.480-0.490 metres. In its double form, it was used in Egypt, Assyria, Persia, Asia Minor, Greece, Africa, and Sardinia. *Inductive Metrology, or the Recovery of Ancient Measures from the Monuments* was first published in London, in 1877 by Hargrove Saunders.

29.   See Taylor, pp. 65-68.

30.   See p. 91, including footnote 2, and pp. 115 and 116 of Nafis Ahmad's *Muslim Contribution to Geography.*

31.   P. 49, *Geography in the Middle Ages.*

32.   The 18 uncia cubit is the Roman cubit, equal to 3/2 the Roman foot of 0.2963 m. The 15 uncia cubit is the Roman palmipes, equal to 5/4 the Roman foot. 4 of the larger cubits make the 6 foot pace. 4 of the smaller cubits make the 5 foot pace. (See Table 5, Chapter 2.)

33.   Note 87 (pp. 81-83), *The Letter and Chart of Toscanelli,* by Henry Vignaud, first published in 1902, reprinted by Books for Libraries Press, Freeport, N.Y., 1971.

34.   Pp. 54, 56, 60, 61, 64, *Islamische Masse und Gewichte .*

35.   P. 65, *Admiral of the Ocean Sea*, by Samuel Eliot Morison; Little Brown & Co., Boston, 1942.   Dr. Erwin Raisz of Harvard's Institute of Geographical Exploration was Admiral Morison's cartographical and geographical advisor.   The specific reference was that " . . . 56 2/3 Arabian miles . . . works out . . . (to be) . . 66.2 nautical miles . . "   Thus, Morison/Raisz rate the Arabian mile at 66.2 x 1852 meters ÷ 56 2/3 = 2163.6 metres.   The black cubit, then, must have been equal to 2163.6 m ÷ 4000 = 0.5409 metres, quite close to Hinz' 0.5404 metres.

36.   Pp. 29, 30, *The Evolution of Weights and Measures* .   The authors refer to Boeckh, *Metrologische Untersuchungen* (Berlin, 1838), pp. 246, 250-3, for the statement by Jomard. E.F. Jomard, 1777-1862, French cartographer, was author of *Monuments de La geographie*, 1842, and an expert metrologist.

37.   The cubit or elle originated as the length of the forearm, from elbow to the tip of the middle finger.   It was thus in all of the ancient systems of metrology and usually equalled 1 1/2 feet (where both units were employed).

38.   J. L. E. Dreyer (1852-1926) was born and educated in Copenhagen (Phd, Univ. of Copenhagen). At the age of 30, he was appointed Director of Armagh Observatory, in which position he remained for 34 years.   He had a distinguished career as an astronomical observer and author (stellar catalogues and ancient and medieval astronomy).   He won the Gold Medal of the Royal Astronomical Society in 1916 and was its President from 1923 to 1925.

39.   Pp. 249, 250, 257, 258, *A History of Astronomy* .

40.   Both quotations by Christman (a 1590 A.D. Frankfurt professor and early translator, from the Arabic, of Alfragan's *Chronologica et Astronomica Elementa*) are found on p. 66, J. P. J. Delambre's *Histoire de l'Astronomie du Moyen Age* , V. Courcier, Paris, 1819; republished by Johnson Reprint Corp., N.Y., 1965.

41.   Footnote 91, pp. 84 and 85, *The Letter and Chart of Toscanelli*.   Edmadeddin Ismael Aboulfeda (Abulfeda) of Damascus, 1273-1331, wrote a *Chronicle* and a *Geography* , published in Europe in 1754.   The latter was translated into French and published in Paris, in 1840, by Joseph Toussaint Reinaud, 1795-1867, French Orienialist and writer on maps.

42.   Pp. 50, 51, *Geography in the Middle Ages.*   See also pp. 13, 19, 24, 52, *Muslim Contribution to Geography.*

43.   For a fuller treatment on Al-Biruni (Abu Raihan Muhammad b. Ahmad) see *Muslim Contribution to Geography,* pp. 15, 17, 18, 20, 36-42, 63-89, 109, 113-118, 130, 150, 156.

44.   P. 250 (footnote 1), *A History of Astronomy* . . .   The original source appears to have been H. Ethe's translation of Zakerina Ben Muhammed Ben Mahmud El Kazwini's *Kosmographie*, p. 195, Leipzig, 1868.

45.   Pp. 46-48, *Muslim Contribution to Geography.*

46.   Pp. 56, 57, *Geography in the Middle Ages.*

47. P. 242, *A History of Astronomy*.

48. P. 116, *Muslim Contribution to Geography*.

Chapter 5

1. Pp. 86, 87, 92, *Geography in the Middle Ages*, by G. H. T. Kimble.

2. P. 93b, Nordenskiold's *Facsimile-Atlas*.

3. P. 92 and footnote no. 5 on p. 242, *Geography in the Middle Ages*.

4. Footnote 1, p. 9, *Geography in the Middle Ages*.

5. P. 64, Morison's *Admiral of the Ocean Sea, A Life of Christopher Columbus*.

6. See pp. 3-23, *Marco Polo and the Mapmakers*, Chapter V of R. A. Skelton's *Explorer's Maps*, F. A. Praeger, New York, 1958.

7. The court of the Great Khan during the period of Plano Carpini's and Rubruqui's visits was at Karakoram, Mongolia.

8. Since the printing press had not yet been invented, "publication" was accomplished rather laboriously by copying the manuscript. A popular account, story, or relation such as Marco Polo's would have teams of copiers, employed at various locations, visited in turn by a "reader" bearing the original manuscript. As copiers became readers, and as translations into other languages took place, some departures from the original version of the manuscript inevitably developed.

9. We are reasonably safe in taking Yule's Italian mile at 1482 metres. In the *Book of Ser Marco Polo*, distance of the return sea journey from Zayton to Chamba (Vietnam), Locac (Maylaya), Java the Less (Sumatra), Seilan (Sri Lanka), India, and Hormuz (Iran) are usually, but not always, given in miles. The land journeys, however, are usually estimated in days and sometimes months and years--although occasionally in miles. Sometimes, specific elements of a journey are described both in days and miles. Occasionally, no bearings are given for legs of a journey and sometimes the bearings given are wrong. In the main, however, sufficient information is imparted for a diligent reader to reconstruct the itineraries of the Polo's (Marco and his Elders) and to form some concepts as to the east-west length of Asia.

   It is in the conversion from days of a journey to Italian miles that Yule may have gone astray, for 25 miles per day on a sustained basis for camels, horses, or mules, over generally harsh terrain, may be too high by as much as 50%. 15 miles per day on a long journey over mountain and desert terrain appears to be a more reasonable allowance.

10. The ancients in referring to the *Pillars of Hercules* were alluding to two peaks at the entrance to the Mediterranean, Gibraltar (Jabal Tariq, Calpe), on the European side, and Jebel Musa, on the Moroccan side.

11. See Chapter II, Book III of any edition of *The Book of Ser Marco Polo*, or *The Travels of Marco Polo*. In Yule's 3rd edition of the former title, revised and edited by Henry Cordier, 1903-1920, and reprinted in 1975 by Philo Press, Amsterdam, this segment is found in Volume III, page 253.

12. Lest there be any doubt in the reader's mind as to the length of the "Italian" mile Yule attributes to Polo, we quote from p. 592, Appendix K, Volume II. "For measures Polo uses the palm rather than the foot. I do not find a value of the Venice palm, but over Italy that measure varies from 9 1/2 inches to something over 10. The Genoa palm is stated at 0.725 inches." Yule then provides a table titled "Old Venice Measures of Length" with the following relationships:

4	fingers	=	1	handbreadth
4	handbreadths	=	1	foot
5	feet	=	1	pace
1000	paces	=	1	mile
4	miles	=	1	league

This is identical to the ancient Roman system in which--if the finger (or digit) is taken as 1/54 metre (0.0185185 m.)--the foot becomes 0.2963 m., (the Palesgic foot), the mile equals 1481.48 m., and the league 5.925 kilometres. In this system, 1.5 feet equalled the cubit of 0.4444 metres.

There was another system which came into use in Italy later which was based upon the palm rather than the foot. The palm, here, was the length of the hand from finger tip to, and including, the wrist joint. In this system, 2 palms equalled a brachium, and 3000 brachia made the Roman or Italian mile. Yule's attribution of 9.75 (English) inches, or 0.247 m., to the Genoa palm, would have yielded a brachium of 0.494 metres and a mile of 1482 metres (the most popular Roman mile). However, the indicated variation in the length of the palm "from 9 1/2 inches to something over 10," thorughout Italy, means that the Italian mile on this basis could have varied from 1448 metres to over 1524.

While some writers have indicated that they consider the brachium and the cubit one and the same, the origins of these arm measures are different. The cubit, which appears to be identical to the ell or elle, has from the most ancient times signified the length of the lower arm from the longest finger-tip to and including the elbow. The brachium, however, originally meant the length of the upper arm from elbow to shoulder. The Roman cubit was, as has been indicated, 1.5 feet or 24 fingers or digits and appears to have been pretty well pegged around 0.444 metres. The brachium, was quite variable throughout Italy, the smallest reported by Kennelly measuring 0.3476 metres and the largest 0.7443 m. (See pp. 108-124 *Vestiges of Pre-Metric Weights and Measures,* by Arthur E. Kennelly, Bureau of International Research of Harvard University and Radcliffe College, published by Little and Ives, New York, 1928.)

13. See p. 54, *Introductory Notices* , Vol.I, *The Book of Ser Marco Polo*, Henry Yule & Henri Cordier, Philo Press, Amsterdam, 1975.

Chapter 6

1. See pp. 95-97, *Geography in the Middle Ages,* by G. H. T. Kimble. Bjorn Landstrom, in his *Columbus, . . .* , Macmillan Co., N.Y., 1966, p. 15, indicated Mandeville was a Liege physician by the name of Jean de Bourgogne. *Tooley's Dictionary, of Mapmakers* , Alan R. Liss, N.Y., 1979, p. 417, gives Mandeville's lifespan as (ca. 1300-1372).

2. Pp. 69, 70, 72, *The World of Measurement,* by H. Arthur Klein, Simon and Schuster, New York, 1974.

3. Pp. 105-111, 142, and 143, *Inductive Metrology,* by W. M. Flinders Petrie.

4. Pp. 169-171, *Historical Metrology,* by A. E. Berriman, E. P. Dutton, New York, 1953.

5. P. 6, *The Discovery of America,* by G. R. Crone, Weybright and Talley, New York, 1969.

6. Ibid.

7. Footnote 87, pp. 81-83, *Toscanelli and Columbus,* by Henry Vignaud, first published in 1902, republished by Books for Libraries Press, Freeport, N.Y., 1971.

8. Footnote 107, pp. 100, 101, Ibid.

9. Ibid footnote 8.

10. See Morison's *Admiral of the Ocean Sea,* pp. 92, 95.

11. See pp. 246, 247, Vol. II., of Alexander Von Humboldt's *Cosmos,* translated from the German by E. C. Otte, Harper & Bros., New York, 1852.

12. See Morison's, *Admiral of the Ocean Sea,* pp. 93, 94.

13. See Morison, p. 94.

14. See p. 38, A. E. Nordenskiold's *Facsimile-Atlas,* orginially published in Stockholm in 1889, translated from the Swedish by Johan Adolf Ekelof (Royal Swedish Navy), and Clements R. Markham, republished in 1973 by Dover Publications, New York.

15. Quoting Morison (p. 171, *Admiral of the Ocean Sea*), "In the great days of sail, before man's inventions and gadgets had given him false confidence in his power to conquer the ocean, seamen were the most religious of all workers on land and sea. The mariner's philosophy . . (was taken) . . from the 107th Psalm: 'They that go down to the sea in ships and occupy their business in great waters; these men see the works of the Lord, and his wonders in the deep. For at His word the stormy wind ariseth which lifteth up the waves thereof . . .' "
    On pp. 171-173, Morison describes the prayers, public and private, said regularly throughout the day by Columbus and his crew.
    On pp. 80-81, 98-9, and 158-9, Morison describes the benevolent role played by the Franciscan frairs of La Rabida during the most trying period in Columbus' life, the years

1485-92, when he was struggling to sell his "Project of the Indies" to the Spanish monarchs, Isabella and Ferdinand.

Clearly, it was Columbus' complete familiarity with those elements of Scripture which supported his theses, his ability to quote d'Ailly and other churchmen at length, and his deep belief that the Lord would bless his project, that ultimately won the approval of the equally religious Queen Isabella and, through her, her husband Ferdinand.

16. Footnote 1, p. 250, and p. 258, *A History of Astronomy . . . ,* by J. L. E. Dreyer. Ulug Begh, who died in 1449, was Tamerlane's grandson and one of the last great patrons of Oriental astronomy.

Chapter 7

1. Nicholas of Cusa, 1401-1464, was a prominent writer on astronomical topics in his day. Without endorsing Aristarchus' heliocentric theory, he believed that the heavenly parade was at least partially due to the earth's motion. Regiomontanus, profiled in Chapter 1, is credited by Jean Fernel (see p. 383, J. P. J. Delambre's *Historie de l' Astronomie du Moyen Age,* V. Courcier, Paris, 1819; republished by Johnson Reprint Corp., N.Y., 1965) with reducing Eratosthenes' length of a degree from 700 stades to 640, or - as he interpreted the stade - from 87 1/2 Italian miles to 80, a significant improvement. Both Nicholas of Cusa and Regiomontanus were friends and admirers of Toscanelli.

2. First published in 1902, reprinted in 1971 by Books for Libraries Press, Freeport, N.Y.

3. All references to "miles", in the analysis of Toscanelli's Geographic Concepts, are intended to mean Roman or Italian miles of (approx.) 1480-2 metres, unless specifically indicated otherwise.

4. See pp. 301-307, *Christopher Columbus, His Life, His Work, His Remains,* by John Boyd Thatcher, first published in 1902 but reprinted by AMS Press, Inc. and Kraus Reprint Corp., New York, in 1967.

5. The remark in brackets is generally believed to be Columbus' comment. Las Casas, understanding this, has put it into brackets to distinguish it from Toscanellis' own words. Even so, it is strangely placed!

6. The great 18th/19th century naturalist and geographer, Alexander Von Humboldt, put Toscanellis' estimate of the sea distance from Portugal to China at 52° longitude. See his *Cosmos, A Sketch of a Physical Description of the Universe,* translated from the German by E. C. Otte, Vol. II, p. 268, Harper & Bros., New York, 1850.

7. Despite the rather clear intent of the wording, "And from the city of Lisbon straight toward the west . . . " not all reviewers accept the notion that Toscanelli meant the voyage westward from Lisbon would be made on the parallel of Lisbon or on a course the mean parallel of which was that of Lisbon's. The reason for this is explained in the text which follows (see "Toscanelli's Degree" and Figure 1, this chapter)).

8. P. 194, *"Toscanelli and Columbus. The Letter and Chart of Toscanelli,"* by Henry Vignaud.

9. The harbor entrance for Lisbon is actually at 38º 40' 33" N. latitude and approximately 9º 20' west longitude. The harbor entrance for Quinsay, modern Hangzhous (Hangchau) is at 30º 15' N. latitude and about 121º E. longitude. Thus, the harbor entrance of Quinsay is really almost 230° west of Lisbon by sea and 130° east of Lisbon by land, just the opposite of Toscanelli's view.

10. G. Uzielli, author *Vita e i tempi di P. dal P. Toscanelli*, Rome, 1894. Contemporary and correspondent of Henry Vignaud.

11. Pp. 370-372, Volume I, of Thatcher's *Christopher Columbus* . . .

12. Pp. 173-177, and 193-196, of Henry Vignaud's *Toscanelli and Columbus.*

13. Ibid, p. 200: "Now, the navigator's mile was the Roman or ordinary Italian mile, of which it took four to make a league. Everyone is agreed on this point. The Italians, who were always in considerable numbers in the Portuguese service had spread this system of reckoning since the 13th century, and Columbus knew no other."

14. See reference note 12, Chapter 5, for a discussion of the medieval Italian mile based on the foot and pace, and the palm and brachium.

15. P. 108, *Vestiges of Pre-metric Weights and Measures*, by Arthur E. Kennelly.

16. P. 302, *The Great Pyramid*, by Piazzi Smyth, first published in 1880 in the U.K.; re-published in 1978 by Bell Publishing Co., New York. All of the 8 braccios were cloth measures.

17. P. 111, *Vestiges of Pre-Metric Weights and Measures.*

18. P. 24, *Admiral of the Ocean Sea, A Life of Columbus*, by Samuel Eliot Morison.

19. Reference note 5, Chapter 3, describes a unit in pre-metric France called the ligne (line) which was equal to 1/144th the pied (foot) and was evaluated at 2.2558 mm, or 0.00226 metres. The Florentine filium at 1/12th the uncia (or 1/144th the Roman pes or Italian piede), measuring from 2.06 to 2.22 mm, may have been its counterpart in Tuscany during Toscanelli's time.

20. In 1669-70, the Abbe Jean Picard determined the length of a degree along the meridian of Paris, and in that general latitude, to be 57,060 toises de Paris (about 111.07 kilometres). This information was promptly disseminated to the rest of Europe by the French Academy of Sciences.

21. Pp. 75-78, Morison's *Admiral of the Ocean Sea* . . . Fig. 4 (herein), map of "Ocean Section of Behaim's Globe of 1492" is taken from pp. 66-68 of this book.

22. See pp. 174, 175, 180-182, 191, 204-206, 212, *Toscanelli and Columbus.*

23. Doppelmayer's work was titled, *Historiche Nachricht von den nurnbergischen Mathematicis und*

Kunstlern, Nurnberg, 1730. See Figure 40 and pp. 72b and 73a, *Facsimile-Atlas,* by A. E. Nordenskiold, originally published in English in Stockholm in 1889; republished by Dover Publications, New York, in 1973.

24. Ibid, pp. 73b, 74a (Nordenskiold).

25. Von Humboldt, who has seen Behaim's globe, says Behaim put the coast of China only 100° west of the Azores, or about 119° 40' west of Cape St. Vincent. Ptolemy (see our Figure 3b, Chapter 5) had put Cape St. Vincent about 3° 10' west of the mouth of the Tagus. Thus, allowing for the distance inland from the Chinese coast to the port, Von Humboldt appears to support a Behaim estimate of about 123°-124° from Lisbon to Kinsai. See Von Humboldt, p. 269.

26. P. 206, *Toscanelli and Columbus.*

27. See Chapter 8, *Spanish Ruling at Badajoz,* and in particular p. 136, of *The Diplomatic History of America,* by Henry Harisse, B. F. Stevens, Publisher, London, 1897; or William C. Brown Reprint Library, Dubuque, Iowa. See also pp. 15-16, Benjamin Keen's translation of *The Life of the Admiral Christopher Columbus By His Son Ferdinand,* Rutgers University Press, New Brunswick, 1959.

Chapter 8

1. Henry Harisse's *Diplomatic History of America,* was first published by B.F. Stevens, London, in 1897 and reprinted by William C. Brown Reprint Library, Dubuque, Iowa. (Date uncertain) See pp. 91-97 and footnotes 97-101, inclusive.

2. Since we do not know precisely where Ferrer placed the zero meridian (some point in the Canaries) we follow Harisse in using Greenwich to locate Fogo and the Line of Demarcation.

3. In note 106, p. 191, Harisse indicates Bandini's *Vita de Vespucci,* p. 72, as the source for the statement that the Portuguese pilots of that time (1500) " . . . gave 16 2/3 leagues to a degree on the equator . . . "

    The Cantino map of 1502, Figure 2, contained the Line of Demarcation, 370 leagues from the island of Fogo in the Cape Verdes. As explained in the "Extended Comment", accompanying the map,an "order-of-magnitude" scaling results in 16.3 leagues to the degree at Equator, a rough confirmation of the 16 2/3 equivalency.

    After the Treaty of Tordesillas, another consideration weighed heavily on the minds of the Spanish and Portuguese authorities responsible for setting the "official" relationship between the league and the degree (such as would be required for the development of sea charts). That was: which relationship between league and degree would result in a greater amount of newly discovered, or newly conquered, land falling into the Spanish or Portuguese spheres of influence. This became a two-edged sword as time went on and Spain developed interests in Asia and Oceania, and as Portugal developed them in the New World, for the "Line" became a great circle, extending across both poles from the Atlantic to (at first) Asia and (then) the Pacific, as the length of the Eurasian continent was found to be much less than supposed.

4.   The contradictory statements are not at all atypical for a committee-type product, put together by people with different views, each insistent upon having some input to the final position.

5.   See p. 202, *Explorers' Maps*, by R. A. Skelton , Frederick A. Praeger, New York, 1958.

6.   See pp. 382-385, *Histoire de l'Astronomie du Moyen Age* , by J. B. J. Delambre, V. Courcier, Paris, 1819, republished by Johnson Reprint Corp., New York, 1965.

7.   See p. 599, *Histoire de l'Astronomie Moderne,* by J. B. J. Delambre, Paris, 1821, republished by Johnson Reprint Corp., New York, 1969.

8.   Pp. 150 and 153, *The Diplomatic History of America.*

9.   See p. 196, *Explorers' Maps.*

10.   The voyages to the Solomons and the Marquesas could not have followed a route along the Equator, because of the strong Equatorial counter-current. It is most likely, because of the latitudes of Callao and the eastern-most of the Solomons that the course followed was at about 10° S. Therefore, the "1700 leagues/85 degrees" reported by Skelton would lead to a length for the degree at the Equator of $20/ \cos 10° = 20/0.9848 = 20.31$ leguas. We do not think Gallego was using any such standard, but that of 20 leguas to the degree at the Equator, and ignoring the shortening of the degree (1 1/2%) at latitude 10° S. Figure 20 illustrates the "Principal Ocean Currents of the World."

11.   " . . The telescope was probably invented in 1608 by Hans Lippershey, a Dutchman, but there is evidence that a telescopic device was used in England in the latter years of the 16th century. Galileo heard about the invention and built a telescope based on the known principles of optics. Galileo's fame lies in his use of the telescope to observe the heavens, not its invention." So says George Reed in "Galileo Galilei . . . , Man and Myth," in Griffith Observer, January, 1985, Vol. 49, No.1, p. 6; published by The Griffith Observatory, Los Angeles. The consensus among historians of astronomy is that Galileo developed the instrument which he used to such enormous advantage, regardless of its reputed origin elsewhere.

12.   See pp. 225-240 of Lloyd A. Brown's *The Story of Maps,* Dover Pubs., N.Y., 1979, for a fuller account of the extraordinary effort mounted by the Harrisons and the British Admiralty to develop a reliable chronometer for use at sea.

13.   Good coverage of the development of the metric system can be obtained in the following references:

    •   Berriman, A. E., *Pre-Metric Metrology in France (Chapter X)* and"*Chronological History of the Metric System* (Chapter XI), *Historical Metrology,* 1st Ed., E. P. Dutton & Co., Inc., New York, N.Y.,1953.

    •   Brown, L. A., *The Story of Maps,* Dover Pubs, Inc., New York, N.Y. 1979, first published by Little, Brown & Co., Boston, Mass. in 1949, pp. 24-34, 47-80, 180-202, 208-225, 240-264, 286-295.

- Crone, G. R., *Maps and Their Makers*, Dawson-Archon Books, Hamden,Conn., 5th Ed., 1978, 43-55, 85-91.

- Delambre, J. B. J., *Histoire de l'Astronomie Moderne*, Paris, 1821,Johnson Reprint Corp., New York, 1965, pp. 92-119.

- Fischer, I. K., *The Figure of the Earth - Changes in Concept*, Geophysical Surveys, Vol. 2, No. 1, 1975, 3-54.

- Hallock, W. and Wade, H. T., *The Evolution of Weights and Measures and The Metric System*, The Macmillan Co., New York, N.Y., 1906, pp 36-63.

- Kish, G., La Carte, *Image des Civilisations*, Seuil, Paris, 1980, pp 38-41, 191-192, 268-280.

- Mason, A. H. and Swindler, W. F., *Mason and Dixon; Their Line and Its Heritage*, American Heritage, 15 February 1964, 23-29, 93-96.

- Wilford, J. N., *The Mapmakers*, Alfred A. Knopf, New York, N.Y., 1981, pp. 93-127, 174-189, 216-220.

Appendix A

1. G.H.S. Bushnell, *The First Americans*, McGraw-Hill, New York, 1978, p. 10; published by Thames and Hudson, London, 1968.

2. Ivan Van Sertima, *They Came Before Columbus*, Random House, 1977. The ruins of Tres Zapotes (in southeastern Vera Cruz) and La Venta (western Tabasco) are both on the Isthmus of Tehuantepec, close to the Gulf of Mexico.

3. R.A. Jairazbhoy, *Ancient Egyptians and Chinese in America*, George Pryor Associated Publishers, London, 1974.

4. Charles G. Leland, *Fusang, or The Discovery of America* by Chinese Buddhist Priests in the Fifth   Century, Curzon Press, London, 1875; New Impression, 1973.

5. Dr. Marshall McKusick, *The Bimini Underwater Discoveries* , The Explorers' Journal, New York, March 1980, pp. 42,43; and *Atlantic Voyages to Prehistoric America*, Illinois University Press, Carbondale, 1980.

6. Barry Fell, *America B.C.*, Quadrangle/N.Y. Times Book Co., 1976.

7. Tristan Jones, *B.C. Mariners Crossed the Pacific*, The Explorers Journal, March, 1977. The article is adapted from his book, *The Incredible Voyage: A Personal Odyssey*, Sheed, Andrews and McMeel, Shawnee Mission, Kansas, 1977.

8. Thor Heyerdahl, *Early Man and the Ocean*, Vintage Books, New York, 1978.

9. Figure 20, Chapter 8, *Principal Ocean Currents of the World* identifies the Canary, North Equatorial (Atlantic), and Gulf Stream - used by Columbus; North Equatorial (Pacific) - used by Saavedra; Kurishio (Japanese), North Pacific, and California - used by Urdanetta; and South Equatorial (Pacific) - used by Mendana.

10. G.R. Crone, *The Discovery of America*, Weybright and Talley, New York, 1969, pp. 15, 16.

11. Ibid, pp. 16-25.

12. Not the cartographer who was to accompany Columbus on his second voyage of exploration and discovery and subsequently become famous for his *Map of the World*. See reference note 14 below.

13. Arguments continue to be advanced that the landing occurred at other islands in the Bahama group, among them Eleuthera, East Caicos, Cat Island, Samaña and Rum Cays, Grand Turk, Mayaguana, and Concepcion. Arne B. Molander makes a convincing case for Egg/Royal Islands just west of the northern tip of Eleuthera, 160 nautical miles northwest of Morison's choice, Watling Island. (See *Columbus Landed Here - Or Did He ?* , Americas, 1981, and, *A New Approach to the Columbus Landfall* by Molander. The latter article was included in *In the Wake of Columbus - Islands and Controversy*, edited by Louis DeVorsey, Jr. and John Parker, Wayne State University Press, Detroit, 1985.)
    A strong argument is advanced for Samana Cay by Joseph Judge and Luis Marden in a series of illustrated articles under cover title *Columbus and the New World*, National Geographic Magazine, November 1986. See Figure 4b for the Morison (Central Route), Molander (Northern Route) and Judge-Marden (Southern Route) landfalls and routes through the Bahamas, as well as the location of the other landfalls which have also been suggested.See article by John Noble Wilford in New York Times, 13 Oct 1985.

14. Some Columbian biographers indicate that cartographer Juan de la Cosa's association with Columbus was limited to the 2nd voyage. Not so, says the director of the Museo Naval (Madrid), Sr. Ricardo Cerezo. De la Cosa accompanied Columbus on the first two voyages.

15. Enough gold dust to fill one large hawk's bill, or 25 pounds of cotton.

16. About 25 miles east of Isabela, on the north coast.

17. Isabela Nueva was the new town his brother Bartolomeo had founded on the River Ozama, already re-named Santo Domingo, unbeknownst to Columbus.

18. The map illustrating *The Age of Discovery* in Rand McNally's *World Atlas, Family Edition*, Chicago, 1972, shows Vespucci making 3 voyages of exploration to the New World. The earliest voyage, in 1497-1498, supposedly explored the entire Caribbean coast of Central America, the Gulf of Mexico coasts of Mexico and the U.S., and the Atlantic coast of the U.S. as far north as Cape Hatteras. However, this voyage is indicated as "conjectural"! Both Morison and Landstrom indicate Vespucci's earliest voyage to have been in 1499-1500, but that it was made with Alonso Hojeda. However, the descriptions of the voyages made by each in

1499-1500, i.e. where each went, what each did, and the timing of the events, make it unlikely that a joint Hojeda-Vespucci voyage occurred. This is also the view of the Rev. J. F. Bannon, S.J. (See pp. 55-56, *Latin America, an Historical Survey,* Bruce Publ. @ Milwaukee, 1963).

Appendix B

1. Ronald Edward Zupko, *British Weights and Measures, A History from Antiquity to the 17th Century*, pp. 6-7.

2. George H. T. Kimble, *Geography in the Middle Ages*, p. 4, Methuen, London, 1938.

3. W. M. Flinders Petrie, *Inductive Metrology*, see "Addenda"; Hargrove Saunders, London, 1877.

4. H. Arthur Klein, *The World of Measurements*, pp. 70, 72; Simon & Schuster, New York, 1974.

5. Republished in 1979 under title of *The Complete Encyclopedia of Illustration*, J. G. Heck, Editor, by Park Lane, New York. The map section referred to are pp. 163-214. See, in particular, Plate 154, *Europe Before the French Revolution of 1789.*

6. A. E. Berriman, *Historical Metrology*, p. 136, E. P. Dutton, New York, 1953.

7. See illustration on p. 247 of Lloyd A. Brown's *Story of Maps*, Dover Publications, New York, 1979.

8. Henry Vignaud, *The Letter and Chart of Toscanelli*, pp. 196-200; Douglas Phillips-Birt, *A History of Seamanship*, p. 172, Doubleday, Garden City, N.Y., 1971; Samuel Eliot Morison, *Admiral of the Ocean Sea, A Life of Christopher Columbus*, pp. 65, 190, 191, Little Brown, Boston, 1942; John Boyd Thatcher, *Christopher Columbus*, pp. 306, 315, AMS Press, New York, 1967; Bjorn Landstrom, *Columbus, the Story of Don Cristobal Colon, Admiral of the Ocean Sea and His Four Voyages Westward to the Indies,*" MacMillan, New York, 1966; see Frontispiece.

9. *A Kino Guide, A Life of Eusebio Francisco Kino, Arizona's First Pioneer and A Guide to His Missions and Monuments*, Southwestern Mission Research Center, Tucson, 1976.

10. Published by Landmark Enterprises, Rancho Cordova, California, 1977

11. Full title *Equivalencias entre Las Pesas Y Medidas Usadas Antiguamente en Las Diver Provincias de Espana Y Las Legales del Sistems Metrico-Decimal,* published by the Direccion General del Instituo Geografico Y Estadistico, Madrid, 1886.

INDEX
(Including Glossary)

Abulfeda (Edmadeddin Ismael Aboulfeda of Damascus, 1273-1331), Arab author of *Chronicle* and *Geography*, claimed Arabian and Roman miles were identical; his works were published in Europe, *Geography* translated into French by Joseph Toussaint Reinaud and published in Paris, in 1840.
Chapter 4, pp. 122, 124, 125 (Table 3) . Reference Note 41.

Acharya, Bhaskara, fl. c. A.D. 1150, Indian astronomer and writer who estimated the diameter of the earth to be 1600 yojans and the distance of the moon from the earth's center as 64.5 times the earth's radius.
Chapter 4, p. 133. Reference Notes 47, 48.

Ahmad, Nafis, Pakistani authority on history of Islamic contributions to astronomy, geography and related sciences. Author *Muslim Contribution to Geography* , Ashraf Press, Lahore, Pakistan, 1972.
Chapter 1, Reference Note 27. Chapter 4, pp. 112, 113, 118-119, (Table 2), 123,125 (Table 3), 132, 133. Reference Notes 14, 19, 30, 42, 43, 45, 48.

Al-Biruni, see Degree Measurement.

Alfragan, name corrupted by d'Ailly from Al-Farghani. Born Ahmed Ben Kebir, this 9th century astronomer and writer became identified with the Arab measurement of a degree of latitude in the Syrian desert because he described it in his *Chronologica et Astronomica Elementa*. This work also declared Ptolemy's degree to be too large, in the ratio of 66 2/3 to 56 2/3. His views were widely reported. Columbus found them in d'Ailly's *Imago Mundi* and adopted them.
Chapter 1, pp. 34, 37, 38, 39. Chapter 4, pp. 120, 122, 124, 125 (Table 3), 129, 134. Reference Note 40. Chapter 5, pp. 137, 139, 158. Chapter 6, pp. 161, 164. Chapter 7, pp. 167, 168,181, 183, 184. Chapter 8, pp. 190, 191.

altitude is the angular distance of a celestial body above (or below) the horizon, measured along the great circle passing through the body and the zenith.

Ananias of Shirak (Siracki). 7th century Armenian writer credited with a table of measures which has been interpreted differently by Hans von Mzik, Aubrey Diller and Eva Taylor.
Chapter 4, pp. 114-119 (incl. Tables 1 and 2), 129. Reference Notes 24, 25, 29.

Anaximander, Greek philosopher, 611-547 B.C., born at Miletus, Asia Minor. Student of Thales, but differed from his mentor as to the shape of the earth. An able astronomer and geographer, he taught obliquity of the ecliptic, invented a celestial globe, and introduced the gnomon and sun dial into Greece.
Chapter 3, p. 74.

Anaximenes, Greek philosopher, fl. 6th cent. B.C., became a student of Anaximander, but rejected his views as to the shape and constitution of the earth in favor of his own theory. The first to record in prose a systematic description of the world as then known to the Greeks.
Chapter 3, p. 74.

Antilia, a fictitious island shown on many 14th and 15th century portolanos (sea charts).
Chapter 1, pp. 2, 3, 34, 36-38. Chapter 7, pp. 166, 167, 169 (Fig. 1), 179, (Fig. 3), 180 (Fig. 4), 181.

Apianus, Petrus, 1531-1589, German cosmographer, cartographer, and mathematician. Produced world map in 1520 showing length of Eurasian continent about 235° of longitude at latitude 36° N, and appears to have intended to make a degree of longitude at equator equal to about 62.5 Italian miliaria (Roman miles). Author of a cosmographic work which became a textbook in several European countries. Taught mathematics at famed academy at Ingolstadt. In a 1545 edition of his popular cosmography, he reduced the length of his degree to 60 miliaria, the mariners' measure also employed by C.C.
Chapter 8, pp. 203, 210 (Figure 8), 217, 220 (Figure 10).

Archimedes, 287-212 B.C., born in Syracuse, Sicily, became the most famous of ancient mathematicians. First to propound the theory of the lever, of hydrostatics, of methods of determining parabolic areas, of the surface and volume of spheres, and of the "Archimedean" screw for raising water. Is credited with estimating the earth's circumference at 300,000 stades based upon a "double-measurement" between Lysimachia and Syene. Some, however, believe it may have been the work of Dikaearchus. (See Dikaearchus, under "Degree Measurements".)
Chapter 3, pp. 76, 77 (Figure 1), 104 (Table 3). Chapter 4, p. 109.

Aristotle (Aristoteles), 384-322 B.C., Hellenic philosopher born in Stagira, Macedonia. Went to Athens in 367 B.C., spending 17 years in association with Plato. In 343 B.C. he was recalled to Macedonia by King Philip to educate his son Alexander. In 334 B.C. Alexander passed into Asia "to subdue the world" and Aristotle returned to Athens to found his Peripatetic school at the Lyceum. He spent 12 years there, lecturing and absorbed in scientific research (supported by gifts from Alexander). Famous for his treatises *Meteorologica* , *Decaelo* , *Timaeus.* In *Decaelo* , he observed that certain mathematicians (probably referring to Eudoxus of Cnidus, 409-356 B.C.) had estimated circumference of earth to be 400,000 stades.
Chapter 3, pp. 75, 76, 104 (Table 3), Reference Note 9. Chapter 4, p. 109.

Aryabhata (fl. ca. A.D. 476), Indian astronomer of some repute who anticipated Copernicus. Made a geodetic estimate which set degree at 9.163 yojans and earth's circumference at 3298.672 yojans. This translates to 69.64 miles (statute) for the degree and 25, 069.91 miles for the earth's circumference according to Colebrook and Dreyer/Stahl. However, Ahmad gives Aryabhata's estimates as 92.158 statute miles for the degree and 33,177 miles for the earth's circumference.
Chapter 4, pp. 112, 113, 133. Reference Notes 13-18, incl.

Bacon, Roger, 1214-1294, English philospher, geographer and scientist. Wrote *Opus Majus* which, inter alia, sought to establish the nearness of Western Europe to Eastern Asia. This was copied by d'Ailly for his *Imago Mundi* , where Columbus found it and established it as the spirit, if not the letter, of his cosmography. Bacon invented the *Meridian Projection* (see Chapter 5, Figure 1) and drew a World Map and a *Map of Travels of Rubruck to Tartary, 1253-55*.

> Chapter 3, p. 88. Chapter 4, pp. 122, 125 (Table 3). Chapter 5, pp. 135,136 (Figure 1), 137, Reference Notes 1, 2. Chapter 6, p. 160.

Barley grain, used as a unit of length, usually 4 lengths of barley at 4.6 mm each, or 6 widths, at 3.1 mm, made the digit (approx. 18.5 mm.).

> Chapter 2, pp. 59, 60, Reference Note 20. Chapter 8, p. 211, Reference Note 6.

Barnard, Lewis, American industrialist (measuring devices), author "The Measurement of Length: Yesterday, Today, and Tomorrow", in the "Report of the 50th National Conference on Weights and Measures, 1965", National Bureau of Standards Miscellaneous Publication No. 272, Issued April 1, 1966, U.S. Govt. Printing Office, Washington, D.C.

> Chapter 2, pp. 44, 46 (Figures 1b, 1d, 1e), 50 (Figures 2a, 2b), 57 (Figure 3b), 58 (Figure 4), 69, Reference Note 43.

Behaim, Martin, 1459-1506, Nurenberg cosmographer, contemporary of C.C., with remarkably similar cosmographic ideas as Toscanelli and C.C. Famous for his globe, copies of which exist in several libraries or museums.

> Chapter 1, pp. 34, 35, 38, 39. Chapter 7, pp. 176-181, incl. Figures 3 and 4, 185-188 (Table 2).

Berriman, A.E., author *Historical Metrology,* E.P. Dutton, N.Y. 1953; excellent source on the history of weights and measures of the major civilizations of the world. Not, however, particularly helpful on itinerary units of linear measurement.

> Chapter 2, pp. 44-46 (Figures 1a, 1c), 47-49, 52, 53, 55, 57 (Figure 3a), 59, 63, 64 (Table 2), 66-70, 72,73 (Table 5), Reference Notes 3, 5, 8, 9, 11, 16, 31, 37, 39d, 41. Chapter 4, p. 111 Reference Notes 8, 18. Chapter 6, p. 160. Reference Note 4. Chapter 8, Reference Note 13a. Appendix B, pp. 286, 304-305. (Table 2).

Brown, Lloyd A., late Librarian, Peabody Institute, Baltimore, and Curator of Maps, Clements Library, Univ. of Michigan. His *Story of Maps*, Little, Brown and Co., Boston, 1949, re-published by Dover Publications, New York, 1979, is at once a history of geography, geodesy, cartography, navigation and elements of astronomy.

> Chapter 8, Reference Notes 12, 13. Appendix B, Reference Note 7.

Brahmagupta, Indian astronomer and mathematician, fl. ca. A.D. 628, who estimated earth's circumference to be the equivalent of 50,936 statute miles-according to Ahmad. This would put a degree equal to 141.49 miles - somewhat over twice its actual length.

> Chapter 4, p. 113, Reference Note 19.

Bunbury, E.H., 19th century British historian and author whose superlative work *A History of Ancient Geography* was first published (in two volumes) in London in 1879. The second edition, 1883, was re-published by Dover Publications, New York, in 1959. The work is as much a history of the classical era as it is a major source book on ancient geography.

Chapter 3, pp. 86, 90, 96 (Fig. 5, part 3), 97,99, 104 (Table 3). Reference Notes 2, 19, 26, 32.
Chapter 4, p. 114. Reference Note 22.

Capella, Mineo Felice Marziano (Martianus Capella), Carthaginian who achieved some fame in 3rd century Rome as an astronomer and geographer. Estimated earth's circumference at 31,500 Roman miles and a degree, at 87.5 Roman miles. Said Eratosthenes' degree had been 406,010 stades, and Ptolemy's 62.5 Roman miles.

Chapter 4, pp. 108, 109, Reference Notes 1-3, incl.

Cathay, the name by which Marco Polo and other 13th century travelers referred to a part of China north and usually west of Mangi. It is, however, variously placed by 15th and 16th century as well as modern geographers.

Chapter 1, pp. 2, 3, 34, 36-38. Chapter 5, pp. 139, 144 (Figure 2a). Chapter 7, pp. 166, 178 (Figure 3), 180 (Figure 4). Appendix A, p. 246.

Chipanju (Cipanju, Zipangu, etc), the name by which travelers such as Marco Polo referred to Japan. It was shown thus on 15th and 16th century maps.

Chap. 1, pp. 2, 3, 34, 36-38. Chap. 5, pp. 139, 144 (Fig. 2a), 154. Chap. 7, pp. 166, 167, 169 (Fig. 1), 175 (Fig. 2b), 176, 178-179 (Fig. 3), 180 (Fig. 4), 181, 184, 187 (Table 2, part 2). Appendix A, p. 246.

Cleomedes, disciple of Posidonius from Lysimachia, a Greek settlement in Asia Minor close to Sea of Marmara and south of Black Sea. An astronomer who wrote a treatise on the cyclical motion of celestial bodies, the moon's phases, eclipses, the planets, and atmospheric refraction. It was in this treatise that he describes Eratosthenes' and Posidonius' measurements of the earth's circumference, i.e. this was the basic source for information on these events. While there is no precise data on his life, it is likely that he flourished in 1st century B.C.

Chapter 3, pp. 78, 79, 86, 88, 104 (Table 3). Reference Note 9.

Columbus, Bartolomeo, brother of C.C. Though younger by about 1-2 years, preceded him into the chartmaking profession in Lisbon where there was a sizable Genoese colony. Shared C.C.'s cosmographic views and tried to interest Henry VII of England and Charles VIII of France in backing a westward voyage to the Indies. Failed. Was working in Fontainbleu as chartmaker when news of C.C.'s discovery broke. Commanded a re-supply fleet to Hispaniola after C.C.'s second voyage, arriving in 1494 to nurse his brother back to health. Was made Adelantado of Hispaniola when C.C. returned to Spain in 1496 to obtain the logistics to continue exploration, pacification, and colonization. Established new headquarters and port city at Santo Domingo. Following C.C.'s return, continued turmoil on Hispaniola resulted in his imprisonment (with brothers Christopher and Diego) and return to Spain in 1500. Participated effectively in C.C.'s 4th voyage of exploration to Central America and Panama. Was C.C.'s chief counselor and executive officer.

Chapter 1, p. 5, 9. Appendix A, pp. 267, 268, 276, incl. Figure 11a, 279, 283.

Columbus, Diego (Giacomo), youngest brother (by some 17 years) of C.C. Accompanied the Great Discoverer on his second voyage, 1493. Appointed president of a council to govern Hispaniola while C.C. was off exploring Cuba, Jamaica and part of Hispaniola. Was both a poor administrator and a poor sailor. Was jailed by Bobadillo in 1500 along with his brothers Christopher and Bartolomeo, and returned to Spain in irons. Was with C.C. at his death in 1506.

Appendix A, pp. 257, 266, 267.

Columbus, Domenico, c. 1418-1496, father of C.C., born in Genoa. At age of 11 became an apprentice weaver. By early manhood had advanced to master weaver. Married Susanna Fontanarossa about 1445. She bore him 5 children - Cristoforo (C.C.), Bartolomeo, Giacomo (Diego), Giovanni, and Bianchinetta. Family lived in Genoa until 1471 by which time Domenico had advanced to master clothier and sons Cristoforo and Bartolomeo had left home to try pursuits other than carding in which they had been trained. Family moved to Savona in 1471, where Domenico continued as a maker of cloth. Around 1473-4, both Susanna and Giovanni died. Shortly thereafter, what was left of the family still at home moved back to Genoa. Bianchinetta married and Giacomo left to follow his brothers. Domenico lived out his life in Genoa, supported by remittances from his sons.

Chapter 1, p. 6.

Columbus, Dona Felipa Perestrello e Moniz, wife of Christopher (married in 1479), mother of their son Diego, died in Porto Santo around 1482. Came from a prominent Portuguese family with excellent maritime connections. Father had been a squire to Prince Henry the Navigator, later Captain-General of Porto Santo.

Chapter 1, pp. 11, 13.

Columbus, Ferdinand, born out of wedlock to C.C. and Beatriz Enriquez de Harana in 1488. At age of 12 accompanied his father on 4th voyage. In 1509, accompanied his half-brother Don Diego, the 2nd Admiral and new governor to Hispaniola. Collected his father's library containing the annotated works which underlay C.C.'s cosmography, later known as *Biblioteca Columbina*. Before his death, he completed a biography of his father known as the "Historie"; first translation published in Venice in 1571. This work is considered one of the 4 major sources (along with Las Casas, Oviedo, and Peter Martyr) on C.C.'s life and accomplishments. An English translation by Benjamin Keen, *The Life of the Admiral Christopher Columbus by His Son Ferdinand* , Rutgers Univ. Press, 1959, is a major source for C.C.C.

Chapter 1; pp. 3, 4, 7, 37. Chapter 7, p. 183. Reference Note 27. Chapter 8, p. 211 Appendix A, p. 279.

Cosmographer: One who practices the science of cosmography, which describes and maps the main features of the heavens and the earth, including astronomy, geography, and geology. As used in this book, one with well developed views on the size and shape of the earth and the distribution of its lands and seas. With reference to the size of the earth, one who - preferably - defines the units of linear measurement employed in his pronouncements, oral or written. A term used most frequently in the Middle Ages, but applied in this book to personalities from antiquity to the Renaissance.

The following, listed separately in this Index, are considered to qualify as cosmographers: Acharya, Ala-Ed-Din Al Kusgi, Al-Biruni, Alfragan, Al-Idrisi, Ananias, Anaximander, Anaximenes, Apianus, Archimedes, Aristotle, Aryabhata, Bacon, Barentzoon, Behaim, Brahmagupta, Capella, Christman, Cleomedes, Columbus (Christopher and son Ferdinand), D'ailly (D'Aliaco), Dikaearchus, (de)Enciso, Eratosthenes, Eudoxus, Fernel, Ferrer, Herodotus, Hipparchus, Hoching-Tien, Isidore, Khurdadbih,

Macrobius, Marinus, Pliny, Posidonius, Ptolemy, Regiomontanus, Rustah, Sacrobosco, Simplicius, Strabo, Sylvani, Toscanelli, Y-Hang. (Some of the foregoing are listed under "Degree Measurements", others under "Great Circle Circumference (of Earth) Estimates, 330 B.C. to A.D. 1670."

Crone, G.R., British author: *The Discovery of America,* Waybright and Talley, New York, 1969; *Maps and Their Makers, an Introduction to the History of Cartography,* Archon Books, Hamden, Conn., first published in 1953, fifth ed. 1978. Formerly Librarian and Map Curator, Royal Geographical Society. Crone's bibliographies are especially helpful to the researcher.
     Chapter 6, p. 161. Ref. Notes 5, 6. Chapter 8, Ref. Note 13. Appendix A, p. 244. Ref. Note 10.

culmination = the instant a heavenly body crosses the meridian of the observer; more precisely, the passage through the point of greatest altitude in the diurnal (daily) path. For circumpolar stars or the moon, upper culmination, or transit, is the celestial body's crossing closer to the observer's zenith. Lower culmination is the crossing farther from the zenith.
     Chapter 3, pp. 76, 77 (Figure 1), 78, 79, 80 (Figures 2a, 2b), 81 (Figure 2c), 86, 87 (Figure 3), 88. Chapter 4, pp. 121, 122, incl. Figure 2. Chapter 7, pp. 182 (Figure 5), 183. Chapter 8, p. 217.

D'Ailly, Pierre (D'Aliaco, Pedro in Spanish), d. 1422, Cardinal of Cambrai, France, cosmographer, author of comprehensive world geography titled *Tractatus de Imagine Mundi (Imago Mundi),* and the single individual whose ideas impressed C.C. the most.
     Chapter 1, pp. 34, 37, 39. Reference Note 39. Chapter 3, p. 88. Chapter 4, pp. 124, 125 (Table 3), 129. Chapter 5, p. 136. Chapter 6, pp. 159-164, incl. Figure 1. Reference Notes 5, 7, 8, 9. Chapter 7, p. 184.

declination = the angular distance, of a heavenly body on the celestial sphere, north or south of the celestial equator. It is measured along the Hour circle or meridian passing through both poles and the heavenly body.
     Chapter 1, pp. 12, 19, 20, 22 (Fig. 4a), 23 (Fig. 4b), 24 (Fig. 5), 25 (Table 1).

DEGREE MEASUREMENTS (in chronological order)

Dikaearchus, 330-296 B.C., philosopher, astronomer, and geographer of Messene in the Pelopennesus, is believed by some to have made the "double-measurement" between Lysimachia, Thrace, and Syene, Egypt, credited to Archimedes. The difference in latitude of sites was determined from the difference in declination of the stars gamma Draco and delta Cancer, believed to pass through the zenith of Lysimachia and Syene, respectively. Surface distance between the two sites (believed to be on same meridian) estimated. Result: 833.33 stades per degree or 300,000 st. for meridional circumference.
     Chapter 3, pp. 76, 77 (Figure 1). Reference Note 7.

Eratosthenes, 276-194 B.C., astronomer/geographer of Cyrene and Alexandria, conducted "double-measurement" between Alexandria and Syene (believed to be on same meridian). Difference in the latitude of end sites was determined from the difference in zenith distances of the sun at the 2 sites at noon on the day of the summer solstice. Distance between sites was determined by pacing. Result: 700 stades per degree or 252,000 stades for meridional circumference. Another measurement conducted on day of winter solstice confirmed previous result. A skiotheran was employed to give direct readings of sun's zenith distance at noon.
     Chapter 1, pp. 35, 39. Chapter 3, pp. 78, 79, 80 (Figures 2a, 2b), 81 (Figure 2c), 82.

Posidonius, 135-50 B.C., philosopher/astronomer of Apameia and Rhodes. Double measurement between Rhodes and Alexandria. Difference in latitude of the sites determined from the difference in the altitude of bright star Canopus, at culmination, at each of the sites. The surface distance between the sites was estimated. Result: 666 2/3 stades per degree/240,000 stades per meridional circumference - according to Cleomedes; 500 stades per degree/180,000 stades per meridional circumference -according to Strabo. See
 Chapter 3, pp. 86, 87 (Figure 3), 88, 89.

Simplicius, 5th century writer born in Cilicia who wrote a critique on Aristotle's *Decaelo* which is highly quoted, claims to have measured a degree by the following method. Employing a diopter, he found two (un-named) stars whose declinations differed by exactly one degree. Then, using same diopter, he found two sites on a common meridian such that the two stars passed through the zeniths of the two sites (i.e. the southern star through the zenith of the southern site and . . . .). Measurement of the distance between the two sites showed it to be 500 stades. Thus, he concluded a degree of latitude measured 500 stades and a meridional great circle 180,000 stades. See Chapter 4, pp. 111,112.

Hoching-Tien, 5th century Chinese astronomer, made a double measurement between the villages of Tong-Feng and Tong King at noon on the day of the summer solstice, employing 8 foot tall gnomons. The difference in latitude between the two sites (believed to be on a common meridian) determined from the length of shadow each gnomon cast (see reference for details) turned out to be 8°13. The distances between the two gnomons was 1000 lys (lis). Thus, the degree was 123 lys and a meridional great circle 44, 280 lys.
 See Chapter 4, p. 113, 118-119 (Table 2).

Y-Hang, a Buddhist priest with astronomical competence, in 721 A.D. directed a large geodetic project in which the latitudes of a number of sites within China proper and Cochin China were to be determined. Two methods were employed at each site, observation of the altitude of the pole star - corrected for circumpolarity, and observations of the sun's altitude at noon - corrected for the declination of the sun for that day. During the course of their work, the field parties were to measure the surface distance between two sites on the same meridian for which they had determined latitudes. A degree was thus found to have a length of 351 lys and 80 paces.
 See Chapter 4, pp. 113, 114, 118, 119 (Table 2).

The Arab Measurement at Sinjar, 830 A.D., At the behest of the Caliph Al-Ma'mun, 2 double measurments were carried out simultaneously from a point in the Syrian desert at about 35° N. latitude. Having noted the altitude of the North Star at upper culmination, one party set out, proceeding due north, the other due south - each measuring the distance travelled. When the northern party was able to detect an altitude of the North Star (at upper culmination) of one degree more than the original reading, and the southern party one degree less, they stopped and recorded the distance each party had travelled. One result was 56 2/3 Arabian miles, the other 56 miles. By consensus the correct measurement was set at 56 2/3 Arabian miles for the degree, making a meridional great circle 20,400 Arabian miles.
 See Chapter 4, pp. 120-129, incl. Fig. 2, and Tables 3 and 4.

Al-Biruni, ca. 1038, confirmed (generally) the results of the Sinjar measurement by a non-astronomical double-measurement.
 Chapter 4, p. 131 (Figure 3). See also Chapt. 4, pp. 129 - 132.

Fernel, Jean, b. 1485 at Montdidier, France, became a prominent physician with a penchant for astronomy. During the period 1526-28, he carried out a double-measurement between two points on the Paris-Amiens road where that road ran "true north and south". The difference in latitude between his end observation points was obtained from the difference in the sun's altitude at noon at the two sites, four days apart. Allowance being made for the difference in the sun's declination on these days, the difference in latitude was found to be 1° 2' 35". The surface distance between his end points was determined by counting the number of revolutions of his 6 (Italian) foot, 6 digit, carriage wheels (equivalent to a circumference of 4.0055 Italian paces. Fernel's result for the length of a degree of latitude was 68,095 1/4 paces, equivalent to 68.09525 Italian miles. The circumference of a meridional great circle was thus taken as 24,514.2857 Italian miles.

    See Chapter 8, pp. 216, 217. Reference Note 6.

Picard, Jean, 1620-1682, French abbe, astronomer, geodesist, working under the auspices of the French Academy of Sciences, made a double measurement of a degree on the Paris-Amiens road, employing triangulation to determine the distance between his end points. His result was 57,060 toises = 111,212 metres.

    Appendix B, p. 296.

Delambre, Jean Baptiste Joseph, 1749-1822, Parisian astronomer, geodesist and author: *Base du Systeme metrique*, 1810; *Histoire de l'Astronomie Ancienne*, 1812; "Histoire de l'Astronomie du Moyen Age", 1819; *Histoire de l'Astronomie Moderne*, 1821. Participated in measurements of the length of a degree of latitude at different segments of the meridian of Paris, including triangulation (with P.F.A. Mechain, in 1792-1798, between Dunkirk and Montjouy (Barcelona), a distance of 1,075.1 Km. The difference in latitude of these endpoints being 9.6738 degrees, the mean length of a degree over this span turned out to be 111,131 metres. Measurements were actually made in toises, for the Mechain-Delambre effort, together with similar efforts in the Arctic and near the Equator, had as their objectives the determination of the shape of the earth (spherical, oblate, or prolate) and the length of a quadrant of the Meridian of Paris (from N. Pole to Equator), one-ten millionth of which would define the length of one metre.

    Chap 2, Ref. Note 20. Chap. 3, p. 78. Ref. Notes 9, 10, 12. Chap. 4, pp.108, 112, 113 . Ref. Notes 2, 3, 11, 20, 21, 40. Chap. 7, Ref. Note 1. Chap. 8, p. 217. Ref. Notes 6, 7, 13.

End of DEGREE MEASUREMENTS

Diller, Aubrey, former Indiana Univ. professor and writer on history of science, esp. geography and metrology. Considered an authority on ancient and medieval metrology. See *Ancient Measurements of the Earth*, ISIS, Feb. 1949; *Geographic Latitudes in Eratosthenes, Hipparchus, and Posidonius*, KLIO, 27, 1934.

    Chap. 2, p. 62, Ref. Notes 29, 30. Chap. 3, pp. 75, 88, 90, 99, 103 (Table 2), 104 (Table 3), 105. Ref. Notes 3, 16, 30. Chap. 4, pp. 114-120, incl. Tables 1 and 2, 125. Ref. Notes 24, 25, 27.

Drabkin, I.E., professor of mathematics, College of City of New York, who with the late Morris R. Cohen (prof. of philosophy, C.C.N.Y. and Univ. of Chicago) wrote *A Source Book in Greek Science*, Mc-Graw-Hill, 1948. Drabkin exhibits the same degree of sophistication about early geodesy and metrology as Aubrey Diller. (See *Posidonius and the Circumference of the Earth*, ISIS, 1943.)

    Chapter 3, pp. 82, 88, 99, 102-103 (Table 2), 104 (Table 3). Reference Notes 9, 22, 27, 31.

Dreyer, J.L.E., 1852-1926, born and educated in Copenhagen; Director, Armagh Observatory for 34 years; author of several works on astronomy, his *History of Planetary Systems from Thales to Kepler,* Cambridge Univ. Press, 1906 is considered a classic in its field. In 1953, revised slightly by W.H. Stahl of New York University, the book was re-issued by Dover Publications, New York, re-titled as *A History of Astronomy from Thales to Kepler.* It is of particular value to students of the history of geodesy for its coverage of efforts to determine the dimensions of the earth. Dreyer's treatment of metrology is sophisticated.

Chapter 1, Reference Note 29. Chapter 3, pp. 75, 76, 86, 98, 99, 102-103 (Table 2), 104 (Table 3). Reference Notes 4, 7, 14, 15, 27. Chapter 4, pp. 112, 118-119 (Table 2), 123, 125 (Table 3), 133. Reference Notes 12, 13, 38, 39, 44, 47. Chapter 6, p. 164. Ref. Note 16.

Earth, shape of: throughout this book earth is treated as a sphere by each of the philosophers, mathematicians, astronomers, geodesists, geographers, and seamen who attempted to measure the length of a degree of latitude or estimate the circumference of a great circle of the earth's surface.

Equinox = either of the points on the celestial sphere at which the ecliptic intersects the celestial equator; also the time at which the sun passes thru either of these intersection points. On these days (usually 20 or 21 March and 22 or 23 Sept) days and nights are of equal length, the declination of the sun at noon is a minimum for the year (very close to zero) and the zenith distance of the sun at noon is very close to equalling the latitude of the site.

Chapter 1, pp. 20, 24 (Figure 5), 25 (Table 1), 26 (Figure 6a). Chapter 3, pp. 81 (Figure 2c), 93 (Figure 4). Reference Note 25.

ERATOSTHENES, 276-194 B.C., Alexandrian philosopher, mathematician, astronomer, geographer and head of the famed Alexandria Museum and Library. Considered to be the first mathematical geographer, he has been called 'the father of geodesy' for his celebrated "measurement " of the earth's meridional circumference. This effort, despite several minor errors, was brilliantly conceived and yielded a result not definitely improved on for almost two millenia. Likewise, his estimate of the length of the Eurasian continent - despite the sometimes gross unreliability of his sources - would not be improved on for almost the same period.

Double-measurement of the length of a degree of latitude, conducted between Syene (Aswan) and Alexandria, Egypt: Chapter 3, pp. 78, 79, 80 (Figures 2a, 2b), 81 (Figure 2c).
Interpreting Eratosthenes' Result: Chapter 3, pp. 79, 82
Eratosthenes' Geographic Views: Chapter 3, pp. 82, 83, 84 (Figure 2d), 85.
Comparing Eratosthenes' and Ptolemy's Views: Chapter 3, pp. 84 (Figure 2d), 94 - 96 (Figure 5), 98 - 106, incl. Tables 1, 2, and 3. Chapter 4, pp. 118-119 (Table 2). Chapter 5, pp. 150-151 (Table 1). Chapter 7, pp. 185-188 (Table 2).

Misc. References to Eratosthenes:
Chapter 1, p. 35, 39. Chapter 3, pp. 79, 82, 85, 86, 88-91. Chapter 4, pp. 109, 111, 103-115. Chapter 5, p. 137. Chapter 8, pp. 191, 192, 212, 216.

Eudoxus, 409-356 B.C., born at Cnidus, Asia Minor, considered by Cicero (Roman orator, statesman, and man of letters, 106-43 B.C.) to be the greatest astronomer up to Cicero's time. Dreyer rates Eudoxus as the first scientific astronomer. Had an observatory in Cnidus, but studied at Athens and in Egypt. Credited by Dreyer and other astronomy-historians as the likely source of Aristotle's estimate of 400,000

stades for the earth's circumference.
Chapter 3, p. 75.

EXPLORATION:

Garcilaso de la Vega, el Ynca, descendant of Royal Incas, chronicler of Spanish conquests on the mainlands of America, including *Histoire des Yncas . . ., Histoire de la Conquete de Floride, La Florida del Ynca,* et al. An authority on Hernando de Soto, and well versed on Spanish operations in America up to about 1600. Apparently, also well versed in Spanish metrology. Wrote during end of 16th and beginning of 17th centuries.
Chapter 1, p. 8. Reference note 8. Appendix B, pp. 293, 296, 304-305 (Table 2).

geocentric coordinates = the latitude and longitude of a point on the earth's surface relative to the center of the earth.

Geodesy: The branch of science concerned with measuring, or determining the shape of, the earth or a large part of its surface, or with locating exactly points on its surface. Geodesists in this book are most easily found under "Degree Measurement", in Index.

geodetic coordinates = the latitude and longitude of a point on the earth's surface determined from the geodetic vertical (normal to the specified spheroid).

geographic coordinates: Every point on the surface of the earth, considered as a sphere, can be located by a system of spherical coordinates based on two systems of circles, meridians and parallels. Meridians of longitude are (imaginary) great circles on the surface of the earth, passing through both poles, generally evenly spaced, and related to a zero meridian - since the late 19th century, passing through Greenwich, England. Parallels of latitude, generally evenly spaced around the meridians and perpendicular thereto at the points of intersection, are related to the zero parallel, the equator. Meridians of longitude are numbered from zero degrees at the equator to 90 degrees north latitude at the North Pole and 90 degrees south latitude at the South Pole. The equator divides the earth into two hemispheres, north and south. The zero and 180 degree meridians can, likewise, be considered to divide the earth into eastern and western hemispheres.
Hipparchus is credited with originating the method of fixing terrestrial positions by means of circles of latitude and longitude. Marinus conceived the idea of a zero meridian which he placed at the Fortunate

(Canary) Islands, believed by him to lie 2 1/2 degrees west of the Sacred Promontory (Cape St. Vincent). Ptolemy, however, wrote a *Geography* (based largely on Marinus, Hipparchus and, through Strabo, Eratosthenes.) This was the first effort to put geography on a scientific basis. He also was the first to describe how to construct a map depicting the earth's spherical surface (up to a hemisphere) on a plane. Today, there are many different projections for depicting terrestrial surfaces of various sizes and shapes, and in various locations, on flat maps with (hopefully) minimum distortion.

Geography: The descriptive science dealing with the surface of the earth, its division into continents, and countries, and the climate, plants, animals, natural resources, inhabitants, etc. The physical features, especially the surface features of a region, area, or place. Geographers of note whose works are described or illustrated in this book are Herodotus, Eratosthenes, Strabo, Marinus, Ptolemy, Alfragan, Marco Polo (perhaps more traveler than formal geographer but who nevertheless influenced 14th, 15th and 16th century cartography enormously), Toscanelli, Martellus, de la Cosa, Contarini, Waldseemuller, Ruysch, Sylvani, Stobnicza, Apianus, Ribeiro, Finaeus, Grynaeus, Mercator (Gerardus), Ortelius, Mercator(Rumold), De Judaeis, Barentzoon and Nordenskiold. Some of the foregoing might more accurately be described as cartographers - those who make maps and charts.

Gillings, Richard, J., author, *Mathematics in the Time of the Pharaohs,* first published by MIT Press, Cambridge, Mass., 1972, re-published by Dover Publications, New York, 1982.
    Chapter 2, pp. 59, 60, 69, Reference Notes 17, 18, 19, 21, 28, 42.

Gnomon, a vertical rod used, first by the Babylonians and later the Egyptians and Greeks, for a variety of purposes by measuring the length of the rod's shadow, cast by the sun. At any time of day, the ratio of the sun's shadow to the rod's length equals the tangent of the sun's zenith distance (angle), and the sun's altitude equals 90° - sun's zenith distance. At noon, with a knowledge of the sun's declination, the latitude of a site is given by the zenith distance (angle) plus the declination. The declination of the sun on any day of the year, at noon, can be determined by noting the shadow's length each day for a year, at noon, and calculating the zenith distance. The change in the zenith distance from noon on any day to noon on the next day is precisely equal to the change in declination. The sun's declination on each of the equinocti is zero. It is a maximum on the day of the summer solstice, a minimum on the day of the winter solstice.
    See Chapter 1, p. 24 (Figure 5), p. 25 (Table 1) and Chapter 3, p. 93 (Figure 4).

To avoid the use of trigonometry, or whatever approximation was used before its invention by Hipparchus, the vertical gnomon was set in the center of a hemispherical bowl (called a scaphe or skiotheran by the Greeks) in which the height of the gnomon equalled the radius of the bowl and the bowl's interior was inscribed with concentric rings. The rod's shadow thus gave a direct reading of the zenith distance.
    See Chapter 3, p. 81 (Figure 2c). Reference Notes 8 and 12.

Gomara, Francisco Lopez de, 1510-60, author *Tode la Tierra de las Indias, Primera y segunda parte de le historia general de las Indias . . . hasta el ano de 1551. Con la conquista de Mexico y dela Nueva Espana,* Medina del Campo, 1553. Authority on Coronado expedition. Produced woodcut *Map of the World, 1552.*
    Chapter 1, p. 8, Reference Note 8.

Gosselin, Pascal Francois Joseph, 1751-1830, b. Lille, became Geographer of Paris. His *Geography des Anciens,* (1797)-1813 considered authoritative.
    Chapter 3, pp. 76, 79, 102, 103, (Table 2). Reference Notes 6, 23. Chapter 4, p. 109.

GREAT CIRCLE CIRCUMFERENCE (of Earth) ESTIMATES, 330 B.C. to A.D. 1670, IN CHRONOLOGICAL ORDER, IN UNITS STATED BY THE ESTIMATOR.
(For equivalence in modern units, see reference pages given for each estimate.)

Aristotle (Eudoxus?), c. 330 B.C.: 400, 000 stades. Ref.: Chapter 3, pp. 75, 76, 104 (Table 3).

Archimedes (Dikaearchus?), c. 250 B.C.: 300,000 stades. Ref: Chapter 3, pp. 76, 104 (Table 3). Note: Capella makes Archimedes result 406,010 stades; ref. Chapter 4, p. 109.

Eratosthenes, c. 220 B.C.: 252,000 stades. Ref.: Chapter 3, pp. 78, 104 (Table 3). Note: Capella makes Eratosthenes measurement 406,010 stades; ref. Chapter 4, p. 109.

Hipparchus, c. 150 B.C.: 252, 000 stades. Ref.: Chapter 3, p. 86.

Posidonius, c. 85 B.C.: 240,000 stades according to Cleomedes; Ref. Chap. 3, pp. 86, 104 (Table 3).Note: Strabo makes this 180,000 stades; ref. Chap. 3, pp. 88, 89, 104 (Table 3).

Strabo, c. A.D. 5: 252,000 stades or 31,500 Roman miles. Ref: Chap. 3, pp. 89, 104 (Table 3).

Pliny, c. A.D. 70: 252,000 stades or 31,500 Roman miles. Ref: Chap. 3, pp. 90, 104 (Table 3).

Marinus, c. A.D. 100: 180,000 stades or 22,500 Rom. mi. Ref: Chap. 3, p. 104 (Table 3).

Ptolemy, c. A.D. 150: 180,000 stades. Ref: Chap. 3, pp. 98, 104 (Table 3).

Capella, c. A.D. 250: 31,500 Roman miles. Ref.: Chap. 4, pp. 108, 109, 118-119 (Table 2).

Macrobius, c. A.D. 400: 252,000 stades or 25,200 Rom. mi. Ref: Chap. 4, pp. 109-110, 118-119 (Table 2).

Simplicius, c. A.D. 450: 180,000 stades. Ref: Chap. 4, pp. 111, 112, 118-119 (Table 2).

Aryabhata, c. A.D. 475: 3300 yojana. Ref: Chap. 4, pp. 112, 113, 118-119 (Table 2).

Hoching-Tien, c. A.D. 450: 44,280 lys (lis). Ref.: Chap. 4, 113, 118-119 (Table 2).

Brahmagupta, c. A.D. 630: 50,936 English (statute)mi. Ref: Chap. 4, pp. 113, 118-119 (Table 2).

Isidore, c. A.D. 630: 252,000 stades or 31,500 Roman miles. Ref: Ch. 4, pp. 114, 118-119 (Table 2).

Ananias, c. A.D. 650: According to Diller -
 a) Persian system: 252,000 stades or 25,200 Rom. mi.
 b) "By Air" system (Greek): 180,000 stades or 24,000 Rom. mi.
 c) According to Taylor - 180,000 stades or 25, 714 Roman miles. Ref. Chap. 4, pp. 114-120, incl. Tables 1 and 2.

Y-Hang, c. A.D. 720: 351 lys and 80 paces for the degree. (The relationship between the ly (li) and the pace is highly conjectural, hence the reluctance to multiply reported length of a degree by 360.)

Alfragan, c. 9th century A.D., Al Battani, c. A.D. 900, Ibn Yunus, c. A.D. 1000, Al Biruni, c. 1025, and Abulfeda, c. A.D. 1320 reported the measurement in the Syrian desert, c. A.D. 830 as 20,400 Arabian miles. Ref. Chap. 4, pp. 120-128, incl. Figure 2 and Table 3.

Ibn Rustah, c. 900 A.D. puts the result of this meas. at 24,000 Arabian miles. Ref. Chapt. 4, p. 122.

Ibn Khurdadhbih, c. A.D. 850: 9000 parasangs. Ref.. Chap. 4, p. 129.

Al Biruni, c. 1025: 20,160 Arabian miles or 24,713 English (statute) miles, according to Ahmad. Ref. Chapt. 4, pp. 129-132, incl. Figure 3. According to Kazwini, Al Biruni estimated 20,400 Arabian miles or 6800 parasangs. Ref. Chap. 4, p. 132.

Bhaskara Acharya, c. A.D. 1150: 5027 yojans (yojana), according to Dreyer, ref. Chap. 4, p. 133. According to Ahmad, Acharya's estimate was equivalent to 48,714 English (statute) miles. Ref. Chap. 4, p. 133.

Al-Idrisi, c. 1154: 22,900 miles (no specification as to type of mile). Ref. Chap. 4, pp. 132, 133.

Sacrobosco, c. 1240: 252,000 stadia or 25,200 Rom. mi. Ref. Chap. 5, p. 137.

Dante, c. 1300: 20,400 Roman miles. Ref. Chap. 5, p. 139.

Jean d'Outremeux ("Sir John Mandeville"), c. 14th century A.D.: 31,500 English miles (not necessarily statute). Ref. Chap. 6, pp. 159, 160.

Pierre d'Ailly, c. 1400: 20,400 Italian (Roman) miles. Ref. Chap. 6, pp. 160,161.

Ala-Ed-Din Al Kusgi, c. 1425: 8000 parasangs or 24,000 Arabian miles. Ref. Chap. 6, p. 164.

Regiomontanus, c. 1460: 230,400 stades or 28,800 Italian miles. Ref. Chap. 8, p. 216.

Paolo Toscanelli, c. 1460: Several possibilities proposed -

    a) 22,500 Italian miles; d'Avezac and Ruge, Chap. 7, pp. 167, 173 (Table 1).

    b) 24,000 Italian miles; Wagner, Chap. 7, pp. 167, 173 (Table 1).

    c) 20,400 Italian miles; Columbus and Vignaud, Chap. pp. 167, 173 (Table 1).

    d) 21,600 Italian miles; possibility, see Chapt. 7, p. 168.

    e) 24,360 Italian miles; Vignaud, Chap. 7, pp. 168, 170, 173 (Table 1).

    f) 24,360 Florentine miles or 27, 216 Italian miles; Uzielli and Thatcher, Chap. 7, pp.171-173, Table 1.

Christopher Columbus, c. 1482: 20,400 Italian miles. Ref. Chap. 7, pp. 181-188, incl. Fig. 5 and Table 2.

Martin Behaim, c. 1492: Two possibilities proposed -

    a) 20,400 Italian miles; Morison, Chap. 7, p. 177.

    b) 22,500 Italian miles; this reviewer, Chap. 7, p. 177.

Jaime Ferrer, c. 1494: 252,000 stades, 31,500 Italian miles, or 7875 leagues. Ref. Chap. 8, p. 191-193

Bernardi Sylvani, c. 1511: 22,500 Italian miles. Ref. Chapter 8, pp. 203, 206-207 (Figure 6), 208.

Martin Fernandez de Enciso, c. 1518: 6000 leagues. Ref. Chap. 8, pp. 192-195, incl. Table 2.

Petrus Apianus, c. 1520: Apparently 22,500 Italian miles. Ref. Chap. 8, pp. 203, 210 (Figure 8), 211.

Thomas Duran, Sebastian Cabot, and Juan Vespuccius at Badajoz Conference of 1523-4:
    Testimony was contradictory and would allow of at least 3 conclusions -

    a) 6300 leagues, 25,200 Italian miles, or 201,600 stades,

    b) 6300 leagues, 22,500 Italian miles, or 180,000 stades,

    c) 5625 leagues, 22,500 Ital. miles, or 180,000 stades.

    Ref. Chap. 8, pp. 211-214, incl. Table 3.

Jean Fernel, c. 1525: 21,600 Ital. miles. Ref. Chap. 8, p. 216.

Jean Fernel, c. 1528: 24,514.3 Ital. miles. Ref. Chap. 8, p. 217.

Alonso de Chaves, c. 1537: 25,200 Ital. miles. Ref. Chap. 8, p. 217.

Petrus Apianus, c. 1545: 21,600 Miliaria (Italian miles) or 5400 German Miles.
    Ref. Chap. 8, p. 218.

Gonzalo Fernandez de Oviedo, c. 1545: 6300 leagues. Ref. Chap. 8, p. 218.

Christman, c. 1590:

    a) 24, 480 Roman miles (based on his translation from Alfragan's Arabic);

    b) 16,303 Roman miles (based on translation from Arabic of *Abraham, son of* Chia).

    Ref. Chapter 4, pp. 123, 125, (Table 3).

Willem Barentzoon, c. 1595: 21,600 Italian miles, 5400 German miles, or 6300 Spanish leguas.
    Ref. Chap. 8, pp. 233-235, incl. Figure 17.

Alvaro de Mendana, c. 1595: 7200 Spanish leguas. Ref. Chap. 8, p. 228.

Jean Picard, c. 1670: 20,541,600 toises de Paris. Ref. Chap. 7, Reference Note 20.

End of GREAT CIRCLE ESTIMATES

Hallock, William (late professor of physics, Columbia Univ.), who with Wade, Herbert T. (late editor for physics and applied science, *The New International Encyclopedia*) authored *Outlines of the Evolution of Weights and Measures,* the Macmillan Co., New York, 1906, a major metrological source of C.C.C.

Chapter 2, pp. 63, 64 (Table 2), 72-73 (Table 5), Reference Notes 5, 7, 11, 22, 25, 27, 35, 36. Chapter 4, pp. 123, 127 (Table 3). Reference Notes 16, 36. Chapter 8, Reference Note 13.

Harisse, Henry, 19th century French scholar and prolific author on Columbian and contemporary voyages to New World, incl. *Colomb et Toscanelli*, Paris, 1893, *Encore La Biblioteque Colombine*, Paris, 1897, and *The Diplomatic History of America*, London, 1897.
    Chapter 1, p. 2, 3, 5. Chapter 7, Reference Note 27. Chapter 8, pp. 191 - 195, incl. Tables 1 and 2, 211-214, incl. Table 3. Ref. Notes 1, 2, 3, 8. Appendix B, pp. 287, 288, 304 - 305 (Table 2).

Heath, Thomas L., knighted British astronomer whose proficiency in translating from the classical languages, as well as modern French and German, have made his *Greek Astronomy*, E.P. Dutton, New York, 1932, and the earlier *Aristarchus of Samos, the Ancient Copernicus*, Clarendon Press, London, 1913, veritable treasure troves of information on Greek, and to a lesser degree, Egyptian and Babylonian astronomy, philosophy, and geodesy.
    Chapter 3, Reference Notes 9 and 11.

Herodotus, native of Halicarnassus (a Greek city in Asia Minor), b. ca. 484 B.C., lived for a time in Samos (island of the Aegean off Ionia), later in an Athenian colony at Thurii, in southern Italy. A keen observer, he travelled widely and recorded what he saw and learned. His "History" in 9 books combines a general history of the Greeks and the Barbarians (to the Greeks all foreigners whose language was not Greek; to the Romans, all who spoke neither Latin nor Greek) with the history of the wars of the Greeks and Persians. Has earned the well merited appelation, "The Father of History". Died in Thurii, date unknown.
    Chapter 2, pp. 52, 69, Referene Note 41. Chapter 3, p. 76.

Hinz, Walther J., author *Islamiche Masse und Gewichte, Ungerechnet ins Metrische System*, E.J. Brill, Leiden, 1955. Excellent source on the very wide ranging medieval Islamic systems of weights and measures.
    Chapter 2, Reference Note 4. Chapter 4, pp. 112, 122, 123, 125-128 (incl. Tables 3 and 4). Reference Notes 17, 34.

Hipparchus, one of the greatest of classical Greek astronomers, born at Nicea in Bythnia around 190 B.C., flourished 160-125 B.C. Carried out his observations at Rhodes and Alexandria. Famous for his catalogue of 1080 stars, the invention of trigonometry, the method of fixing terrestrial positions by means of circles of latitude and longitude, the introduction of the 360° circle vice the Babylonian 60° circle, and the method of determining longitude by timing lunar eclipses. A severe critic of Eratosthenes, he produced no geographic work of his own and accepted Eratosthenes' measurement of 252,000 stades for the earth's circumference.
    Chapter 3, pp. 85, 86, 90-92, 98. Reference Note 9.

Homer, Greek epic poet, flourished sometime between 1200 and 800 B.C. Considered by Strabo to be the father of geography.
    Chapter 3, p. 75. Reference Note 2.

Hoching-Tien, see Degree Measurement.

horizon = a plane perpendicular to the line from an observer to the zenith.

Judge, Joseph: See Marden, Luis.

Keen, Benjamin, translator (par excellence!) of *The Life of the Admiral Christopher Columbus by His Son Ferdinand*, Rutgers University Press, New Brunswick, N.J., 1959. This has become one of the prime biographical sources in the English language on the Great Discoverer. The following illustrations have been reproduced (some with minor changes) from Professor Keen's book - with his permission.
    Chapter 1: Figure 1a; Appendix A: Figures 1, 4a, 6, 7, 8, 9, 10, 11, and 12.

Kennelly, Arthur E., author *Vestiges of Pre-Metric Weights and Measures*, J. J. Little, and Ives Co., N.Y., 1928.
    Chapter 4. Reference Note 26.
    Chapter 7, pp. 170, 171. Reference Notes 15, 17.

Kimble, George H.T., British author of *Geography in the Middle Ages*, Methuen & Co., London, 1938; one of the few writers on geographical topics with a sophisticated approach to the metrologic nuances of his subject.
    Chapter 3, Reference Notes 13, 17. Chapter 4, pp. 111, 120, 122, 125 (Figure 3), 129, 133. Reference Notes 7, 9, 23, 31, 42, 46. Chapter 5, p. 137. Reference Notes 1, 3, 4. Chapter 6, Reference Note 1.

Kinsay (Kinsai, Quinsay, etc), Chinese port city reported by Marco Polo to be in Mangi. Later, postulated by Toscanelli as one likely terminus of a westward ocean crossing from Lisbon. Believed by some investigators to be Hangzhous (Hangchau).
    Chapter 1, pp. 3, 34-37. Chapt. 5, pp. 135, 142 (Fig. 2a), 147 (Fig. 2b), 149, 150-151 (Table 1). Chapter 7, pp. 166, 167, 169 (Figure 1), 177, 179, 180 (Figure 4), 181. Ref. Note 9.

Klein, H. Arthur, author World of Measurements, Simon and Schuster, New York, 1974.
    Chapter 2, Reference Notes 5, 7, 11, 15. Chapter 6, p. 160. Reference Note 2. Appendix B, pp. 285, 286, 304-305 (Table 2). Reference Note 4.

La Cosa, Juan de, Basque chartmaker who signed as able seaman on the Nina on C.C.'s 2nd voyage of expl. and disc. (Madrid's Museo Naval insists he was with C.C. on his first two voyages.) La Cosa was with C.C. when he explored Cuba's south coast. Accompanied Alonso de Hojeda on his 1499-1500 voyage to the Spanish Main, Hispaniola, and the Bahamas. Produced planisphere around 1500 which was the first map to show parts of the New World. Died 1509.
    Chapter 8, pp. 197-198, incl. Figure 1, 203, 237, 238 (Table 4). Appendix A, pp. 257, 266, 274. Reference Notes 12, 14.

La Cosa, Juan de, owner and master of the Santa Maria, C.C.'s flagship on his 1st voyage. Ship ran aground on Christmas Eve on N. coast of Haiti at a place C.C. appropriately named Navidad. Instead of obeying Columbus' orders, after grounding, to take ship's anchor in a boat to a position somewhat astern of the ship in deep water, drop the anchor - secured to ship by a heavy line - and return to assist in windlassing (winching) ship off the shoal, La Cosa deserted the ship, rowing off to the Nina. He never again participated in any of C.C.'s voyages of explorations and discovery.
    Appendix A, p. 247. Reference Note 12.

Landstrom, Bjorn, native of Finland, residing and working in Sweden, author of *Columbus, the Story of Don Cristobal Colon, Admiral of the Ocean Sea,* MacMillan Co., New York, 1966; *Ships,* and *Sailing Ships,* Doubleday, Garden City (the latter reprinted in 1978); and other works; highly regarded as an illustrator and authority on the history of sailing and sailing ships.

    Chapter 1, p. 1, 14. Reference Note 2. Chapter 6, Reference Note 1. Appendix A, p. 249 (Figure 3a), p. 250 (Figure 3b), p. 251 (Figure 3c), p. 258 (Figure 5). Appendix B, p. 287. Reference Note 8.

Las Casas, Bartolome de, 1474-1566, Spanish missionary and historian, author of *Historia de las Indias,* generally acknowledged as the most authentic single source on Spanish activities in the New World during the period of the 4 Columbus voyages and for some time thereafter.

    Chapter 1, pp. 2-6, 8. Reference Note 8. Chap. 6, p. 161. Reference Note 9. Chap. 7, p. 166 Reference Note 5.

latitude, terrestrial = angular distance on the Earth measured north or south of the equator along the meridian of a geographic location.

longitude, terrestrial = angular distance measured along the earth's equator from the zero meridian to the meridian of a geographic location. (Since about 1884, Greenwich Observatory, London, has served as the zero meridian. Prior to then many geographic sites were utilized, depending upon the cartographer and the country in which his atelier was located.)

Macrobius, Ambrosius Aurelis Theodosius (sometimes called Theodosius), 399-423 A.D., Roman grammarian, philosopher and geographer. His *Interpretatio in Somnium Scipionis,* an eassy on metaphysical and cosmographical topics, and his other works were widely read and highly regarded for hundreds of years after his time. Preferred Eratosthenes to Ptolemy.

    Chapter 3, pp. 79, 98, 102, 103 (Table 2). Chapter 4, pp. 109, 110 (Figure 1),111, 118-119 (Table 2), 124. Reference Note 4. Chapter 8, pp. 191, 192.

Mangi, the name by which Marco Polo and other travelers referred to South China.
Chap. 1, pp. 2, 3, 34. Chap. 5, p. 139. Chap. 7, pp. 166, 169 (Figure 1). Appendix A, p. 246.

Marden, Luis, National Geographic staffer, who with Joseph Judge (Senior Associate Editor, Nat'l Geog.) authored essays in the November, 1986 issue of this magazine under the symposium title *Columbus and the New World* and essay titles *The Island of Landfall* (Judge and James L. Stanfield) and *Tracking Columbus Across the Atlantic*(Marden). The conclusion of the Marden-Judge team is that Columbus' San Salvador was Samana Cay, some 70 nautical miles SE of Morison's choice, Watling Island. See Appendix A, Figure 4b, herein.

Marinus of Tyre, gifted geographer and cartographer of the end of the first century or beginning of 2nd A.D., first to introduce map projections. While his work was criticized by Ptolemy, it served as a prototype for the latter's *Geographia.* Adopted Strabo's version of Posidonius, 180,000 stades as the circumference of a great circle of the earth. Decided the length of the Eurasian continent was no less than 225 degrees of longitude. A major pillar of Columbus' cosmography.

    Chapter 1, pp. 34, 35, 38, 39. Chapter 3, pp. 90-92, 98, 104 (Table 3), 105, 106. Reference Note 32.Chapter 5, 150-151 (Table 1). Chapter 7, pp. 184, 185-188 (Table 2).

Martin, Fernao; canon of Cathedral of Lisbon, confidante of Portuguese monarch, Alonso V, and correspondent of Toscanelli.

Chap. 1, pp. 2-5, 34.

Mercator, Gerardus (Gerhard Kremer), was born in the Flemish town of Rupelmonde in 1512 and died in Duisberg, Germany in 1594. Studied under Gemma Frisius at Louvain. After marraige in 1536, he turned to map drawing, engraving in copper, the manufacture of astronomical instruments, and surveying - becoming quite adept at each pursuit. His first known map was that of the Holy Land, done in 1537. The next year he produced the world map shown in Chapter 8, Figure 13. In 1541 he produced a large terrestrial globe and published a pamphlet on the use of Italic letters in map print. Produced various cosmographical instruments for Emperor Charles V which were highly regarded. Spent 4 months in prison on a charge of heresy, but his many friends obtained his release. In 1551, he finished his large celestial globe, a pendant for his globe, a manual for the use of both, and 2 small related tracts. In 1552, he moved to Duisberg where he soon produced 2 elaborate globes for Emperor - 1 celestial and 1 terrestrial. His map of Europe was produced in 1554. For the next several years he produced several globes, sold at Nurnberg, and an elaborate map of England engraved on copper. In 1568, he published his comprehensive *Chronologia,* much praised by contemporary scholars. In 1569, he published his large epochal map of the world on an 'increasing' cylindrical projection - one of the most original and valuable cartographical works ever published, but little understood or appreciated by his contemporaries. A short description of the Mercator projection follows.

Meridians are parallel, equally spaced, and perpendicular to the equator and all parallels - the distance between which increase with latitude. In fact, both meridian and parallel spacing are increased with latitude (as compared to the spacing on a globe grid), the expansion being equal to the secant of the latitude.

Such a projection is a conformal chart showing true angles and true distance. A rhumb line (a line which intersects all meridians at the same angle) plots as a straight line on a Mercator chart. The disadvantage of a Mercator projeciton is the distortion at high latitudes, the secant of 90°, for instance, being infinity. The projeciton was originally designed for, and is still used in, navigation.

In 1578, Mercator published a new edition of Ptolemy's Geography which was well received and often reprinted in the next 150 years. The Rumold Mercator world map of 1587, Chapter 8, Figure 15, was originally part of an atlas prepared by his father but with-held from publication because of the preemptive introduction of Ortelius' *Theatrum,* in 1570. (The world map, Figure 14, Chapter 8, is part of Ortelius' atlas.)

Sons Arnold, Barthelemy, and Rumold and grandsons Joannes, Gerard, and Michael all assisted Gerardus senior, or continued as cartographers after his death.

Metrology : The science of weights and measures. This book concerns itself with a limited phase of this subject - ancient, medieval, and renaissance linear metrology. The authors most knowledgeable or most helpful in their coverage of this subject are: Eratosthenes, Strabo, Pliny, Macrobius, Isidore, Alfragan, Abulfeda, Christman, Ferrer, Fernel, Apianus, Picard, Delambre, Garcilaso de la Vega, Gosselin, Petrie, Smyth, Reinaud, Yule, Bunbury, Vignaud, Harisse, Von Humboldt, Nordenskiold, Hallock and Wade, Berriman, Ahmad, Hinz, Sarton, Barnard, Dreyer, Diller, Drabkin and Cohen, Gillings, Kennelly, Kimble, Klein, Morison, and Taylor. All of the foregoing are referenced, as appropriate, in the text and have individual entries in this Index.

Molander, Arne B, senior staff engineer at TRW, Inc., specialist in analysis of navigational systems for U.S. Department of Defense, experienced amateur yachtsman who since 1971 has been re-examining available data on the course followed by Columbus on his first trans-Atlantic crossing, the likely first landfall, and the route taken by Columbus through the Bahamas to Cuba. Molander argues, convincingly, in publications (see Appendix A, Figure 4b and reference note 13) and in correspondence with this author that Columbus entered the Bahamas just north of the northern tip of Eleuthera and made his landfall at Egg/ Royal Islands about 9 nautical miles SW of the northern tip of Eleuthera and 160 nautical miles northwest (approx.) of Watling Island, S.E. Morison's choice (and that of Benjamin Keen), see Figures 1a and 1b, Chapter 1 and Figures 4a and 4b, Appendix A. Molander's landfall is about 226 nautical miles NW of Samana Cay, the landfall choice of Luis Marden and Joseph Judge (see Index entry for Luis Marden). Molander's choice of a route through the Bahamas and the islands visited by Columbus en-route to Cuba differs, therefore, from Morison-Keen and Marden-Judge.

Morison, Samuel Eliot, 1887-1976, distinguished professor of history (Harvard and Oxford), author of 40 volumes and 100 articles of an historical, biographical, nautical thrust, frequently combining all three, as in *Admiral of the Ocean Sea, A Life of Christopher Columbus* , Little Brown & Co., Boston, 1942; *The European Discovery of America: The Northern Voyages, A.D. 500-1600* , Oxford University Press, 1971; *The Southern Voyages, A.D. 1492-1616,* copyright Samuel Eliot Morison, reprinted by Oxford Univ. Press, 1974; *Sailor Historian, the Best of Samuel Eliot Morison,* edited by Emily Morison Beck, Houghton Mifflin Co., Boston, 1977. Saw military service as a private of Infantry in W.W. I. and as a LCdr. to RAdm. in U.S. Navy chronicling *History of U.S. Naval Operations in World War II,* in 15 vols., 1947-62. An avid and skillful "salt water sailor", he retraced Columbus' first voyage and those of several other European discoverers of America. A major source for this volume.

Chap. 1, pp. 6, 7, 12, 14, 16 (Figure 1a), 17 (Figure 1b), 19, 28, 29 (Figure 7b), 32-38. Reference Notes 1, 7, 22, 24, 25, 28, 30, 32, 35. Chap. 4, pp. 119, 121 (Table 3). Reference Note 35. Chap. 5, pp. 135, 144-145 (Table 1). Ref. Note 5. Chap. 6, p. 156. Ref. Notes 10-13, incl., 15. Chap. 7, pp. .165, 170, 171, 174, 175, 179-182 (Table 2), Reference Notes 18, 21. Appendix B, pp. 281, 298-299 (Table 2). Ref. Note 8.

Nordenskiold, Adolf Erik, 1832-1901, born and reared in Finland of Swedish parents, spent the majority of his lifespan after age of 25 in or working out of Sweden. A mineralogist of some note, he became a highly successful arctic explorer, engaging in 10 expeditions, one of which (in 1878-80) circumnavigated Asia for the first time. Dissatisfied with the manner in which the history of cartography had been presented up to his time, he published his *Facsimilie - Atlas to the Early History of Cartography with Reproductions of the Most Important Maps Printed in the XV and XVI Centuries,* Stockholm, 1889. The 1973 reproduction of this masterpiece by Dover Publications, New York, is a principal pillar on which "C.C.C." rests.

Maps reproduced from Nordenskiold's Atlas are listed hereunder: Chapter 3, Figure 5: The World of Ptolemy. Chapter 4, Fig. 1: A Map of the World from the 1483 Edition of Macrobius. Chapter 5, Figure 1a: A Map of the World by Sacrobosco (John of Holywood). Chapter 5, Figure 3a: Marinus' and Ptolemy's Concepts of the Prime Meridian, the . . . Mediterranean, and its Eastern and Western Ends. Chapter 5, Figure 3b: Marinus' and Ptolemy's Concepts of the Sacred Promontory . . ., the Pillars of Hercules, and the Port of Lisbon. Chapter 5, Figure 3c: Ptolemy's Concept of the Eastern End of the Habitable World. Chapter 6, Figure 1: Map of the World, from Ymago Mundi, by Petrus de Aliaco. Chapter 7, Figure 3: Martin Behaim's Globe of 1492, from J.G. Doppelmayer. Chapter 8, Figure 5: The Conical Projection of Johannes Ruysch . . . 1508. Chapter 8, Figure 6: Sylvani's Map of the World, 1511. Chapter 8, Figure 7: Johannes Stobnicza's Homeother Projection . . . 1512. Chapter 8, Figure 8: The Cordiform Projection of Pietrus Apianus of 1520. Chapter 8, Figure 10: Petrus Apianus' . . . (Cosmographic Map) . . . of 1545. Chapter 8, Figure 11: Finaeus' Double

Cordiform Map . . . of 1531. Chapter 8, Figure 12: Grynaeus' 1532 Map of the World . . . Chapter 8, Figure 13: Gerardus Mercator's Double Cordiform Map of 1538. Chapter 8, Figure 14: Abraham Ortelius' . . . Map. . . of 1570. Chapter 8, Figure 15: Rumold Mercator's World Map of 1587. Chapter 8, Figure 16: Cornelius de Judaeis' World Map on a Polar Projection . . . 1593. Chapter 8, Figure 17: Chart of the Mediterranean Sea by Willem Barentzoon . . .1595.

Other references to Nordenskiold's Atlas:

Oviedo y Valdez, Gonzalo Fernandez de, 1478-1557, Spanish author *La Historia Natural y General de las Indias*, Seville, 1535; reprinted Salamanca, 1547; Madrid (in 4 volumes), 1851-55. The 4th volume was not in earlier editions. Spent 34 years in different parts of the Caribbean. His account of Columbus' voyages drew on oral sources. Good knowledge of navigation.

parallax = the difference in the apparent direction of an object as seen from two different locations; conversely, the angle at the object that is subtended by the line joining two designated points. (See Chap. 3, Ref. Note 12.)

Petrie, W.M. Flinders, 19th century Egyptologist and metrologist, author *Inductive Metrology, or the Recovery of Ancient Measures from the Monuments,* Hargrove Saunders, London, 1877, a major source for C.C.C.

Plato, 427-347 B.C., Greek philosopher born at Athens or the island of Aegina. Spent most of his life in Athens immersed in philosophy. Studied under Socrates and after his death he devoted himself to the mission of his martyred teacher, the record of which is contained in his collecton of "Dialogues". In ethics and politics, Plato was the first thinker to offer a good account of the principles that form and govern conduct and character. "Justice" is the virtue of the good citizen and his idea of good is to be realized in the life of the commonwealth. In his *Phaedo* he mentions Socrates' views as regards the "current" controversy as to the shape and size of the earth, this subject being developed further in his "Timaeus", but no dimensions are suggested.

Pliny (Caius Plinius Secundus) commonly known as Pliny the Elder, 23-79 A.D., born at Verona or Comum, in northern Italy. Held various political posts during lifetime, including procurator of Spain. Killed at Misenum during famous eruption of Visuvius, 79 A.D., which buried Herculaneum and Pompeii. Best known for his "Natural History", an enormous essay, in 37 books, towards a description of the Universe. Despite its many deficiencies, it did provide a political and statistical description of the geography of the many countries making up the Roman Empire of his day. From a metrological stand-point, his contributions are valuable for the relationships it provided between the Greek and Roman units utilized in itinerary and geodetic measurement from Eratosthenes' time to his own. His interpretations were quoted, if not parroted, for centuries after his time. An important, if imperfect, link in the history of ancient geography.

Chapter 2, p. 53. Chapter 3, pp. 79, 82, 86, 89, 90, 102-103 (Table 2), 104 (Table 3), 105. Reference Note 20. Chapter 4, pp. 109, 114.

Plutarch (Plutarchus), 48-122 A.D., Greek biographer and philosopher who spent much time in Rome. His fame stems from his work *Parallel Lives*, a collection of biographies of notable men in pairs, one Greek and one Roman. The work is important, historically, for it contains much information from older authorities whose output is now lost.

Chapter 3, p. 88.

Polo, Marco, 1254-1324, the Venetian traveler in Asia whose book extolled the wonders and wealth of Mangi, Cathay, the brillant Mongol court of Kubla Khan at Cambaluc, the busy seaports of Kinsay and Zaitun, the fabulous island of Chipanju, 1500 miles east of the Chinese mainland, and, indirectly, tended to confirm and even extend Marinus' already over-stretched east-west length of the Eurasian continent.

Introduction to *The Book of Ser Marco Polo* , Chap. 5, p. 140.
Original trip, 1260-1266, of Nicolo and Maffeo Polo to court of Kubla Khan, return with Marco, 1271-1275, and 17 year service of the 3 Polos for the great Mongol monarch; Chap. 5, p. 141. Marco Polo's Travels, Yule's Summary of: Chap. 5, pp. 142, 144-145 (Figure 2a, which is an outline map illustrating how Yule believed Marco Polo envisaged Eurasia). Marco Polo's Impact on C.C., per Morison, Chap. 5, p. 139. Skelton's assessment of general impact of Marco Polo's book; Chap. 5, p. 138, 141. Marco Polo's impact on 15th century cosmography, Chap. 5, pp. 148-151, incl. Table 1. An assessment of Polo's book, Chap. 5, pp. 154, 158. Reference Notes 6-9, 11, 13. See also Ch.1, pp. 2, 34-39; Chap. 6, p. 159; Chap. 7, pp. 181, 184, 186-188 (Table 2). Appendix A, p. 246.

Polybius, Greek historian, born in Megalopolis, Arcadia (center of Peloponnesus) about 210 B.C. Deported to Italy after Macedonian conquest. Formed lifelong friendship with important Roman families. After Roman conquest of Greece, he was successful in obtaining favorable terms for his defeated countrymen. Highly honored in Greece, he wrote a history of Rome in 40 books. Important to our story because he related the stade to the Roman mile, 8 1/3 to 1.

Chapter 2, pp. 61, 70, Reference Note 26. Chapter 3, pp. 82, 98. Ref. Note 21.

Posidonius, 135-50 B.C., born in Apameia (Syria), later member of famed Stoic school of philosophy at Rhodes. An astronomer of some note in his day, he undertook to check Eratosthenes' determination of the length of a degree of latitude employing the stars, instead of the sun, for his astronomical measurements. His results started a controversy which persists to this day. His measurment of a degree of latitude:

Chapter 3, pp. 76, 86-89, incl Figure 3. Ref. Note 31. Interpretations and Effects of Posidonius' measurment: Chapter 3, pp. 89, 90, 91, 98, 100 (Table 1), 104 (Table 3). His geographic views: Chapter 3, pp. 88,106. Chapter 5, pp. 150, 151, (Table 1). Chapter 7, pp. 179-181 (Table 2). Misc. references: Chapter 3. Ref. Note 22. Chapter 4, pp. 115, 118-119 (Table 2).

Ptolemy (Claudius Ptolemaeus), fl. middle of 2nd Century, Alexandrian whose great astronomical work was translated into Arabic in the 8th century and named the Almagest (the best), and whose *Geographik Syntaxis* (Atlas of the World) was the first effort to put geography on a scientific basis. His manuscript was first translated (from Greek) into Latin by Jacopo Angelo in 1406. First edition with maps, Bologna 1477. First German edition with 5 new maps added, Ulm 1482. Not believed to have drawn any of the maps in the original MS. Confirms his geographic atlas borrows heavily from Marinus (as his astronomical work does from Hipparchus). Reduced longitudinal extent of Eurasian continent from Marinus' 225° to 180°, but left no details as to what lay beyond the 180th meridian. Adopted Posidonius' standard of 500 stades to the degree, without defining his stade. Maps based upon Ptolemy's geography (most of which include information from Marco Polo's book, or from C.C.'s and later voyages of exploration) are listed hereunder:

Ch. 3, Fig. 5: The World of Ptolemy. Chapter 5, Figures 3a, 3b, 3c: See Nordenskiold. Ch. 7, Figs. 3 & 4: Martin Behaim's Globe of 1492. Ch. 8, Figures 5-8, 11-16: See Nordenskiold. His Geographic and Geodetic Views: Ch. 3, pp. 92-98 (incl. Fig. 4 and 5). Interpretations of His Geodetic Views, including a discussion of the length in metres of his stade and degree: Ch. 3, pp. 98-101 (incl. Table 1), 104 (Table 3). Misc. Refs. to Ptolemy: Ch. 1, pp. 34, 35, 36, 39. Ref. Notes 17, 26,31. Ch. 3, pp. 88, 90. Ref. Note 32. Ch. 4, pp. 109, 111-113, 115, 118 (Table 2), 120, 124, 125 (Table 3), 134. Ch. 5, pp. 148-151, incl. Table 1, 153 (Figure 3a), 155 (Figure 3b), 156-157 (Figure 3c), 158 (Figure 3d). Ch. 6, pp. 160, 161. Ch. 7, pp. 167, 176, 183-188, (Table 2). Ch. 8, pp. 201, 202 (Figure 4), 206-207 (Figure 6), 216.

Pythagoras, Greek philosopher, born c. 582 B.C. at Samos. Founded a school at Crotona, Italy, c. 529 B.C., where his students were trained in gymnastics, mathematics, and music, were vegetarians, and believed in immortality and transmigration of the soul. Pythagoras is regarded as the founder of Geometry and the first to conceive of the earth as a sphere.

Chapter 3, p. 75.

Quinn, David B., History professor, University of Liverpool, author of the exceptionally complete, informative, and accurate *North America from Earliest Discovery to First Settlements: the Norse Voyages to 1612*, Harper and Row, New York, 1977.

Chapter 1, pp. 36, 37. Reference Note 36.

refraction, astronomical = the change in the direction of travel (bending) of a light ray as it passes obliquely through the atmosphere. As a result of refraction, the observed altitude of a celestial object is greater than the geometric altitude. Refraction is greatest at the horizon and zero at the zenith.

Chapter 3, pp. 86, 87. Reference Notes 12, 27, 31. Chapter 4, p. 132

Regiomontanus (John Muller, better known as Johannes de Monte Regio, and, after his death, as Regiomontanus), 1436-1476. Born in Konigsberg, Franconia, studied under astronomer George Peurbach, became proficient in Greek, translated Ptolemy into Latin, published astronomical ephemerides used by Portuguese, Spanish and Italian navigators, including C.C. Credited with reducing length of a degree from Eratosthenes' 700 stades (which he interpreted as 87 1/2 Roman ((Italian))miles) to 640 stades, or 80 miles - an improvement.

Chap. 1, pp. 31, 32, Ref. Note 29. Chap. 7, p. 165 Ref. Note 1. Chap. 8, p. 216. Ref. Note 6.

Reinaud, Joseph Toussaint (1795-1867), French Orientalist and writer on maps, translated from Arabic to French Abulfeda's "Geography", published in Paris, 1840; evidently agreed with Abulfeda that the Arabian and Roman miles were equal in length.

Chapter 4, pp. 122, 125 (Table 3). Reference Note 41.

Rustah, Ibn (c. 10th century A.D.), Persian geographer from Isfahan who interpreted the Arab measurement of a degree at Sinjar in 830 A.D. as 66 2/3 miles, presumably Arabian.

    Chapter 4, p. 122. Reference Note 30.

Sacrobosco, Joannes (John of Holywood), d. 1256, transplanted English cosmographer and author (*Tractatus de Sphaera*) who taught at University of Paris and, indirectly, influenced Cardinal d'Ailly and C.C.

    Chapter 1, p. 11. Reference Note 17. Chapter 5, pp. 137, 138 (Figure 1a).

Sarton, George, d. 1956, humanist, professor (Harvard Univ.), and author of the epic "A History of Science", envisaged as requiring 8 or 9 volumes. Only 2 had been completed by the time of his death: the first volume, the monumental *Introduction to the History of Science*, Carnegie Institution of Washington, 1927, 1931, 1947-48; and the second *A History of Science, Hellenistic Science and Culture in the Last Three Centuries B.C.*, Harvard Univ. Press, Cambridge, 1959. A bibliography of his writings was published in a *George Sarton Memorial Issue*, ISIS, Sept. 1957.

    Chapter 3, Reference Notes 9, 23. Chapter 4, p. 112. Ref. Note 15.

Simplicius, see Degree Measurement.

Skelton, R.A., British author of several works on the history of exploration and of cartography, among them *Explorers' Maps*, F.A. Praeger, New York, 1958 and *The Cartography of the First Voyage*, an appendix to *The Journal of Christopher Columbus*, translated by Cecil Jane, Clarkson N. Potter, New York, 1960.

    Chapter 1, pp. 37, 38. Reference Note 37. Chapter 5, pp. 139, 140, 142, 143. Ref. Note 6. Chapter 8, pp. 216, 234. Ref. Notes 5, 9.

Skiotheran, see Gnomon

Smyth, Piazzi, 19th century Astronomer Royal of Scotland who became convinced - along with certain other religious zealots - that the Great Pyramid of Gizeh had been planned by ancient Hebrews acting under Divine Guidance. Further, many astronomic, geodetic, and geographic mysteries were incorporated into the dimensions of the huge structure. Smyth's book, detailing the foregoing, was first published in 1880 in Great Britain. In 1978, it was republished by Bell Publishing Co., New York, under the title *The Great Pyramid, Its Secrets and Mysteries Revealed.* The book is, inter alia, replete with ancient and medieval metrologic data.

    Chapter 2, Reference Note 32. Chapter 3, Reference Note 1. Chapter 7, p. 171. Reference Note 16.

Socrates, Greek philosopher, born c. 470 B.C. became a sculptor but his ideas on philosophy soon made him famous and controversial in Athens. Despite serving with distinction in the war with Sparta (c. 432 B.C.), his ideas later got him into trouble. In 399 B.C. he was jailed and later given the death penalty for corrupting youth and introducing new divinities in place of those recognized by the state. Commit suicide in prison by dirnking hemlock. His most famous student was Plato who is responsible for most of what is known about Socrates, including his concept of the earth as "... a sphere at rest in the center of a swirl . ... " (see Plato).

    Chapter 3, p. 75.

Solstice = either of the two points on the ecliptic at which the apparent longitude of the sun is 90° or 270°; also the time at which the sun is at either point. Usually occurs on 21 June and 22 Dec. when the sun is respectively farther north and south (at noon) than on any other day (at noon).

Chapter 1, pp. 20, 24 (Fig. 5), 25 (Table 1); Chapter 3, 78-81, incl. Figures 2a, 2b, 2c.

Strabo, c. 63 B.C.- c. A.D. 19 (Tooley's *Dictionary of Mapmakers*, R.V. Tooley, Allen R. Liss, New York, 1979, says c. 50 B.C. - A.D. 25), Greek geographer and historian, settled in Rome about 29 B.C. and worked there the rest of his life. Travelled widely in the Mediterranean world. His *Geographica*, in 17 books, is most important source on ancient geography up to time of Ptolemy, especially on Eratosthenes and Posidonius. First translation, Treviso, 1480. Basle, 1539 and others followed. Early English translation, used in C.C.C., is H.C. Hamilton and W. Falconer, *The Geography of Strabo, Literally Translated*, Henry G. Bohn, London, 1857. This work is especially valuable because it challenges many Straboisms on the authority of many mathematical geographers since Strabo's time.

Ch. 2, p. 61, Ref. Note 24. Ch. 3, pp. 78, 79, 82, 83, 88-91, 98, 99, 101 (Table 1, part 2), 102, 104 (Table 3), 105,106. Ref. Notes 6, 9, 13, 18, 21, 23, 24, 28, 29, 33, 34. Ch. 4, pp. 109, 113, 114. Ch. 5, p. 150 (Table 1).

Taylor, E.G.R., British writer on cosmographic, geographic, cartographic, and navigational topics. Handles metrologic aspect with sophistication. Her *The Haven Finding Art,* London, 1956, has been called 'the clearest study of the art of navigation to the time of James Cook' (when the seagoing chronometer received its first extensive utilization in the determination of longitude). Her *Some Notes on Early Ideas of the Form and Size of the Earth,* Geographical Journal, London, Jan. 1935, is referenced and discussed in C.C.C.

Chapter 4, pp. 114-119, including Tables 1 and 2. Reference Notes 7, 24, 29.

Thales, b. 640 B.C. in Miletus, Asia Minor, founded Ionian school of philosophy, is considered the pioneer among the Greeks in the sciences of geometry and astronomy, was first of record to suggest a scientific, rather than mythological explanation of the universe.

Chapter 3, p. 74.

Thatcher, John Boyd, author of *Christopher Columbus, His Life, His Work, His Remains*, first published in 1902, reprinted by AMS Press, Inc. and Kraus Reprint Corp., New York, 1967. Of all the Columbus biographers, Thatcher and Vignaud appear to have possessed the deepest understanding of medieval/renaissance metrology.

Chapter 7, pp. 168, 170, 171, 173 (Table 1), Reference Notes 4, 11. Appendix B, pp. 287, 304-305, (Table 2). Reference Note 8.

Tooley, R.V., author *Tooley's Dictionary of Mapmakers,* Alan R. Liss, New York, 1979. Excellent biographical source on cosmographers, geographers, cartographers, geodesists, philosophers and astronomers with cosmographic leanings.

Chapter 4, Reference Note 1. Chapter 6, Reference Note 1.

Toscanelli, Paolo dal Pozzo, 1397-1482; Florentine physician, astronomer, cosmographer and sometime correspondent with Fernao Martin and C.C.

Contents of letter to Martin; Chap. 1, pp. 2, 3-5, Reference Note 5; Chap. 7, pp. 166, 167. Biography; Chap. 7, p. 165. Degree of Latitude, length of; Chap. 7, pp. 167, 168, 170-173, incl. Table 1. Cosmography; Chap. 1, pp. 34-39. Chap. 7, pp. 166, 167, 169 (Fig. 1), 170-174, incl.

Table 1, and Fig. 2a, 176, 186-188 (Table 2). Reference Notes 1, 3, 5-10, 12, 16, 19, 22. Misc., Chapter 2, p. 48.

Vignaud, Henry, 1830-1922, French-American scholar, friend of Henry Harisse and severe critic of many Columbus claims. Author of *La Lettre et La Carte de Toscanelli,* Paris, 1901; *Le Vrai Christophe Colomb et La Legende,* Paris, 1921; 3 other volumes on Columbus and one on Amerigo Vespucci.
Chapter 1, pp. 3-5, Reference Notes 6, 26. Chap. 4, pp. 122, 124, 125 (Table 3). Reference Notes 33, 41.Chapter 6, p. 161. Ref. Notes 7, 8, 9. Chapter 7, pp. 165, 167, 168, 170, 172, 173 (Table 1), 176, 177, 181. Reference Notes 8, 10, 12, 13, 22, 26. Appendix B, pp.287, 304-305 Table 2. Reference Note 8.

Von Humboldt, Alexander, 1769-1859; German scientist, explorer, and writer.
Chapter 1, pp. 1, 4. Reference Note 1. Chapter 4, p. 109. Chapter 6, pp. 161, 162. Ref. Note 11. Chapter 7, p. 177. Reference Notes 6, 25.

Y-Hang, see Degree Measurment

Yule, Henry, 19th century British Army Colonel with considerable knowledge of eastern Asia stemming from his duty in India and travels in adjoining and nearby Asian countries. Has written extensively on the geography and customs of these countries. Author of *The Book of Ser Marco Polo,* first published in 1871. The third edition, revised and edited by Henry Cordier, distinguished French Orientalist, and geographer, extends the thrust of Yule's original work. This was not merely a translation from early manuscripts, but an interpretation, profusely augmented by maps, illustrations, and notes intended to make clearer to the reader where the Polo's went and what they experienced. The third edition, first published in 1903, was reprinted in two volumes by John Murray, London, in 1975.
Chapter 5, pp. 140, 142, 144-145 (Figure 2a), 148, 150-151 (Table 1), 154, 158. Reference Notes 9, 11-13, incl.

Zaitun (Zaytun, Zaiton), Chinese port city which, along with Kinsay, were reported by Marco Polo as major centers of commerce in Mangi. Toscanelli later considered this port as another terminus for a possible westward ocean crossing from Lisbon. Believed by some investigators to be Quanzhou, by others -Changchow.
Chapter 1, p. 34. Chapter 5, pp. 139, 144-145 (Figure 2a), 146-147 (Figure 2b), 148, 149, 150-151 (Table 1). Chapter 7, pp. 167, 169 (Fig. 1), 180 (Figure 4).

zenith distance = angular distance on the celestial sphere measured along the great circle from the zenith to the celestial object = 90° - altitude of object above horizon. The zenith is, in general, the point on the celestial sphere directly over the observer. The astronomical zenith is the extension to infinity of a plumb line (assuming no extraordinary aberrations in the earth's gravitational attraction at the site). The geocentric zenith is defined by the line from the center of the earth through the observer. The geodetic zenith is the normal to the geoid at the observers location. (See references cited for each of the "Degree Measurements").